W9-AFJ-189

Graduate Texts in Mathematics 135

Graduate Texts in Mathematics

1 TAKEUTI/ZARING. Introduction to Axiomatic Set Theory. 2nd ed.
2 OXTOBY. Measure and Category. 2nd ed.
3 SCHAEFFER. Topological Vector Spaces.
4 HILTON/STAMMBACH. A Course in Homological Algebra.
5 MAC LANE. Categories for the Working Mathematician.
6 HUGHES/PIPER. Projective Planes.
7 SERRE. A Course in Arithmetic.
8 TAKEUTI/ZARING. Axiometic Set Theory.
9 HUMPHREYS. Introduction to Lie Algebras and Representation Theory.
10 COHEN. A Course in Simple Homotopy Theory.
11 CONWAY. Functions of One Complex Variable. 2nd ed.
12 BEALS. Advanced Mathematical Analysis.
13 ANDERSON/FULLER. Rings and Categories of Modules.
14 GOLUBITSKY/GUILEMIN. Stable Mappings and Their Singularities.
15 BERBERIAN. Lectures in Functional Analysis and Operator Theory.
16 WINTER. The Structure of Fields.
17 ROSENBLATT. Random Processes. 2nd ed.
18 HALMOS. Measure Theory.
19 HALMOS. A Hilbert Space Problem Book. 2nd ed., revised.
20 HUSEMOLLER. Fibre Bundles. 2nd ed.
21 HUMPHREYS. Linear Algebraic Groups.
22 BARNES/MACK. An Algebraic Introduction to Mathematical Logic.
23 GREUB. Linear Algebra. 4th ed.
24 HOLMES. Geometric Functional Analysis and Its Applications.
25 HEWITT/STROMBERG. Real and Abstract Analysis.
26 MANES. Algebraic Theories.
27 KELLEY. General Topology.
28 ZARISKI/SAMUEL. Commutative Algebra. Vol. I.
29 ZARISKI/SAMUEL. Commutative Algebra. Vol. II.
30 JACOBSON. Lectures in Abstract Algebra I. Basic Concepts.
31 JACOBSON. Lectures in Abstract Algebra II. Linear Algebra.
32 JACOBSON. Lectures in Abstract Algebra III. Theory of Fields and Galois Theory.
33 HIRSCH. Differential Topology.
34 SPITZER. Principles of Random Walk. 2nd ed.
35 WERMER. Banach Algebras and Several Complex Variables. 2nd ed.
36 KELLEY/NAMIOKA et al. Linear Topological Spaces.
37 MONK. Mathematical Logic.
38 GRAUERT/FRITZSCHE. Several Complex Variables.
39 ARVESON. An Invitation to C^*-Algebras.
40 KEMENY/SNELL/KNAPP. Denumerable Markov Chains. 2nd ed.
41 APOSTOL. Modular Functions and Dirichlet Series in Number Theory. 2nd ed.
42 SERRE. Linear Representations of Finite Groups.
43 GILLMAN/JERISON. Rings of Continuous Functions.
44 KENDIG. Elementary Algebraic Geometry.
45 LOÈVE. Probability Theory I. 4th ed.
46 LOÈVE. Probability Theory II. 4th ed.
47 MOISE. Geometric Topology in Dimentions 2 and 3.

continued after index

Steven Roman

Advanced Linear Algebra

With 26 illustrations in 33 parts

Springer-Verlag
New York Berlin Heidelberg London Paris
Tokyo Hong Kong Barcelona Budapest

Steven Roman
Department of Mathematics
California State University at Fullerton
Fullerton, CA 92634 USA

Mathematics Subject Classifications (1991): 15-01, 15A03, 15A04, 15A18, 15A21, 15A63, 16D10, 54E35, 46C05, 51N10, 05A40

Library of Congress Cataloging-in-Publication Data
Roman, Steven.
Advanced linear algebra / Steven Roman.
p. cm. -- (Graduate texts in mathematics . 135)
Includes bibliographical references and index.
ISBN 0-387-97837-2
1. Algebras, Linear. I. Title. II. Series.
QA184.R65 1992
512'.5--dc20 92-11860

Printed on acid-free paper.

Production managed by Karen Phillips; manufacturing supervised by Robert Paella.
Camera-ready copy prepared by the author.
Printed and bound by R.R. Donnelley & Sons, Harrisonburg, VA.
Printed in the United States of America.

9 8 7 6 5 4 3 2 1

ISBN 0-387-97837-2 Springer-Verlag New York Berlin Heidelberg
ISBN 3-540-97837-2 Springer-Verlag Berlin Heidelberg New York

To Donna

Preface

This book is a thorough introduction to linear algebra, for the graduate or advanced undergraduate student. Prerequisites are limited to a knowledge of the basic properties of matrices and determinants. However, since we cover the basics of vector spaces and linear transformations rather rapidly, a prior course in linear algebra (even at the sophomore level), along with a certain measure of "mathematical maturity," is highly desirable.

Chapter 0 contains a summary of certain topics in modern algebra that are required for the sequel. *This chapter should be skimmed quickly and then used primarily as a reference.* Chapters 1-3 contain a discussion of the basic properties of vector spaces and linear transformations.

Chapter 4 is devoted to a discussion of modules, emphasizing a comparison between the properties of modules and those of vector spaces. Chapter 5 provides more on modules. The main goals of this chapter are to prove that any two bases of a free module have the same cardinality and to introduce noetherian modules. However, the instructor may simply skim over this chapter, omitting all proofs. Chapter 6 is devoted to the theory of modules over a principal ideal domain, establishing the cyclic decomposition theorem for finitely generated modules. This theorem is the key to the structure theorems for finite dimensional linear operators, discussed in Chapters 7 and 8.

Chapter 9 is devoted to real and complex inner product spaces. The emphasis here is on the finite-dimensional case, in order to arrive as quickly as possible at the finite-dimensional spectral theorem for normal operators, in Chapter 10. However, we have endeavored to state as many results as is convenient for vector spaces of arbitrary dimension.

The second part of the book consists of a collection of independent topics, with the one exception that Chapter 13 requires Chapter 12. Chapter 11 is on metric vector spaces, where we describe the structure of symplectic and orthogonal geometries over various base fields. Chapter 12 contains enough material on metric spaces to allow a unified treatment of topological issues for the basic Hilbert space theory of Chapter 13. The rather lengthy proof that every metric space can be embedded in its completion may be omitted.

Chapter 14 contains a brief introduction to tensor products. In order to motivate the universal property of tensor products, without getting too involved in categorical terminology, we first treat both free vector spaces and the familiar direct sum, in a universal way. Chapter 15 is on affine geometry, emphasizing algebraic, rather than geometric, concepts.

The final chapter provides an introduction to a relatively new subject, called the umbral calculus. This is an algebraic theory used to study certain types of polynomial functions that play an important role in applied mathematics. We give only a brief introduction to the subject – emphasizing the algebraic aspects, rather than the applications. This is the first time that this subject has appeared in a true textbook.

One final comment. Unless otherwise mentioned, omission of a proof in the text is a tacit suggestion that the reader attempt to supply one.

Steven Roman *Irvine, Ca.*

Contents

CHAPTER 0

Preliminaries

In this chapter, we briefly discuss some topics that are needed for the sequel. This chapter should be skimmed quickly and then used primarily as a reference.

Contents: ***Part 1: Preliminaries.*** *Matrices. Determinants. Polynomials. Functions. Equivalence Relations. Zorn's Lemma. Cardinality.* ***Part 2: Algebraic Structures.*** *Groups. Rings. Integral Domains. Ideals and Principal Ideal Domains. Prime Elements. Fields. The Characteristic of a Ring.*

Part 1 Preliminaries

Matrices

If F is a field, we let $\mathcal{M}_{m,n}(F)$ denote the set of all $m \times n$ matrices whose entries lie in F. When no confusion can arise, we denote this set by $\mathcal{M}_{m,n}$, or simply by $\mathcal{M}$. The set $\mathcal{M}_{n,n}(F)$ will be denoted by $\mathcal{M}_n(F)$ or $\mathcal{M}_n$.

We expect that the reader is familiar with the basic properties of matrices, including matrix addition and multiplication. If $A \in \mathcal{M}$, the (i,j)-th entry of A will be denoted by $A_{i,j}$. The identity matrix of size $n \times n$ is denoted by I_n.

Definition The **transpose** of $A \in \mathcal{M}_{n,m}$ is the matrix A^T defined by

$$(A^T)_{i,j} = A_{j,i}$$

A matrix A is **symmetric** if $A = A^T$ and **skew-symmetric** if $A^T = -A$. □

Theorem 0.1 **(Properties of the transpose)** Let $A, B \in \mathcal{M}$. Then
1) $(A^T)^T = A$
2) $(A+B)^T = A^T + B^T$
3) $(rA)^T = rA^T$, for all $r \in F$
4) $(AB)^T = B^T A^T$, provided that the product AB is defined
5) $\det(A^T) = \det(A)$. ∎

Recall that there are three types of *elementary row operations*. Type 1 operations consist of multiplying a row of A by a nonzero scalar (that is, an element of F). Type 2 operations consist of interchanging two rows of A. Type 3 operations consist of adding a scalar multiple of one row of A to another row of A.

If we perform an elementary operation of type k ($=1,2$ or 3) to an identity matrix I_n, we get an **elementary matrix** of type k. It is easy to see that all elementary matrices are invertible.

If A has size $m \times n$, then in order to perform an elementary row operation on A, we may instead perform that operation on the identity I_m, to obtain an elementary matrix E, and then take the product EA. Note that we must multiply A *on the left* by E, since multiplying on the right has the effect of performing *column* operations.

Definition A matrix R is said to be in **reduced row echelon form** if
1) All rows consisting only of 0s appear at the bottom of the matrix.
2) In any nonzero row, the first nonzero entry is a 1. This entry is called a **leading entry**.
3) For any two consecutive rows, the leading entry of the lower row is to the right of the leading entry of the upper row.
4) Any column that contains a leading entry has 0s in all other positions. □

Here are the basic facts concerning reduced row echelon form.

Theorem 0.2 Two matrices A and B in $\mathcal{M}_{m,n}$ are **row equivalent** if one can be obtained from the other by a series of elementary row operations. We denote this by $A \sim B$.
1) Row reduction is an equivalence relation. That is,
   a) $A \sim A$
   b) $A \sim B \Rightarrow B \sim A$
   c) $A \sim B,\ B \sim C \Rightarrow A \sim C$.
2) Any matrix A is row equivalent to one and only one matrix R that is in reduced row echelon form. The matrix R is called the **reduced row echelon form of** A. Furthermore, we have

$$A = E_1 \cdots E_k R$$

where E_i are the elementary matrices required to reduce A to reduced row echelon form.

3) A is invertible if and only if R is an identity matrix. Hence, a matrix is invertible if and only if it is the product of elementary matrices. ∎

Determinants

We assume that the reader is familiar with the following basic properties of determinants.

Theorem 0.3 Let A be an $n \times n$ matrix over F. Then det(A) is an element of F. Furthermore,

1) $\det(AB) = \det(A)\det(B)$, for any $B \in \mathcal{M}_n(F)$.
2) A is nonsingular (invertible) if and only if $\det(A) \neq 0$.
3) The determinant of an upper triangular, or lower triangular, matrix is the product of the entries on its main diagonal.
4) Let A(i,j) denote the matrix obtained by deleting the ith row and jth column from A. The **adjoint** of A is the matrix *adj*(A) defined by
$$(adj(A))_{i,j} = (-1)^{i+j}\det(A(i,j))$$
If A is invertible, then
$$A^{-1} = \frac{1}{\det(A)} adj(A)$$
∎

Polynomials

If F is a field, then F[x] denotes the set of all polynomials in the variable x, with coefficients from F. If $p(x) \in F[x]$, we say that p(x) is a polynomial *over* F. If
$$p(x) = a_0 + a_1 x + \cdots + a_n x^n$$
is a polynomial, with $a_n \neq 0$, then a_n is called the **leading coefficient** of p(x), and the **degree** deg p(x) of p(x) is n. We will set the degree of the zero polynomial to $-\infty$. A polynomial is **monic** if its leading coefficient is 1.

Theorem 0.4 (Division algorithm) Let $f(x) \in F[x]$ and $g(x) \in F[x]$, where $\deg g(x) > 0$. Then there exist unique polynomials q(x) and r(x) in F[x] for which
$$f(x) = q(x)g(x) + r(x)$$
where $r(x) = 0$ or $0 \leq \deg r(x) < \deg g(x)$. ∎

If $p(x)$ divides $q(x)$, that is, if there exists a polynomial $f(x)$ for which

$$q(x) = f(x)p(x)$$

then we write $p(x) \mid q(x)$.

Theorem 0.5 Let $f(x)$ and $g(x)$ be polynomials over F. The **greatest common divisor** of $f(x)$ and $g(x)$, denoted by $\gcd(f(x),g(x))$, is the unique monic polynomial $p(x)$ over F for which

1) $p(x) \mid f(x)$ and $p(x) \mid g(x)$
2) if $r(x) \mid f(x)$ and $r(x) \mid g(x)$, then $r(x) \mid p(x)$.

Furthermore, there exist polynomials $a(x)$ and $b(x)$ over F for which

$$\gcd(f(x),g(x)) = a(x)f(x) + b(x)g(x)$$ ∎

Definition Let $f(x)$ and $g(x)$ be polynomials over F. If $\gcd(f(x),g(x)) = 1$, we say that $f(x)$ and $g(x)$ are **relatively prime.** In particular, $f(x)$ and $g(x)$ are relatively prime if and only if there exist polynomials $a(x)$ and $b(x)$ over F for which

$$a(x)f(x) + b(x)g(x) = 1$$ □

Definition A nonconstant polynomial $f(x) \in F[x]$ is **irreducible** if whenever $f(x) = p(x)q(x)$, then one of $p(x)$ or $q(x)$ must be constant. □

The following two theorems support the view that irreducible polynomials behave like prime numbers.

Theorem 0.6 If $f(x)$ is irreducible and $f(x) \mid p(x)q(x)$, then either $f(x) \mid p(x)$ or $f(x) \mid q(x)$. □

Theorem 0.7 Every nonconstant polynomial in $F[x]$ can be written as a product of irreducible polynomials. Moreover, this expression is unique up to order of the factors and multiplication by a scalar. □

Functions

To set our notation, we should make a few comments about functions.

Definition Let $f:S \to T$ be a function (map) from a set S to a set T.

1) The **domain** of f is the set S.
2) The **image** or **range** of f is the set $im(f) = \{f(s) \mid s \in S\}$.
3) f is **injective** (one-to-one), or an **injection**, if $x \neq y \Rightarrow f(x) \neq f(y)$.

4) f is **surjective** (onto T), or a **surjection**, if $im(f) = T$.
5) f is **bijective**, or a **bijection**, if it is both injective and surjective. □

If $f:S \to T$ is injective, then its inverse $f^{-1}:im(f) \to S$ exists and is well-defined. It will be convenient to apply $f:S \to T$ to subsets of S and T. In particular, if $X \subset S$, we set $f(X) = \{f(x) \mid x \in X\}$ and if $Y \subset T$, we set $f^{-1}(Y) = \{s \in S \mid f(s) \in Y\}$. Note that the latter is defined even if f is not injective.

If $X \subset S$, the **restriction** of $f:S \to T$ is the function $f|_X:X \to T$. Clearly, the restriction of an injective map is injective.

Equivalence Relations

The concept of an equivalence relation plays a major role in the study of matrices and linear transformations.

Definition Let S be a nonempty set. A binary relation $\sim$ on S is called an **equivalence relation** on S if it satisfies the following conditions.

1) **(reflexivity)**
$$a \sim a$$
for all $a \in S$.
2) **(symmetry)**
$$a \sim b \;\Rightarrow\; b \sim a$$
for all $a, b \in S$.
3) **(transitivity)**
$$a \sim b,\ b \sim c \;\Rightarrow\; a \sim c$$
for all $a, b, c \in S$. □

Definition Let $\sim$ be an equivalence relation on S. For $a \in S$, the set
$$[a] = \{b \in S \mid b \sim a\}$$
is called the **equivalence class** of a. □

Theorem 0.8 Let $\sim$ be an equivalence relation on S. Then

1) $b \in [a] \;\Leftrightarrow\; a \in [b] \;\Leftrightarrow\; [a] = [b]$
2) For any $a, b \in S$, we have either $[a] = [b]$ or $[a] \cap [b] = \emptyset$. ∎

Definition Let S be a nonempty set. A **partition** of S is a collection $\{A_1, \ldots, A_n\}$ of *nonempty* subsets of S, called **blocks**, for which

1) $A_i \cap A_j = \emptyset$, for all i,j
2) $S = A_1 \cup \cdots \cup A_n$. □

The following theorem sheds considerable light on the concept of an equivalence relation.

Theorem 0.9

1) Let $\sim$ be an equivalence relation on S. Then the set of distinct equivalence classes with respect to $\sim$ are the blocks of a partition of S.
2) Conversely, if $\mathcal{P}$ is a partition of S, the binary relation $\sim$ defined by
$$a \sim b \Leftrightarrow a \text{ and } b \text{ lie in the same block of } \mathcal{P}$$
is an equivalence relation on S, whose equivalence classes are the blocks of $\mathcal{P}$.

This establishes a one-to-one correspondence between equivalence relations on S and partitions of S. ∎

The most important problem related to equivalence relations is that of finding an *efficient* way to determine when two elements are equivalent. Unfortunately, in most cases, the definition does not provide an efficient test for equivalence, and so we are led to the following concepts.

Definition Let $\sim$ be an equivalence relation on S. A function $f:S \to T$, where T is any set, is called an **invariant** of $\sim$ if
$$a \sim b \Rightarrow f(a) = f(b)$$
A function $f:S \to T$ is a **complete invariant** if
$$a \sim b \Leftrightarrow f(a) = f(b)$$
A collection $f_1, \ldots, f_k$ of invariants is called a **complete system of invariants** if
$$a \sim b \Leftrightarrow f_i(a) = f_i(b) \text{ for all } i = 1, \ldots, n$$
□

Definition Let $\sim$ be an equivalence relation on S. A subset $C \subset S$ is said to be a set of **canonical forms** for $\sim$ if for every $s \in S$, there is *exactly* one $c \in C$ such that $c \sim s$. □

Example 0.1 Define a binary relation $\sim$ on $F[x]$ by letting $p(x) \sim q(x)$ if and only if there exists a nonzero constant $a \in F$ such that $p(x) = aq(x)$. This is easily seen to be an equivalence relation. The function that assigns to each polynomial its degree is an invariant, since
$$p(x) \sim q(x) \Rightarrow \deg(p(x)) = \deg(q(x))$$
However, it is not a complete invariant, since there are inequivalent

polynomials with the same degree. The set of all *monic* polynomials is a set of canonical forms for this equivalence relation. ▯

Example 0.2 We have remarked that row equivalence is an equivalence relation on $\mathcal{M}_{m,n}(F)$. Moreover, the subset of reduced row echelon form matrices is a set of canonical forms for row equivalence, since every matrix is row equivalent to a *unique* matrix in reduced row echelon form. ▯

Example 0.3 Two matrices $A, B \in \mathcal{M}_n(F)$ are row equivalent if and only if there is an invertible matrix P such that $A = PB$. Similarly, A and B are column equivalent (that is, A can be reduced to B using elementary column operations) if and only if there exists an invertible matrix Q such that $A = BQ$.

Two matrices A and B are said to be **equivalent** if there exists invertible matrices P and Q for which

$$A = PBQ$$

Put another way, A and B are equivalent if A can be reduced to B by performing a series of elementary row and/or column operations. (The use of the term equivalent is unfortunate, since it applies to all equivalence relations – not just this one. However, the terminology is standard, so we use it here.)

It is not hard to see that a square matrix R that is in both reduced row echelon form and reduced column echelon form must have the form

$$J_k = \begin{bmatrix} 1 & 0 & & \cdots & & 0 \\ 0 & \ddots & & \cdots & & 0 \\ & & 1 & & & \\ \vdots & & & 0 & & \vdots \\ & & & & \ddots & \\ 0 & 0 & 0 & 0 & 0 & 0 \end{bmatrix}$$

with 0s everywhere off the main diagonal, and k 1s, followed by $n - k$ 0s, on the main diagonal.

We leave it to the reader to show that every matrix A in $\mathcal{M}_n$ is equivalent to exactly one matrix of the form J_k, and so the set of these matrices is a set of canonical forms for equivalence. Moreover, the function f defined by $f(A) = k$, where $A \sim J_k$, is a complete invariant for equivalence.

Since the rank of J_k is k, and since neither row nor column operations affect the rank, we deduce that the rank of A is k. Hence, rank is a complete invariant for equivalence. ▯

Example 0.4 Two matrices $A, B \in \mathcal{M}_n(F)$ are said to be **similar** if there exists an invertible matrix P such that

$$A = PBP^{-1}$$

Similarity is easily seen to be an equivalence relation on $\mathcal{M}_n$. As we will learn, two matrices are similar if and only if they represent the same linear operators on a given n-dimensional vector space V. Hence, similarity is extremely important for studying the structure of linear operators. One of the main goals of this book is to develop canonical forms for similarity.

We leave it to the reader to show that the determinant function and the trace function are invariants for similarity. However, these two invariants do not, in general, form a complete system of invariants. □

Example 0.5 Two matrices $A, B \in \mathcal{M}_n(F)$ are said to be **congruent** if there exists an invertible matrix P for which

$$A = PBP^{T}$$

where P^T is the transpose of P. This relation is easily seen to be an equivalence relation, and we will devote some effort to finding canonical forms for congruence. For some base fields F (such as $\mathbb{R}$, $\mathbb{C}$ or a finite field), this is relatively easy to do, but for other base fields (such as $\mathbb{Q}$), it is extremely difficult. □

Zorn's Lemma

In order to show that any vector space has a basis, we require a result known as Zorn's lemma. To state this lemma, we need some preliminary definitions.

Definition A **partially ordered set** is a nonempty set P, together with a partial order defined on P. A **partial order** is a binary relation, denoted by $\leq$ and read "less than or equal to," with the following properties.

1) (**reflexivity**) For all $a \in P$,

$$a \leq a$$

2) (**antisymmetry**) For all $a,b \in P$,

$$a \leq b \quad \text{and} \quad b \leq a \quad \text{implies} \quad a = b$$

3) (**transitivity**) For all $a,b,c \in P$,

$$a \leq b \quad \text{and} \quad b \leq c \quad \text{implies} \quad a \leq c$$

□

Definition If P is a partially ordered set and if $m \in P$ has the property that $m \leq p$ implies $m = p$, then m is called a **maximal element** in P. ▯

Definition Let P be a partially ordered set and let $a,b \in P$. If there is a $u \in P$ with the property that

1) $a \leq u$ and $b \leq u$, and
2) if $a \leq x$ and $b \leq x$, then $u \leq x$

then we say that u is the **least upper bound** of a and b, and write $u = \text{lub}\{a,b\}$. If there is an element $\ell \in P$ with the property that

3) $\ell \leq a$ and $\ell \leq b$, and
4) if $x \leq a$ and $x \leq b$, then $x \leq \ell$

then we say that ℓ is the **greatest lower bound** of a and b, and write $\ell = \text{glb}\{a,b\}$. ▯

Note that in a partially ordered set, it is possible that not all elements are comparable. In other words, it is possible to have $x,y \in P$ with the property that $x \nleq y$ and $y \nleq x$. A partially ordered set in which every pair of elements is comparable is called a **totally ordered set**, or a **linearly ordered set.** Any totally ordered subset of a partially ordered set P is called a **chain** in P.

Let S be a subset of a partially ordered set P. We say that an element $u \in P$ is an **upper bound** for S if $s \leq u$ for all $s \in S$.

Example 0.6

1) The set $\mathbb{R}$ of real numbers, with the usual binary relation $\leq$ is a partially ordered set. It is also a totally ordered set. It has no maximal element.
2) The set $\mathbb{N}$ of natural numbers, together with the binary relation of divides, is a partially ordered set. It is customary to write $n \mid m$ to indicate that n divides m. The subset S of $\mathbb{N}$ consisting of all powers of 2 is a totally ordered subset of $\mathbb{N}$, that is, it is a chain in $\mathbb{N}$. The set $P = \{2,4,8,3,9,27\}$ is a partially ordered set under $\mid$. It has two maximal elements, namely 8 and 27.
3) Let S be any set, and let $\mathcal{P}(S)$ be the power set of S, that is, the set of all subsets of S. Then $\mathcal{P}(S)$, together with the subset relation $\subseteq$, is a partially ordered set. ▯

Now we can state Zorn's lemma.

Theorem 0.10 **(Zorn's lemma)** Let P be a partially ordered set in which every chain has an upper bound. Then P has a maximal element. ▮

The reader who is interested in looking at an example of the use of Zorn's lemma now might wish to refer to the proof in Chapter 1 that every vector space has a basis.

Cardinality

We will say that two sets S and T have the same *cardinality*, and write

$$|S| = |T|$$

if there is a bijective function (a one-to-one correspondence) between the sets. The reader is probably aware of the fact that

$$|\mathbb{Z}| = |\mathbb{N}| \quad \text{and} \quad |\mathbb{Q}| = |\mathbb{N}|$$

where $\mathbb{N}$, $\mathbb{Z}$ and $\mathbb{Q}$ are the natural numbers, integers, and rational numbers, respectively.

If S is in one-to-one correspondence with a subset of T, we write $|S| \leq |T|$. If S is in one-to-one correspondence with a proper subset of T, and if $|S| \neq |T|$, we write $|S| < |T|$. The second condition is necessary, since, for instance, $\mathbb{N}$ is in one-to-one correspondence with a proper subset of $\mathbb{Z}$, and yet $|\mathbb{N}| \not< |\mathbb{Z}|$.

This is not the place to enter into a detailed discussion of cardinal numbers. The intention here is that the cardinality of a set, whatever that is, represents the "size" of the set, and it happens that it is much easier to talk about two sets having the same, or different, size (cardinality) than it is to explicitly define the size (cardinality) of a given set.

Be that as it may, we associate to each set S a cardinal number, denoted by $|S|$ or *card*(S), that is intended to measure the size of the set. Actually, cardinal numbers are just very special types of sets. However, we can simply think of them as vague amorphous objects that measure the size of sets.

A set is **finite** if it can be put in one-to-one correspondence with a set of the form $\mathbb{Z}_n = \{0,1,\ldots,n-1\}$, for some positive integer n. The cardinal number (or cardinality) of a finite set is just the number of elements in the set. The cardinal number of the set $\mathbb{N}$ of natural numbers is $\aleph_0$ (read "aleph nought"), where $\aleph$ is the first letter of the Hebrew alphabet . Hence,

$$|\mathbb{N}| = |\mathbb{Z}| = |\mathbb{Q}| = \aleph_0$$

Any set with cardinality $\aleph_0$ is called a **countably infinite** set, and any finite or countably infinite set is called a **countable** set.

Since it can be shown that $|\mathbb{R}| > |\mathbb{N}|$, the real numbers are not countable.

If S and T are finite sets, then it is well known that

$$|S| \leq |T| \text{ and } |T| \leq |S| \Rightarrow |S| = |T|$$

The first part of the next theorem tells us that this is also true for infinite sets.

The reader will no doubt recall that the **power set** $\mathcal{P}(S)$ of a set S is the set of all subsets of S. For finite sets, the power set of S is always bigger than the set itself. In fact,

$$|S| = n \Rightarrow |\mathcal{P}(S)| = 2^n$$

The second part of the next theorem says that the power set of *any* set S is bigger than S itself. On the other hand, the third part of this theorem says that, for infinite sets S, the set of all *finite* subsets of S is the same size as S.

Theorem 0.11

1) **(Schröder-Bernstein theorem)** For any sets S and T,

$$|S| \leq |T| \text{ and } |T| \leq |S| \Rightarrow |S| = |T|$$

2) **(Cantor's theorem)** If $\mathcal{P}(S)$ denotes the power set of S, then

$$|S| < |\mathcal{P}(S)|$$

3) If $\mathcal{P}_0(S)$ denotes the set of all *finite* subsets of S, and if S is an *infinite* set, then

$$|S| = |\mathcal{P}_0(S)|$$

Proof. We prove only parts (1) and (2).

1) To prove the Schröder-Bernstein theorem, we follow the proof of Halmos [1960]. Let $f:S \to T$ be an injective function from S into T, and let $g:T \to S$ be an injective function from T into S. We want to show that there is a bijective function from S to T. For this purpose, we make the following definitions. An element $s \in S$ has **descendants**

$$f(s), g(f(s)), f(g(f(s))), \ldots$$

If t is a descendant of s, then s is an **ancestor** of t. We define descendants of t and ancestors of s similarly. Now, by tracing an element's ancestry to its beginning, we find that there are three possibilities – the element may originate in S, or in T, or it may have no originator. Accordingly, we can write S as the union of three *disjoint* sets

$$S_S = \{s \in S \mid s \text{ originates in } S\}$$
$$S_T = \{s \in S \mid s \text{ originates in } T\}$$

and

$$S_\infty = \{s \in S \mid s \text{ has no originator}\}$$

Similarly, we write T as the disjoint union of T_S, T_T and T_∞.

Now, the restriction

$$f|_{S_S}: S_S \to T_S$$

is a bijection. For if $t \in T_S$, then $t = f(s')$, for some $s' \in S$. But s' and t have the same originator, and so $s' \in S_S$. We leave it to the reader to show that the functions

$$g|_{T_T}: T_T \to S_T \text{ and } f|_{S_\infty}: S_\infty \to T_\infty$$

are also bijections. Putting these three bijections together gives a bijection between S and T. Hence, $|S| = |T|$.

2) The inclusion map $\epsilon: S \to \mathcal{P}(S)$ defined by $\epsilon(s) = \{s\}$ is an injection from S to $\mathcal{P}(S)$, and so $|S| \le |\mathcal{P}(S)|$. To complete the proof of Cantor's theorem, we must show that if $f: S \to \mathcal{P}(S)$ is any injection, then f is *not* surjective. To this end, let

$$X = \{s \in S \mid s \notin f(s)\}$$

Then $X \in \mathcal{P}(S)$, and we now show that X is *not* in *im*(f). For suppose that $X = f(x)$ for some $x \in X$. Then if $x \in X$, we have by definition of X that $x \notin X$. On the other hand, if $x \notin X$, we have again by definition of X that $x \in X$. This contradiction implies that $X \notin im(f)$, and so f is not surjective. ∎

Now let us define addition, multiplication and exponentiation of cardinal numbers. If S and T are sets, the **cartesian product** $S \times T$ is the set of all ordered pairs

$$S \times T = \{(s,t) \mid s \in S, t \in T\}$$

Also, we let S^T denote the set of all functions from T to S.

Definition Let κ and λ denote cardinal numbers.

1) The sum $\kappa + \lambda$ is the cardinal number of $S \cup T$, where S and T are any *disjoint* sets for which $|S| = \kappa$ and $|T| = \lambda$.
2) The product $\kappa\lambda$ is the cardinal number of $S \times T$, where S and T are any sets for which $|S| = \kappa$ and $|T| = \lambda$.
3) The power κ^λ is the cardinal number of S^T, where S and T are any sets for which $|S| = \kappa$ and $|T| = \lambda$. □

We will not go into the details of why these definitions make sense. (For instance, they *seem* to depend on the sets S and T, but in fact, they do not.) It can be shown, using these definitions, that cardinal addition and multiplication is associative, commutative and that multiplication distributes over addition.

Theorem 0.12 Let κ, λ and μ be cardinal numbers. Then the following properties hold.

1) **(Associativity)**

$$\kappa+(\lambda+\mu)=(\kappa+\lambda)+\mu \quad \text{and} \quad \kappa(\lambda\mu)=(\kappa\lambda)\mu$$

2) **(Commutativity)**

$$\kappa+\lambda=\lambda+\kappa \quad \text{and} \quad \kappa\lambda=\lambda\kappa$$

3) **(Distributivity)**

$$\kappa(\lambda+\mu)=\kappa\lambda+\kappa\mu$$

4) **(Properties of Exponents)**

a) $\kappa^{\lambda+\mu}=\kappa^{\lambda}\kappa^{\mu}$
b) $(\kappa^{\lambda})^{\mu}=\kappa^{\lambda\mu}$
c) $(\kappa\lambda)^{\mu}=\kappa^{\mu}\lambda^{\mu}$ ∎

On the other hand, the arithmetic of cardinal numbers can seem a bit strange at first.

Theorem 0.13 Let κ and λ be cardinal numbers. Then
1) $\kappa+\lambda=\max\{\kappa,\lambda\}$
2) $\kappa\lambda=\max\{\kappa,\lambda\}$ ∎

It is not hard to see that there is a one-to-one correspondence between the power set $\mathcal{P}(S)$ of a set S and the set of all functions from S to $\{0,1\}$. This leads to the following theorem.

Theorem 0.14
1) If $|S|=\kappa$ then $|\mathcal{P}(S)|=2^{\kappa}$
2) $\kappa<2^{\kappa}$ ∎

We have already observed that $|\mathbb{N}|=\aleph_0$. It can be shown that $\aleph_0$ is the smallest infinite cardinal, that is,

$$\kappa<\aleph_0 \ \Rightarrow\ \kappa \text{ is a natural number}$$

It can also be shown that the set $\mathbb{R}$ of real numbers is in one-to-one correspondence with the power set $\mathcal{P}(\mathbb{N})$ of the natural numbers. Therefore,

$$|\mathbb{R}|=2^{\aleph_0}$$

The set of all points on the real line is sometimes called the *continuum*, and so $2^{\aleph_0}$ is sometimes called the *power of the continuum*, and denoted by c.

Theorem 0.13 shows that cardinal addition and multiplication has a kind of "absorption" quality, which makes it hard to produce larger cardinals from smaller ones. The next theorem demonstrates this more dramatically.

Theorem 0.15

1) Addition and multiplication, applied a finite number of times to the cardinal number $\aleph_0$, do not yield anything more than $\aleph_0$. Specifically, for any nonzero $n \in \mathbb{N}$,

$$n \cdot \aleph_0 = \aleph_0 \quad \text{and} \quad \aleph_0^n = \aleph_0$$

2) Addition and multiplication, applied a countable number of times to the cardinal number $2^{\aleph_0}$ do not yield more than $2^{\aleph_0}$. Specifically, we have

$$\aleph_0 \cdot 2^{\aleph_0} = 2^{\aleph_0} \quad \text{and} \quad (2^{\aleph_0})^{\aleph_0} = 2^{\aleph_0}$$ ∎

Using this theorem, we can establish other relationships, such as

$$2^{\aleph_0} \le (\aleph_0)^{\aleph_0} \le (2^{\aleph_0})^{\aleph_0} = 2^{\aleph_0}$$

which, by the Schröder-Bernstein theorem, implies that

$$(\aleph_0)^{\aleph_0} = 2^{\aleph_0}$$

We mention that the problem of evaluating κ^λ in general is a very difficult one, and would take us far beyond the scope of this book.

We will have use for the following result, whose proof is omitted.

Theorem 0.16 Let $\{A_k \mid k \in K\}$ be a collection of sets, indexed by the set K, with $|K| = \kappa$. If $|A_k| \le \lambda$ for all $k \in K$, then

$$\left| \bigcup_{k \in K} A_k \right| \le \lambda\kappa$$ ∎

Let us conclude by describing the cardinality of some famous sets.

Theorem 0.17

1) The following sets have cardinality $\aleph_0$.
   a) The rational numbers $\mathbb{Q}$.
   b) The set of all *finite* subsets of $\mathbb{N}$.
   c) The union of a countable number of countable sets.
   d) The set $\mathbb{Z}^n$ of all ordered n-tuples of integers.
2) The following sets have cardinality $2^{\aleph_0}$.
   a) The set of all points in $\mathbb{R}^n$.
   b) The set of all infinite sequences of natural numbers.
   c) The set of all infinite sequences of real numbers.

d) The set of all *finite* subsets of $\mathbb{R}$.
e) The set of all irrational numbers. ∎

Part 2 Algebraic Structures

Groups

Definition A **group** is a nonempty set G, together with a binary operation denoted by $*$, which satisfies the following properties.

1) (**associativity**) For all $a,b,c \in G$

$$(a*b)*c = a*(b*c)$$

2) (**identity**) There exists an element $e \in G$ for which

$$e*a = a*e = a$$

for all $a \in G$.

3) (**inverses**) For each $a \in G$, there is an element $a^{-1} \in G$ for which

$$a*a^{-1} = a^{-1}*a = e$$

□

Definition A group G is **abelian**, or **commutative**, if $a*b = b*a$, for all $a,b \in G$. When a group is abelian, it is customary to denote the operation $*$ by $+$, thus writing $a*b$ as $a+b$. It is also customary to refer to the identity as a **zero element**, and to denote the inverse a^{-1} by $-a$, referred to as the **negative** of a. □

Example 0.7 The set $\mathcal{F}$ of all *bijective* functions from a set S to S, is a group under composition of functions. □

Example 0.8 The set $\mathcal{M}_{m,n}(F)$ is an abelian group under addition of matrices. The identity is the zero matrix $\mathbf{0}_{n,m}$ of size $m \times n$.

The set $\mathcal{M}_n(F)$ is not a group under multiplication of matrices, since not all matrices have multiplicative inverses. However, the set of invertible matrices of size $n \times n$ is a nonabelian group under multiplication. □

A group G is **finite** if it contains only a finite number of elements. The cardinality of a finite group G is called its **order** and is denoted by $o(G)$. Thus, for example, Z_n is a finite group, but $\mathcal{M}_{n,m}(\mathbb{R})$ is not finite.

Rings

Definition A **ring** is a nonempty set R, together with two binary operations, called *addition* (denoted by +), and *multiplication* (denoted by juxtaposition), for which the following hold.

1) R is an abelian group under addition
2) (**associativity**) For all $a,b,c \in R$,

$$(ab)c = a(bc)$$

3) (**distributivity**) For all $a,b,c \in R$,

$$(a+b)c = ac + bc \quad \text{and} \quad c(a+b) = ca + cb$$

□

Definition A ring R is said to be **commutative** if $ab = ba$ for all $a,b \in R$. If a ring R contains an element e with the property that

$$ae = ea = a$$

for all $a \in R$, we say that R is a **ring with identity**. The identity e is usually denoted by 1. □

Example 0.9 The set $\mathbb{Z}_n = \{0,1,\ldots,n-1\}$ is a commutative ring under addition and multiplication modulo n

$$a \oplus_2 b = (a+b) \bmod n, \qquad a \odot_2 b = ab \bmod n$$

The element $1 \in \mathbb{Z}_n$ is the identity.

Example 0.10 The set of even integers $E \subset \mathbb{Z}$ is a commutative ring under the usual operations on $\mathbb{Z}$, but it has no identity. □

Example 0.11 The set $\mathcal{M}_n(F)$ is a *noncommutative* ring under matrix addition and multiplication. The identity matrix I_n is the identity for $\mathcal{M}_n(F)$. □

Example 0.12 Let F be a field. The set F[x] of all polynomials in a single variable x, with coefficients in F, is a commutative ring, under the usual operations of polynomial addition and multiplication. What is the identity for F[x]? □

Definition A **subring** of a ring R is a subset S of R that is a ring in its own right, using the same operations as defined on R. □

Applying the definition is not generally the easiest way to show that a subset of a ring is a subring. The following characterization is usually easier to apply.

Theorem 0.18 A nonempty subset S of a ring R is a subring if and only if

1) S is *closed under subtraction*, that is
$$a,b \in S \Rightarrow a-b \in S$$

2) S is *closed under multiplication*, that is,
$$a,b \in S \Rightarrow ab \in S$$
∎

Integral Domains

Definition Let R be a ring. A *nonzero* element $r \in R$ is called a **zero divisor** if there exists a nonzero $s \in R$ for which $rs = 0$. A commutative ring R with identity is called an **integral domain** if it contains no zero divisors. □

Example 0.13 If n is *not* a prime number, then the ring Z_n has zero divisors, and so is not an integral domain. To see this, observe that if n is not prime, then $n = ab$ in $\mathbb{Z}$, where $a,b \geq 2$. But in $\mathbb{Z}_n$, we have
$$a \odot_n b = ab \bmod n = n \bmod n = 0$$
and so a and b are both zero divisors. As we will see later, if n is a prime, then $\mathbb{Z}_n$ is an integral domain. □

Example 0.14 The ring $F[x]$ is an integral domain, since $p(x)q(x) = 0$ implies that $p(x) = 0$ or $q(x) = 0$. □

If R is a ring and $rx = ry$ for $r,x,y \in R$, then we cannot in general, cancel the r's, and conclude that $x = y$. For instance, in $\mathbb{Z}_4$, we have $2 \cdot 3 = 2 \cdot 1$, but we cannot cancel the 2's, to get $3 = 1$. However, it is precisely the integral domains in which we can cancel.

Theorem 0.19 Let R be a commutative ring with identity. Then R is an integral domain if and only if the **cancellation law**
$$r \neq 0 \text{ and } rx = ry \Rightarrow x = y$$
holds in R.

Proof. Suppose that R is an integral domain. Then
$$r \neq 0 \text{ and } rx = ry \Rightarrow r(x-y) = 0 \Rightarrow x - y = 0 \Rightarrow x = y$$
Conversely, suppose that the cancellation law holds and that $ab = 0$. If $a \neq 0$, then we have $ab = a0$, and so $b = 0$. Hence, R is an integral domain. ∎

Ideals and Principal Ideal Domains

Rings have another important substructure, besides subrings.

Definition Let R be a ring. A subset $\mathfrak{I}$ of R is called an **ideal** if
1) $\mathfrak{I}$ is *closed under subtraction*, that is

$$a,b \in R \Rightarrow a-b \in R$$

2) $\mathfrak{I}$ is closed under multiplication by *any* ring element, that is,

$$a \in \mathfrak{I},\ r \in R \Rightarrow ar \in \mathfrak{I} \text{ and } ra \in \mathfrak{I}$$

□

Observe that a subring is closed under multiplication, in the sense that the product of two elements *in the subring* is also in the subring. However, an ideal has a stronger closure property, namely, the product of an element in the ideal and *any element in the ring* is in the ideal.

Example 0.15 Let $p(x)$ be a polynomial in $F[x]$. The set of all multiples of $p(x)$

$$\langle p(x)\rangle = \{q(x)p(x) \mid q(x) \in F[x]\}$$

is an ideal in $F[x]$. □

Definition Let S be a *subset* of a ring R with identity. The set

$$\langle s_1,\ldots,s_n\rangle = \{r_1s_1 + \cdots + r_ns_n \mid r_i \in R,\ s_i \in S\}$$

is an ideal in R, called the **ideal generated** by S. It is the *smallest* (in the sense of set inclusion) ideal of R containing S. □

Note that in the previous definition, we require that R have an identity. This is to insure that, for example, $s \in \langle s\rangle$.

Definition Let R be a ring with identity, and let $a \in R$. The **principal ideal** generated by a is the ideal

$$\langle a\rangle = \{ra \mid r \in R\}$$

□

We will use the following algebraic structure quite a bit in the sequel.

Theorem 0.20 Let R be a ring.
1) The intersection of any collection $\{\mathfrak{I}_k \mid k \in K\}$ of ideals is an ideal.
2) If $\mathfrak{I}_1 \subset \mathfrak{I}_2 \subset \cdots$ is an *ascending* sequence of ideals, each one contained in the next, then the union $\bigcup \mathfrak{I}_k$ is also an ideal.

Proof. To prove (1), let $\mathfrak{J} = \bigcap \mathfrak{I}_k$. Then if $a,b \in \mathfrak{J}$, we have $a,b \in \mathfrak{I}_k$

for all $k \in K$. Hence, $a - b \in \mathfrak{I}_k$ for all $k \in K$, and so $a - b \in \mathfrak{I}$. Hence, $\mathfrak{I}$ is closed under subtraction. Also, if $r \in R$, then $ra \in \mathfrak{I}_k$ for all $k \in K$, and so $ra \in \mathfrak{I}$.

To prove (2), observe that if $a,b \in \bigcup \mathfrak{I}_k$, then $a \in \mathfrak{I}_i$ and $b \in \mathfrak{I}_j$ for some $i, j \in \mathbb{N}$. Hence, if $m = \max\{i,j\}$, we have $a,b \in \mathfrak{I}_m$, and so $a - b \in \mathfrak{I}_m \subset \bigcup \mathfrak{I}_k$. Hence, $\bigcup \mathfrak{I}_k$ is closed under subtraction. Also, if $r \in R$ and $a \in \bigcup \mathfrak{I}_k$, then $a \in \mathfrak{I}_i$ for some $i \in \mathbb{N}$, and so $ra \in \mathfrak{I}_i \subset \bigcup \mathfrak{I}_k$. Thus, $\bigcup \mathfrak{I}_k$ is closed under multiplication by any ring element, and so it is an ideal. ∎

Note that in general the union of ideals is not an ideal. However, as we have just proved, the union of an *ascending* chain of ideals is an ideal.

Definition An integral domain R in which every ideal is a principal ideal is called a **principal ideal domain.** □

Theorem 0.21 The integers form a principal ideal domain. In fact, an ideal $\mathfrak{I}$ in R is generated by the smallest positive integer a that is contained in $\mathfrak{I}$. ∎

Theorem 0.22 The ring $F[x]$ is a principal ideal domain. In fact, any ideal $\mathfrak{I}$ is generated by the unique monic polynomial of smallest degree contained in $\mathfrak{I}$. Moreover, for polynomials $p_1,\ldots,p_n$,

$$\langle p_1,\ldots,p_n\rangle = \langle \gcd\{p_1,\ldots,p_n\}\rangle$$

Proof. Let $\mathfrak{I}$ be an ideal in $F[x]$, and let $m(x)$ be a monic polynomial of smallest degree in $\mathfrak{I}$. First, we observe that there is only one such polynomial in $\mathfrak{I}$. For if $n(x) \in \mathfrak{I}$ is monic, and $\deg n(x) = \deg m(x)$, then

$$b(x) = m(x) - n(x) \in \mathfrak{I}$$

and since $\deg b(x) < \deg m(x)$, we must have $b(x) = 0$, and so $n(x) = m(x)$.

Now, let us show that $\mathfrak{I}$ is generated by $m(x)$. Since $\mathfrak{I}$ is an ideal, and $m(x) \in \mathfrak{I}$, we have

$$\langle m(x)\rangle \subset \mathfrak{I}$$

To establish the reverse inclusion, if $p(x) \in \langle m(x)\rangle$, then dividing $p(x)$ by $m(x)$ gives

$$p(x) = q(x)m(x) + r(x)$$

where $r(x) = 0$ or $0 \leq \deg r(x) < \deg m(x)$. But since $\mathfrak{I}$ is an ideal,

$$r(x) = p(x) - q(x)m(x) \in \mathfrak{I}$$

and so $0 \leq \deg r(x) < \deg m(x)$ is impossible. Hence, $r(x) = 0$, and

$$p(x) = q(x)m(x) \in \langle m(x) \rangle$$

This shows that $\mathfrak{I} \subset \langle m(x) \rangle$, and so $\mathfrak{I} = \langle m(x) \rangle$.

To prove the second statement, let $\mathfrak{I} = \langle p_1(x), \ldots, p_n(x) \rangle$. Then, by what we have just shown,

$$\mathfrak{I} = \langle p_1(x), \ldots, p_n(x) \rangle = \langle m(x) \rangle$$

for the unique monic polynomial $m(x)$ in $\mathfrak{I}$ of smallest degree. In particular, since $p_i(x) \in \langle m(x) \rangle$, we have

$$p_i(x) = a_i(x)m(x)$$

for some polynomial $a_i(x)$, and so $m(x) \mid p_i(x)$, for each $i = 1, \ldots, n$. In other words, $m(x)$ is a common divisor of the $p_i(x)$'s.

Moreover, if $q(x) \mid p_i(x)$, for all i, then each $p_i(x)$ is a multiple of $q(x)$, and so

$$p_i(x) \in \langle q(x) \rangle$$

for all i, which implies that

$$\langle m(x) \rangle = \langle p_1(x), \ldots, p_n(x) \rangle \subset \langle q(x) \rangle$$

In particular, this implies that $m(x) \in \langle q(x) \rangle$, and so $q(x) \mid m(x)$. This shows that $m(x)$ is the *greatest* common divisor of the $p_i(x)$'s, and completes the proof. ∎

Example 0.16 Let $R = F[x,y]$ be the ring of polynomials in two variables x and y. Then R is *not* a principal ideal domain. To see this, observe that the subring $\mathfrak{I}$ of all polynomials with zero constant term is an ideal in R. Also, $x \in \mathfrak{I}$ and $y \in \mathfrak{I}$. Now, suppose that $\mathfrak{I}$ is the principal ideal $\mathfrak{I} = (p(x,y))$. Then there exist polynomials $a(x,y)$ and $b(x,y)$ for which

$$x = a(x,y)p(x,y) \quad \text{and} \quad y = b(x,y)p(x,y) \tag{0.1}$$

But if $p(x,y)$ is a constant polynomial, then $\mathfrak{I} = (p(x,y))$ is all of R, which is not the case. Hence, $deg(p(x,y)) \geq 1$, and so $a(x,y)$ and $b(x,y)$ must both be constants, which implies that (0.1) cannot possibly hold. □

Prime Elements

We can define the notion of a prime element in any integral domain. For $r,s \in R$, we say that r **divides** s (written $r \mid s$) if there exists an $x \in R$ for which $s = xr$.

Definition Let R be an integral domain.

1) An invertible element of R is called a **unit.** Thus, $u \in R$ is a unit if $uv = 1$ for some $v \in R$.
2) Two elements $a,b \in R$ are said to be **associates** if there exists a unit u for which $a = ub$.
3) A nonzero *nonunit* $p \in R$ is said to be **prime** if $p \mid ab \Rightarrow p \mid a$ or $p \mid b$.
4) A nonzero *nonunit* $r \in R$ is said to be **irreducible** if $r = ab \Rightarrow$ either a or b is a unit. □

Theorem 0.23

1) An element $u \in R$ is a unit if and only if $\langle u \rangle = R$.
2) r and s are associates if and only if $\langle r \rangle = \langle s \rangle$.
3) r divides s if and only if $\langle s \rangle \subset \langle r \rangle$.
4) r *properly* divides s (that is, $s = xr$ where x is *not* a unit) if and only if $\langle s \rangle \subsetneq \langle r \rangle$. ∎

In the case of the integers, an integer is prime if and only if it is irreducible. However, this is not the case in general. But it is true for principal ideal domains.

Theorem 0.24 Let R be a principal ideal domain.

1) If $r \in R$ is irreducible, then the principal ideal $\langle r \rangle$ is **maximal**, that is, $\langle r \rangle \neq R$ and there is no ideal $\langle a \rangle$ for which $\langle r \rangle \subsetneq \langle a \rangle \subsetneq R$.
2) An element in R is prime if and only if it is irreducible.
3) Any $r \in R$ can be written as a product

$$r = up_1 \cdots p_n$$

where u is a unit, and $p_1, \ldots, p_n$ are primes. Furthermore, this factorization is unique up to order, and unit element u.

Proof. To prove (1), suppose that r is irreducible, and that $\langle r \rangle \subset \langle a \rangle \subset R$. Then $r \in \langle a \rangle$, and so $r = xa$ for some $x \in R$. The irreducibility of r now implies that a or x is a unit. But if a is a unit, then $\langle a \rangle = R$, and if x is a unit, then $\langle a \rangle = \langle xa \rangle = \langle r \rangle$. This shows that $\langle r \rangle$ is maximal. (We have $\langle r \rangle \neq R$, since r is not a unit.)

To prove (2), assume first that p is prime, and let $p = ab$. Then $p \mid ab$, and so $p \mid a$ or $p \mid b$. We may assume that $p \mid a$. Therefore, $a = xp$, and $p = ab = xpb$. Canceling p's, we get $1 = xb$, and so b is a unit. Hence, p is irreducible. (Note that this argument applies in any integral domain.)

Conversely, suppose that r is irreducible, and let $r \mid ab$. We wish to prove that $r \mid a$ or $r \mid b$. In the terminology of ideals, we assume that $ab \in \langle r \rangle$, where by part (1), $\langle r \rangle$ is maximal, and we want to show that $a \in \langle r \rangle$ or $b \in \langle r \rangle$. But

$$a \notin \langle r \rangle \Rightarrow \langle a,r \rangle = R \Rightarrow 1 = xa + yr, \quad \text{for some } x,y \in R$$

and

$$b \notin \langle r \rangle \Rightarrow \langle b,r \rangle = R \Rightarrow 1 = x'b + y'r, \quad \text{for some } x',y' \in R$$

From this, we get

$$1 = (xa + yr)(x'b + y'r) = xx'ab + xy'ar + yx'br + yy'r^2 \in \langle r \rangle$$

which implies that r is a unit. This contradiction shows that $a \in \langle r \rangle$ or $b \in \langle r \rangle$.

To prove (3), let $r \in R$. If r is irreducible, then we are done. If not, then $r = r_1 r_2$, where neither factor is a unit. If r_1 and r_2 are irreducible, we are done. If not, suppose that r_2 is not irreducible. Then $r_2 = r_3 r_4$, where neither r_3 nor r_4 is a unit. Continuing in this way, we obtain a factorization of the form (after renumbering if necessary)

$$r = r_1 r_2 = r_1(r_3 r_4) = (r_1 r_3)(r_5 r_6) = (r_1 r_3 r_5)(r_7 r_8) = \cdots \tag{0.2}$$

Each step is a factorization of r into a product of nonunits. However, this process must stop after a finite number of steps. To see this, observe that since

$$r_2 \mid r, \quad r_4 \mid r_2, \quad r_6 \mid r_4, \ \ldots$$

the sequence (0.2) gives rise to an *ascending* sequence of ideals

$$\langle r \rangle \subset \langle r_2 \rangle \subset \langle r_4 \rangle \subset \langle r_6 \rangle \cdots$$

Moreover, since none of the r_i's is a unit, the inclusions in this chain are proper. Now, if the factorization process did not stop, we would obtain an *infinite* ascending sequence of such ideals. But, according to Theorem 0.20, the union $\mathcal{U}$ of all of these ideals would be another ideal in R, which must be principal. Suppose that $\mathcal{U} = \langle a \rangle$. Then $a \in \mathcal{U}$ and so $a \in \langle r_{2n} \rangle$ for some n. But this is not possible, since it would imply that

$$\mathcal{U} = \langle a \rangle \subset \langle r_{2n} \rangle$$

which implies that $\langle r_{2n} \rangle = \langle r_{2(n+1)} \rangle = \cdots$, contradicting the fact that the inclusions are proper. ∎

Fields

In a ring, addition is "stronger" than multiplication, in the sense that it must possess more properties. In a field, the two operations have essentially the same strength.

Definition A **field** is a set F, containing at least two elements, together with two binary operations, called *addition* (denoted by $+$) and *multiplication* (denoted by juxtaposition), for which the following hold.

1) F is an abelian group under addition.
2) The set F^* of all *nonzero* elements in F is an abelian group under multiplication.
3) (**distributivity**) For all $a,b,c \in F$,

$$(a+b)c = ac + bc \quad \text{and} \quad c(a+b) = ca + cb$$ ▯

We require that F have at least two elements to avoid the pathological case where $0 = 1$.

Example 0.17 The sets $\mathbb{Q}$, $\mathbb{R}$ and $\mathbb{C}$, of all rational, real and complex numbers, respectively, are fields, under the usual operations of addition and multiplication of numbers. ▯

Example 0.18 The ring $\mathbb{Z}_n$ is a field if and only if n is a prime number. We have already seen that $\mathbb{Z}_n$ is not a field if n is not prime, since a field is also an integral domain. Now suppose that $n = p$ is a prime.

We have seen that $\mathbb{Z}_p$ is an integral domain, and so it remains to show that every nonzero element in $\mathbb{Z}_p$ has a multiplicative inverse. Let $0 \neq a \in \mathbb{Z}_p$. Since $a < p$, we know that a and b are relatively prime. It follows that there exists integers u and v for which

$$ua + vp = 1$$

Hence,

$$ua \equiv (1 - vp) \equiv 1 \bmod p$$

and so $u \odot_p a = 1$ in $\mathbb{Z}_p$, that is, u is the multiplicative inverse of a. ▯

The previous example shows that not all fields are infinite sets. In fact, *finite fields* play an extremely important role in many areas of abstract and applied mathematics.

The Characteristic of a Ring

Let R be a ring. If n is a positive integer, then by $n \cdot r$, we simply mean

$$n \cdot r = \underbrace{r + \cdots + r}_{n \text{ terms}}$$

Now, it may happen that there is a positive integer c for which

$$c \cdot 1 = \underbrace{1 + \cdots + 1}_{c \text{ terms}} = 0$$

For instance, in $\mathbb{Z}_n$, we have $n \cdot 1 = n = 0$. On the other hand, in $\mathbb{Z}$, $c \cdot 1 = 0$ implies $c = 0$, and so no such *positive* integer exists.

Notice that, in any *finite* ring or field, there must exist such a positive integer c, since the infinite sequence of numbers

$$1 \cdot 1,\ 2 \cdot 1,\ 3 \cdot 1,\ \ldots$$

cannot be distinct, and so $i \cdot 1 = j \cdot 1$ for some $i \neq j$. Hence, if $i < j$, we have $(j - i) \cdot 1 = 0$.

Definition Let R be a ring. The *smallest* positive integer c for which $c \cdot 1 = 0$ is called the **characteristic** of R. If no such number c exists, we say that R has **characteristic** 0. The characteristic of R is denoted by $\mathrm{char}(R)$. □

If $\mathrm{char}(R) = c$, then for *any* $r \in R$, we have

$$c \cdot r = \underbrace{r + \cdots + r}_{c \text{ terms}} = (\underbrace{1 + \cdots + 1}_{c \text{ terms}})r = 0 \cdot r = 0$$

Theorem 0.25 Any *finite* ring has nonzero characteristic. Furthermore, any finite field has *prime* characteristic.

Proof. We have already seen that a finite ring has nonzero characteristic. Let F be a finite field, and suppose that $\mathrm{char}(F) = c > 0$. If $c = pq$, where $p,\ q < c$, then $pq \cdot 1 = 0$. Hence, $(p \cdot 1)(q \cdot 1) = 0$, implying that $p \cdot 1 = 0$ or $q \cdot 1 = 0$. In either case, we have a contradiction to the fact that c is the *smallest* positive integer such that $c \cdot 1 = 0$. Hence, c must be prime. ∎

Notice that in any field F of characteristic 2, we have $2a = 0$ for all $a \in F$. Thus, in F, we have

$$2 = 0, \text{ and } a = -a, \text{ for all } a \in F$$

These properties take a bit of getting used to, and make fields of characteristic 2 quite exceptional. (As it happens, there are many important uses for fields of characteristic 2.)

Part 1

Basic Linear Algebra

CHAPTER 1

Vector Spaces

Contents: *Vector Spaces. Subspaces. The Lattice of Subspaces. Direct Sums. Spanning Sets and Linear Independence. The Dimension of a Vector Space. The Row and Column Space of a Matrix. Coordinate Matrices. Exercises.*

Vector Spaces

Let us begin with the definition of our principle object of study.

Definition Let F be a field, whose elements are referred to as **scalars**. A **vector space** over F is a nonempty set V, whose elements are referred to as **vectors**, together with two operations. The first operation, called *addition* and denoted by $+$, assigns to each pair $(\mathbf{u},\mathbf{v}) \in V \times V$ of vectors in V a vector $\mathbf{u}+\mathbf{v}$ in V. The second operation, called *scalar multiplication* and denoted by juxtaposition, assigns to each pair $(r,\mathbf{u}) \in F \times V$ a vector $r\mathbf{v}$ in V. Furthermore, the following properties must be satisfied.

1) **(Associativity of addition)**

$$\mathbf{u}+(\mathbf{v}+\mathbf{w}) = (\mathbf{u}+\mathbf{v})+\mathbf{w}$$

for all vectors $\mathbf{u},\mathbf{v},\mathbf{w} \in V$.

2) **(Commutivity of addition)**

$$\mathbf{u}+\mathbf{v} = \mathbf{v}+\mathbf{u}$$

for all vectors $\mathbf{u},\mathbf{v} \in V$.

3) **(Existence of a zero)**
There is a vector $\mathbf{0} \in V$ with the property that
$$\mathbf{0}+\mathbf{u}=\mathbf{u}+\mathbf{0}=\mathbf{u}$$
for all vectors $\mathbf{u} \in V$.

4) **(Existence of additive inverses)**
For each vector $\mathbf{u} \in V$, there is a vector in V, denoted by $-\mathbf{u}$, with the property that
$$\mathbf{u}+(-\mathbf{u})=(-\mathbf{u})+\mathbf{u}=\mathbf{0}$$

5) **(Properties of scalar multiplication)**
For all scalars r and s, we have
$$\begin{aligned} r(\mathbf{u}+\mathbf{v}) &= r\mathbf{u}+r\mathbf{v} \\ (r+s)\mathbf{u} &= r\mathbf{u}+s\mathbf{u} \\ (rs)\mathbf{u} &= r(s\mathbf{u}) \\ 1\mathbf{u} &= \mathbf{u} \end{aligned}$$
for all vectors $\mathbf{u},\mathbf{v} \in V$. □

Note that the first four properties in the definition of vector space can be summarized by saying that V is an *abelian group* under addition.

Any expression of the form
$$r_1\mathbf{v}_1+\cdots+r_n\mathbf{v}_n$$
where $r_i \in F$ and $\mathbf{v}_i \in V$ for all i, is called a **linear combination** of the vectors $\mathbf{v}_1,\ldots,\mathbf{v}_n$.

Example 1.1

1) Let F be a field. The set $\mathcal{F}$ of all functions from F to F is a vector space over F, under the operations of ordinary addition and scalar multiplication of functions
$$(f+g)(x)=f(x)+g(x)$$
and
$$(rf)(x)=r(f(x))$$

2) The set $\mathcal{M}_{m,n}(F)$ of all $m\times n$ matrices with entries in a field F is a vector space over F, under the operations of matrix addition and scalar multiplication.

3) The set F^n of all ordered n-tuples, whose components lie in a field F, is a vector space over F, with addition and scalar multiplication defined componentwise
$$(a_1,\ldots,a_n)+(b_1,\ldots,b_n)=(a_1+b_1,\ldots,a_n+b_n)$$
and
$$r(a_1,\ldots,a_n)=(ra_1,\ldots,ra_n)$$

When convenient, we will also write the elements of F^n in column form. When F is a finite field F_q with q elements, we use the notation $V(n,q)$, rather than F_q^n. Thus, $V(n,q)$ is the set of all ordered n-tuples, whose components come from the finite field F_q.

4) There are various *sequence spaces* that are vector spaces. The set $Seq(F)$ of all infinite sequences, whose entries lie in a field F, is a vector space, under componentwise operations

$$(s_n)+(t_n)=(s_n+t_n)$$

and

$$r(s_n)=(rs_n)$$

In a similar way, the set c_0 of all sequences of complex numbers that converge to 0 is a vector space, as is the set ℓ^∞ of all *bounded* complex sequences. Also, if p is a positive integer, then the set ℓ^p of all complex sequences (s_n) for which $\sum |s_n|^p<\infty$ is a vector space under componentwise operations. To see that addition is a binary operation on ℓ^p, one verifies *Minkowski's inequality*

$$(\sum |s_n+t_n|^p)^{1/p}\le(\sum |s_n|^p)^{1/p}+(\sum |t_n|^p)^{1/p}$$

which we will not do here. (See the exercises in Chapter 12.) □

Subspaces

Most algebraic structures contain substructures, and vector spaces are no exception.

Definition A **subspace** of a vector space V is a subset S of V that is a vector space in its own right, under the operations obtained by restricting the operations of V to S. □

Since many of the properties of addition and scalar multiplication hold, *a fortiori*, in the subset S, we can establish that a *nonempty* subset is a subspace merely by checking that the subset is *closed* under the operations of V.

Theorem 1.1 A nonempty subset S of a vector space V is a subspace if and only if

1) S is *closed under addition*, that is,

$$\mathbf{u},\mathbf{v}\in S \Rightarrow \mathbf{u}+\mathbf{v}\in S$$

2) S is *closed under scalar multiplication*, that is,

$$r\in F,\ \mathbf{u}\in S \Rightarrow r\mathbf{u}\in S$$

Equivalently, S is a subspace if and only if

3) S is *closed under taking linear combinations*, that is,

$$r,s \in F,\ \mathbf{u},\mathbf{v} \in S \Rightarrow r\mathbf{u} + s\mathbf{v} \in S$$ □

Example 1.2 Consider the vector space $V(n,2)$ of all binary n-tuples. The **weight** $w(\mathbf{v})$ of a vector $\mathbf{v} \in V(n,2)$ is the number of nonzero coordinates in $\mathbf{v}$. For instance, $w(101010) = 3$. Let E_n be the set of all vectors in V of even weight. Then E_n is a subspace of $V(n,2)$.

To see this, note that

$$w(\mathbf{u}+\mathbf{v}) = w(\mathbf{u}) + w(\mathbf{v}) - 2w(\mathbf{u} \cap \mathbf{v})$$

where $\mathbf{u} \cap \mathbf{v}$ is the vector in $V(n,2)$ whose ith component is the product of the ith components of $\mathbf{u}$ and $\mathbf{v}$, taken modulo 2. That is,

$$(\mathbf{u} \cap \mathbf{v})_i = (\mathbf{u}_i \cdot \mathbf{v}_i) \bmod 2$$

Hence, if $w(\mathbf{u})$ and $w(\mathbf{v})$ are both even, so is $w(\mathbf{u}+\mathbf{v})$. Finally, scalar multiplication over F_2 is trivial, and so E_n is a subspace of $V(n,2)$, known as the **even weight subspace** of $V(n,2)$. □

Example 1.3 Any subspace of the vector space $V(n,q)$ is called a **linear code.** Linear codes are among the most important, and most studied, types of codes, because their structure allows for efficient encoding and decoding of information. For a detailed discussion of linear (and other) codes, see Roman [1992]. □

The Lattice of Subspaces

The set $\mathcal{S}(V)$ of all subspaces of a vector space V is partially ordered by set inclusion. The **zero subspace** $\{0\}$ is the smallest element in $\mathcal{S}(V)$, and the entire space V is the largest element.

If $S,T \in \mathcal{L}(V)$, then $S \cap T$ is the largest subspace of V that contains S and T. Hence, in $\mathcal{S}(V)$, the greatest lower bound of S and T is

$$\text{glb}\{S,T\} = S \cap T$$

Similarly, if $\{S_i \mid i \in K\}$ is any collection of subspaces of V, then the intersection

$$\bigcap_{i \in K} S_i$$

is also a subspace of V, and is the greatest lower bound of the collection $\{S_i\}$.

On the other hand, if $S,T \in \mathcal{S}(V)$, then $S \cup T \in \mathcal{S}(V)$ if and only if $S \subset T$ or $T \subset S$. Thus, the union of subspaces is never a subspace in any "interesting" case. We also have the following.

Theorem 1.2 A vector space V over an infinite field F is never the union of a finite number of proper subspaces.

Proof. Suppose that $V = S_1 \cup \cdots \cup S_n$, where we may assume that $S_1 \not\subset S_2 \cup \cdots \cup S_n$. Let $\mathbf{w} \in S_1 - (S_2 \cup \cdots \cup S_n)$, and let $\mathbf{v} \notin S_1$. Consider the infinite set $A = \{\mathbf{w} + r\mathbf{v} \mid r \in F\}$. (This is the "line" through $\mathbf{w}$, parallel to $\mathbf{v}$.) We want to show that each S_i contains at most one vector from the infinite set A, which is contrary to the fact that $V = S_1 \cup \cdots \cup S_n$, and so this will prove the theorem.

Suppose that $\mathbf{w} + r\mathbf{v} \in S_1$ for $r \neq 0$. Then since $\mathbf{w} \in S_1$, we would have $r\mathbf{v} \in S_1$, or $\mathbf{v} \in S_1$, contrary to assumption. Next, suppose that $\mathbf{w} + r_1\mathbf{v}$, $\mathbf{w} + r_2\mathbf{v} \in S_i$, for $i \geq 2$, where $r_1 \neq r_2$. Then

$$S_1 \ni r_2(\mathbf{w} + r_1\mathbf{v}) - r_1(\mathbf{w} + r_2\mathbf{v}) = (r_2 - r_1)\mathbf{w}$$

and so $\mathbf{w} \in S_i$, which is also contrary to assumption. ∎

To determine the smallest subspace of V containing the subspaces S and T, we make the following definition.

Definition Let S and T be subspaces of V. The **sum** $S + T$ is the set of all sums of vectors from S and T, that is,

$$S + T = \{\mathbf{u} + \mathbf{v} \mid \mathbf{u} \in S, \mathbf{v} \in T\}$$

More generally, the **sum** of any collection $\{S_i \mid i \in K\}$ of subspaces is the set of all *finite* sums of vectors from the union $\bigcup S_i$

$$\sum_{i \in K} S_i = \{\mathbf{s}_1 + \cdots + \mathbf{s}_n \mid \mathbf{s}_j \in \bigcup_{i \in K} S_i\} \qquad \square$$

It is not hard to show that the sum of any collection of subspaces of V is a subspace of V, and that

$$\text{lub}\{S,T\} = S + T$$

and, more generally,

$$\text{lub}\{S_i\} = \sum_{i \in K} S_i$$

A partially ordered set in which every pair of elements has a least upper bound and greatest lower bound is called a *lattice.*

Theorem 1.3 The set $\mathcal{S}(V)$ of all subspaces of a vector space V is a lattice under set inclusion, with

$$\text{glb}\{S,T\} = S \cap T \quad \text{and} \quad \text{lub}\{S,T\} = S + T \qquad \square$$

Direct Sums

As we will see, there are many ways to construct new vector spaces from old ones.

Definition Let $V_1,\ldots,V_n$ be vector spaces over the same field F. The **external direct sum** of $V_1,\ldots,V_n$, denoted by $V = V_1 \boxplus \cdots \boxplus V_n$, is the vector space V whose elements are ordered n-tuples

$$V = \{(\mathbf{v}_1,\ldots,\mathbf{v}_n) \mid \mathbf{v}_i \in V_i,\ i = 1,\ldots,n\}$$

with componentwise operations

$$(\mathbf{u}_1,\ldots,\mathbf{u}_n) + (\mathbf{v}_1,\ldots,\mathbf{v}_n) = (\mathbf{u}_1 + \mathbf{v}_1,\ldots,\mathbf{u}_n + \mathbf{v}_n)$$

and

$$r(\mathbf{v}_1,\ldots,\mathbf{v}_n) = (r\mathbf{v}_1,\ldots,r\mathbf{v}_n)$$ □

Example 1.4 The vector space F^n is the external direct sum of n copies of F, that is,

$$F^n = F \boxplus \cdots \boxplus F$$

where there are n summands on the right-hand side. □

This construction can be generalized to any collection of vector spaces, by generalizing the idea that an ordered n-tuple $(\mathbf{v}_1,\ldots,\mathbf{v}_n)$ is just a *function* $f:\{1,\ldots,n\} \to \bigcup V_i$, with the property that $f(i) \in V_i$. One possible generalization is given by the following definition.

Definition Let $\mathcal{F} = \{V_i \mid i \in K\}$ be any family of vector spaces over F. The **direct product** of $\mathcal{F}$ is the vector space

$$\prod_{i \in K} V_i = \left\{ f:K \to \bigcup_{i \in K} V_i \;\middle|\; f(i) \in V_i \right\}$$

thought of as a subspace of the vector space of all functions from K to $\bigcup V_i$. □

The following will prove more useful to us, however.

Definition Let $\mathcal{F} = \{V_i \mid i \in K\}$ be a family of vector spaces over F. The **support** of a function $f:K \to \bigcup V_i$ is the set

$$\mathit{supp}(f) = \{i \in K \mid f(i) \neq \mathbf{0}\}$$

Thus, f has *finite* support if $f(i) = \mathbf{0}$ for *all but* a finite number of $i \in K$. The **external direct sum** of the family $\mathcal{F}$ is the vector space

$$\boxplus_{i \in K} V_i = \left\{ f:K \to \bigcup_{i \in K} V_i \;\middle|\; f(i) \in V_i,\ f \text{ has finite support} \right\}$$

thought of as a subspace of the vector space of all functions from K to $\bigcup V_i$. □

An important special case occurs when $V_i = V$ for all $i \in K$. If we let V^K denote the set of all functions from K to V and $(V^K)_0$ denote the set of all functions in V^K that have finite support, then

$$\prod_{i \in K} V = V^K \quad \text{and} \quad \boxplus_{i \in K} V = (V^K)_0$$

There is also an *internal* version of the direct sum construction.

Definition Let V be a vector space. We say that V is the **(internal) direct sum** of a family $\mathcal{F} = \{S_i \mid i \in K\}$ of subspaces of V if every vector $\mathbf{v}$ in V can be written, in a *unique* way (except for order), as a *finite* sum of vectors from the subspaces in $\mathcal{F}$, that is, if for all $\mathbf{v} \in V$,

$$\mathbf{v} = \mathbf{u}_1 + \cdots + \mathbf{u}_n$$

for some $\mathbf{u}_i \in S_i$, and furthermore, if

$$\mathbf{v} = \mathbf{w}_1 + \cdots + \mathbf{w}_n$$

where $\mathbf{w}_i \in S_i$, then $\mathbf{w}_i = \mathbf{u}_i$ for all $i = 1, \ldots, n$.

If V is the direct sum of $\mathcal{F}$, we write

$$V = \bigoplus_{i \in K} S_i$$

and refer to each S_i as a **direct summand** of V. If $\mathcal{F} = \{S_1, \ldots, S_n\}$ is a finite family, we write

$$V = S_1 \oplus \cdots \oplus S_n$$

If $V = S \oplus T$, then T is called a **complement** of S in V. We will often write S^c to denote a complement of S. □

The reader will be asked in a later chapter to show that the concepts of internal and external direct sum are essentially equivalent. Since the internal version of direct sum will be used more often, we simply refer to it as *the* direct sum. Once we have discussed the concept of a basis, the following theorem can be easily proved.

Theorem 1.4 Any subspace of a vector space has a complement, that is, if S is a subspace of V, then there exists a subspace S^c for which $V = S \oplus S^c$. □

It should be emphasized that a subspace generally has many complements. The reader can easily find examples of this in $\mathbb{R}^2$. The following characterization of direct sums is quite useful.

Theorem 1.5 A vector space V is the direct sum of a family $\mathcal{F} = \{V_i \mid i \in K\}$ of subspaces if and only if

1) $V = \sum_{i \in K} S_i$
2) For each $i \in K$,

$$S_i \cap \Big(\sum_{j \neq i} S_j\Big) = \{0\}$$

Proof. Suppose first that V is the direct sum of $\mathcal{F}$. Then (1) certainly holds, and if

$$\mathbf{v} \in S_i \cap \Big(\sum_{j \neq i} S_j\Big)$$

then

$$\mathbf{v} = \mathbf{0} + \cdots + \mathbf{0} + \mathbf{s}_i + \mathbf{0} + \cdots + \mathbf{0}$$

and

$$\mathbf{v} = \mathbf{s}_1 + \cdots + \mathbf{s}_{i-1} + \mathbf{0} + \mathbf{s}_{i+1} + \cdots + \mathbf{s}_n$$

where $\mathbf{s}_i \in S_i$ for all i. Hence, by the uniqueness of direct sum representations, $\mathbf{s}_i = \mathbf{0}$ for all $i = 1, \ldots, n$, and so $\mathbf{v} = \mathbf{0}$. Thus, (2) holds.

For the converse, suppose that (1) and (2) hold. Then any vector $\mathbf{v}$ is a sum of vectors from the S_i,

$$\mathbf{v} = \mathbf{s}_1 + \cdots + \mathbf{s}_n$$

where $\mathbf{s}_i \in S_i$. If

$$\mathbf{v} = \mathbf{t}_1 + \cdots + \mathbf{t}_n$$

where $\mathbf{t}_i \in S_i$, then

$$(\mathbf{s}_1 - \mathbf{t}_1) + \cdots + (\mathbf{s}_n - \mathbf{t}_n) = \mathbf{0}$$

But if $\mathbf{v}_i = \mathbf{s}_i - \mathbf{t}_i \in S_i$ is nonzero, then $\mathbf{v}_i$ can be written as a sum of vectors from the S_j, with $j \neq i$, which contradicts (2). Hence, $\mathbf{s}_i = \mathbf{t}_i$ for all i, and V is the direct sum of $\mathcal{F}$. ∎

Example 1.5 Any matrix $A \in \mathcal{M}_n$ can be written in the form

$$A = \tfrac{1}{2}(A + A^{\mathsf{T}}) + \tfrac{1}{2}(A - A^{\mathsf{T}}) = B + C \tag{1.1}$$

where A^{T} is the transpose of A. It is easy to verify that B is symmetric, and C is skew-symmetric, and so (1.1) is a decomposition of A as the sum of a symmetric matrix and a skew-symmetric matrix.

Since the sets *Sym* and *SkewSym* of all symmetric and skew-symmetric matrices in $\mathcal{M}_n$ are subspaces of $\mathcal{M}_n$, we have

$$\mathcal{M}_n = \mathit{Sym} + \mathit{SkewSym}$$

Furthermore, if $S + T = S' + T'$, where S and S' are symmetric, and

T and T′ are skew-symmetric, then the matrix

$$U = S - S' = T - T'$$

is both symmetric and skew-symmetric. Hence, *provided that* $char(F) \neq 2$, we deduce that $U = 0$, and so $S = S'$ and $T = T'$. Thus,

$$\mathcal{M}_n = Sym \oplus SkewSym$$

□

Spanning Sets and Linear Independence

A set of vectors *spans* a vector space if every vector can be written as a linear combination of some of the vectors in that set.

Definition The subspace **spanned** (or **generated**) by a set S of vectors in V is the set of all linear combinations of vectors from S

$$\langle S \rangle = span(S) = \{r_1\mathbf{v}_1 + \cdots + r_n\mathbf{v}_n \mid r_i \in F, \mathbf{v}_i \in V\}$$

When $S = \{\mathbf{v}_1, \ldots, \mathbf{v}_n\}$ is a finite set, we use the notation $\langle \mathbf{v}_1, \ldots, \mathbf{v}_n \rangle$, or $span\{\mathbf{v}_1, \ldots, \mathbf{v}_n\}$. A set S of vectors in V is said to **span** V, or **generate** V, if

$$V = span(S)$$

that is, if every vector $\mathbf{v} \in V$ can be written in the form

$$\mathbf{v} = r_1\mathbf{v}_1 + \cdots + r_n\mathbf{v}_n$$

for some scalars $r_1, \ldots, r_n$ and vectors $\mathbf{v}_1, \ldots, \mathbf{v}_n$. □

It is clear that any superset of a spanning set is also a spanning set. Note also that all vector spaces have spanning sets, since the entire space is a spanning set.

Definition The nonempty set S of vectors in V is **linearly independent** if for any $\mathbf{v}_1, \ldots, \mathbf{v}_n$ in V, we have

$$r_1\mathbf{v}_1 + \cdots + r_n\mathbf{v}_n = \mathbf{0} \Rightarrow r_1 = \cdots = r_n = 0$$

If a set of vectors is not linearly independent, it is **linearly dependent.** □

It follows from the definition that any nonempty subset of a linearly independent set is linearly independent.

Theorem 1.6 Let S be a set of vectors in V.

1) S is linearly independent if and only if every vector in the *span* of S has a *unique* expression as a linear combination of the vectors in S.

2) S is linearly independent if and only if no vector in S is a linear combination of the other vectors in S. ▮

The relationship between minimal spanning sets and linear independence is described in the following key theorem.

Theorem 1.7 Let S be a set of vectors in V. The following are equivalent.

1) S is linearly independent and spans V.
2) For every vector $\mathbf{v} \in V$, there is a *unique* set of vectors $\mathbf{v}_1, \ldots, \mathbf{v}_n$ in S, along with a unique set of scalars $r_1, \ldots, r_n$ in F, for which
$$\mathbf{v} = r_1\mathbf{v}_1 + \cdots + r_n\mathbf{v}_n$$
3) S is a *minimal* spanning set in the sense that S spans V, and any proper subset of S does not span V.
4) S is a *maximal* linearly independent set in the sense that S is linearly independent, but any proper superset of S is *not* linearly independent.

Proof. We leave it to the reader to show that (1) and (2) are equivalent. Now suppose (1) holds. Then S is a spanning set. If some proper subset S' of S also spanned V, then any vector in $S - S'$ would be a linear combination of the vectors in S', contradicting the fact that the vectors in S are linearly independent. Hence (1) implies (3).

Conversely, if S is a minimal spanning set, then it must be linearly independent. For if not, some vector $\mathbf{s} \in S$ would be a linear combination of the other vectors in S, and so $S - \{\mathbf{s}\}$ would be a proper spanning subset of S, which is not possible. Hence (3) implies (1).

Suppose again that (1) holds. Then S is linearly independent. If S were not maximal, there would be a vector $\mathbf{v} \in V - S$ for which the set $S \cup \{\mathbf{v}\}$ is linearly independent. But then $\mathbf{v}$ is not in the span of S, contradicting the fact that S is a spanning set. Hence, S is a maximal linearly independent set, and so (1) implies (4).

Conversely, if S is a maximal linearly independent set, then it must span V, for if not, we could find a vector $\mathbf{v} \in V - S$ that is not a linear combination of the vectors in S. Hence, $S \cup \{\mathbf{v}\}$ would be a linearly independent proper superset of S, which is a contradiction. Thus, (4) implies (1). ▮

Corollary 1.8 A finite set $S = \{\mathbf{v}_1, \ldots, \mathbf{v}_n\}$ of vectors in V is a basis for V if and only if
$$V = \langle \mathbf{v}_1 \rangle \oplus \cdots \oplus \langle \mathbf{v}_n \rangle$$
▮

Definition Any set of vectors in V that is linearly independent and spans V is called a **basis** for V. Thus, a set of vectors is a basis for V if and only if it satisfies any (and hence all) of the conditions in Theorem 1.7. □

Example 1.6 The **ith standard vector** in F^n is the vector $\mathbf{e}_i$ that has 0s in all coordinate positions except the ith, where it has a 1. Thus,

$$\mathbf{e}_1 = (1,0,\ldots,0), \quad \mathbf{e}_2 = (0,1,\ldots,0),\ldots, \quad \mathbf{e}_n = (0,\ldots,0,1)$$

The set $\{\mathbf{e}_1,\ldots,\mathbf{e}_n\}$ is called the **standard basis** for F^n. □

The proof that every nontrivial vector space has a basis is a classic example of the use of Zorn's lemma.

Theorem 1.9 Any vector space, except the zero space $\{\mathbf{0}\}$, has a basis.

Proof. Let V be a nonzero vector space, and consider the collection $\mathcal{A}$ of all linearly independent subsets of V. This collection is not empty, since any single nonzero vector forms a linearly independent set. Now, if $I_1 \subset I_2 \subset \cdots$ is a chain of linearly independent subsets of V, then the union

$$U = \bigcup I_i$$

is also a linearly independent set. Hence, every chain in $\mathcal{A}$ has an upper bound in $\mathcal{A}$, and according to Zorn's lemma, $\mathcal{A}$ must contain a maximal element, that is, V has a maximal linearly independent set, which is a basis for V by Theorem 1.7. ∎

Theorem 1.7 makes it easy to prove the following useful result.

Theorem 1.10

1) Any linearly independent set of vectors in V is contained in a basis for V. That is, any linearly independent set can be *extended* to a basis for V.
2) Any spanning set for V contains a basis for V. That is, any spanning set can be *reduced* to a basis for V. ∎

The reader can now show, using Theorem 1.10, that any subspace of a vector space has a complement.

The Dimension of a Vector Space

The next result, with its classical elegant proof, says that if a vector space V has a finite spanning set S, then the size of any linearly independent set cannot exceed the size of S.

Theorem 1.11 Let V be a vector space, and assume that the vectors $\mathbf{v}_1,\ldots,\mathbf{v}_n$ are linearly independent, and the vectors $\mathbf{s}_1,\ldots,\mathbf{s}_m$ span V. Then $n \le m$.

Proof. First, we list the two sets of vectors

$$\mathbf{s}_1,\ldots,\mathbf{s}_m \qquad \mathbf{v}_1,\ldots,\mathbf{v}_n$$

Then we move the last vector $\mathbf{v}_n$ to the front of the first list

$$\mathbf{v}_n,\mathbf{s}_1,\ldots,\mathbf{s}_m \qquad \mathbf{v}_1,\ldots,\mathbf{v}_{n-1}$$

Since $\mathbf{s}_1,\ldots,\mathbf{s}_m$ span V, $\mathbf{v}_n$ is a linear combination of the $\mathbf{s}_i$'s. This implies that we may remove one of the $\mathbf{s}_i$'s, say $\mathbf{s}_j$, from the first list, and still have a spanning set

$$\mathbf{v}_n,\mathbf{s}_1,\ldots,\widehat{\mathbf{s}}_j,\ldots,\mathbf{s}_m \qquad \mathbf{v}_1,\ldots,\mathbf{v}_{n-1}$$

where the hat ^ means that the vector has been removed from the list.

Now we repeat the process, moving $\mathbf{v}_{n-1}$ from the second list to the beginning of the first list

$$\mathbf{v}_{n-1},\mathbf{v}_n,\mathbf{s}_1,\ldots,\widehat{\mathbf{s}}_j,\ldots,\mathbf{s}_m \qquad \mathbf{v}_1,\ldots,\mathbf{v}_{n-2}$$

As before, the vectors in the first list are linearly dependent, since they spanned V before the inclusion of $\mathbf{v}_{n-1}$. However, since the $\mathbf{v}_i$'s are linearly independent, any linear combination of the vectors in the first list must involve at least one of the $\mathbf{s}_i$'s. Hence, we may remove that vector, say $\mathbf{s}_k$, and still have a spanning set

$$\mathbf{v}_{n-1},\mathbf{v}_n,\mathbf{s}_1,\ldots,\widehat{\mathbf{s}}_j,\ldots,\widehat{\mathbf{s}}_k,\ldots,\mathbf{s}_m \qquad \mathbf{v}_1,\ldots,\mathbf{v}_{n-2}$$

It should be clear that this process can be continued until we run out of either the $\mathbf{s}_i$'s or the $\mathbf{v}_i$'s. However, if we run out of the $\mathbf{s}_i$'s before the $\mathbf{v}_i$'s, that is, if $m < n$, then the first list will be a *proper subset* of the $\mathbf{v}_i$'s that spans V, which contradicts the independence of the $\mathbf{v}_i$'s. Therefore, $m \ge n$. ∎

Corollary 1.12 If V has a finite spanning set, then any two bases of V have the same size. ∎

Now let us prove Corollary 1.12 for arbitrary vector spaces.

Theorem 1.13 If V is a vector space, then any two bases for V have the same cardinality.

Proof. We may assume that all bases for V are infinite sets, for if any basis is finite, then V has a finite spanning set, and so Corollary 1.12 applies.

Let $\mathfrak{B}$ be a basis for V. We may write $\mathfrak{B} = \{\mathbf{b}_i \mid i \in I\}$, where I is the index set, used to index the vectors in $\mathfrak{B}$. Note that $|I| = |\mathfrak{B}|$. Now let $\mathfrak{C}$ be another basis for V. Then any vector $\mathbf{c} \in \mathfrak{C}$ can be written as a *finite* linear combination of the vectors in $\mathfrak{B}$, where all of the coefficients are nonzero, say

$$\mathbf{c} = \sum_{i \in U_{\mathbf{c}}} r_i \mathbf{b}_i$$

Here, $U_{\mathbf{c}}$ is a finite subset of the index set I. Now, because $\mathfrak{C}$ is a basis for V, the union of all of the $U_{\mathbf{c}}$'s, as $\mathbf{c}$ varies over $\mathfrak{C}$, must be I, in symbols,

$$\bigcup_{\mathbf{c} \in \mathfrak{C}} U_{\mathbf{c}} = I \tag{1.2}$$

For if all vectors in the basis $\mathfrak{C}$ can be expressed as a finite linear combination of the vectors $\mathfrak{B} - \{\mathbf{b}_k\}$, for some k, then all vectors in V can be expressed in this manner, implying that $\mathfrak{B} - \{\mathbf{b}_k\}$ spans V, which is not the case.

Now, from (1.2), Theorem 0.16 implies that

$$|\mathfrak{B}| = |I| \leq |\mathfrak{C}|\aleph_0 = |\mathfrak{C}|$$

But we may also reverse the roles of $\mathfrak{B}$ and $\mathfrak{C}$, to conclude that $|\mathfrak{C}| \leq |\mathfrak{B}|$, and so $|\mathfrak{B}| = |\mathfrak{C}|$. ∎

Theorem 1.13 allows us to make the following definition.

Definition A vector space V is **finite dimensional** if it is the zero space $\{\mathbf{0}\}$, or if it has a finite basis. All other vector spaces are **infinite dimensional.**

The **dimension** of the zero space is 0, and the **dimension** of any nonzero vector space is the cardinality of any basis for V. If a vector space V has a basis of cardinality κ, we say that V is **κ-dimensional,** and write $\mathit{dim}(V) = \kappa$. □

It is easy to see that if S is a subspace of V, then $\mathit{dim}(S) \leq \mathit{dim}(V)$. Furthermore, if $\mathit{dim}(S) = \mathit{dim}(V) < \infty$ then $S = V$.

Theorem 1.14 Let S and T be subspaces of a finite dimensional vector space V, then

$$\mathit{dim}(S) + \mathit{dim}(T) = \mathit{dim}(S + T) + \mathit{dim}(S \cap T)$$

In particular, if S^c is any complement of S in V, then

$$\mathit{dim}(S) + \mathit{dim}(S^c) = \mathit{dim}(V)$$ ∎

Theorem 1.15 Let V be a vector space.

1) If $\mathcal{B}$ is a basis for V, and if $\mathcal{B} = \mathcal{B}_1 \cup \mathcal{B}_2$ and $\mathcal{B}_1 \cap \mathcal{B}_2 = \emptyset$, then

$$V = span\{\mathcal{B}_1\} \oplus span\{\mathcal{B}_2\}$$

2) Let $V = S \oplus T$. If $\mathcal{B}_1$ is a basis for S and $\mathcal{B}_2$ is a basis for T, then $\mathcal{B}_1 \cap \mathcal{B}_2 = \emptyset$ and $\mathcal{B} = \mathcal{B}_1 \cup \mathcal{B}_2$ is a basis for V. ∎

The Row and Column Space of a Matrix

Let A be an $m \times n$ matrix over F. The rows of A span a subspace of F^n known as the **row space** of A, and the columns of A span a subspace of F^m known as the **column space** of A. The dimensions of these spaces are called the **row rank** and **column rank**, respectively. We denote the row space and row rank by $rs(A)$ and $rr(A)$, and the column space and column rank by $cs(A)$ and $cr(A)$.

It is a remarkable, and useful, fact that the row rank of a matrix is always equal to the column rank, despite the fact that if $m \neq n$, the row space and column space lie in different vector spaces.

To see this, let A be an $m \times n$ matrix. Some subset of the rows of A form a basis for $rs(A)$. Let A' be the submatrix of A containing just these rows. Hence,

$$rr(A') = rr(A) \tag{1.3}$$

and

$$cr(A') \leq cr(A) \tag{1.4}$$

Consider the matrix C obtained by throwing away all columns of A', except those that form a basis for the column space of A'. Thus, C is a matrix of size $rr(A') \times cr(A')$, whose $cr(A')$ columns form a basis for $cs(A')$, which is a subspace of F^r, where $r = rr(A')$. Hence,

$$cr(A') \leq rr(A') \tag{1.5}$$

We propose to show that $cr(A') = rr(A')$.

If $\mathbf{c}_1, \ldots, \mathbf{c}_{k'}$ are the columns of C, then the ith column $\mathbf{a}_i$ of A' has the form

$$\mathbf{a}_i = r_1\mathbf{c}_1 + \cdots + r_{k'}\mathbf{c}_{k'}$$

and so

$$\mathbf{a}_i = C\begin{bmatrix} r_1 \\ \vdots \\ r_{k'} \end{bmatrix} = C\mathbf{r}_i$$

Hence, if R is the $k' \times n$ matrix whose ith column is $\mathbf{r}_i$, we have

$$A' = CR$$

Now, if $cr(A') < rr(A')$, then the matrix C would have linearly dependent rows, and so there would exist a *nonzero* row vector $\mathbf{v}$ for which $\mathbf{v}C = \mathbf{0}$. Hence,

$$\mathbf{v}A' = \mathbf{v}CR = \mathbf{0}$$

and so the rows of A' would be linearly dependent, which is not the case. Therefore, $cr(A') \nless rr(A')$, and so (1.5) implies that

$$cr(A') = rr(A')$$

This, together with (1.3) and (1.4), gives

$$rr(A) = rr(A') = cr(A') \leq cr(A)$$

But, we can apply the same reasoning to the transpose A^{T} of A, to deduce that

$$rr(A^{\mathsf{T}}) \leq cr(A^{\mathsf{T}})$$

and since $rr(A^{\mathsf{T}}) = cr(A)$ and $cr(A^{\mathsf{T}}) = rr(A)$, we have

$$cr(A) \leq rr(A)$$

which, together with the reverse inequality, gives $cr(A) = rr(A)$. (I am indebted to Professor William Gearhart for suggesting the above argument.) Let us summarize.

Theorem 1.16 For any matrix A, we have $rr(A) = cr(A)$. This number is called the **rank** of A, and denoted by $rk(A)$. ∎

Coordinate Matrices

From the point of view of the vector space operations, every n-dimensional vector space is essentially the same. To understand this statement more clearly, we make the following definition.

Definition Let $dim(V) = n$. An **ordered basis** for V is an ordered n-tuple $(\mathbf{v}_1, \ldots, \mathbf{v}_n)$ of vectors, for which the set $\{\mathbf{v}_1, \ldots, \mathbf{v}_n\}$ is a basis for V. □

Thus, the only difference between a basis and an ordered basis is that we impose an order on the vectors in an ordered basis.

Now, let us fix an ordered basis $\mathfrak{B} = (\mathbf{v}_1, \ldots, \mathbf{v}_n)$ for V. For each vector $\mathbf{v} \in V$, there is a *unique* ordered n-tuple $(r_1, \ldots, r_n)$ of scalars for which

$$\mathbf{v} = r_1\mathbf{v}_1 + \cdots + r_n\mathbf{v}_n$$

This allows us to associate to each vector $\mathbf{v} \in V$ a *unique* column matrix of length n as follows

$$(1.6) \qquad \mathbf{v} \to [\mathbf{v}]_{\mathcal{B}} = \begin{bmatrix} r_1 \\ \vdots \\ r_n \end{bmatrix}$$

The matrix $[\mathbf{v}]_{\mathcal{B}}$ is known as the **coordinate matrix** of $\mathbf{v}$ with respect to the ordered basis $\mathcal{B}$. Clearly, knowing the coordinate matrix $[\mathbf{v}]_{\mathcal{B}}$ is just as good as knowing $\mathbf{v}$. (Assuming that we know $\mathcal{B}$.)

Furthermore, performing linear operations on coordinate matrices has essentially the same effect as performing the same operations on the vectors in V. That is,

$$[\mathbf{u}+\mathbf{v}]_{\mathcal{B}} = [\mathbf{u}]_{\mathcal{B}} + [\mathbf{v}]_{\mathcal{B}}$$

and

$$[r\mathbf{v}]_{\mathcal{B}} = r[\mathbf{v}]_{\mathcal{B}}$$

or, more generally,

$$(1.7) \qquad [r_1\mathbf{v}_1 + \cdots + r_n\mathbf{v}_n]_{\mathcal{B}} = r_1[\mathbf{v}_1]_{\mathcal{B}} + \cdots + r_n[\mathbf{v}_n]_{\mathcal{B}}$$

The association (1.6) defines a function

$$\phi_{\mathcal{B}}(\mathbf{v}) = [\mathbf{v}]_{\mathcal{B}}$$

from V to F^n (where we write the elements of F^n as column vectors). Because $\mathcal{B}$ is a basis, it is easy to see that $\phi_{\mathcal{B}}$ is bijective. Moreover, (1.7) is equivalent to

$$\phi_{\mathcal{B}}(r_1\mathbf{v}_1 + \cdots + r_n\mathbf{v}_n) = r_1\phi_{\mathcal{B}}(\mathbf{v}_1) + \cdots + r_n\phi_{\mathcal{B}}(\mathbf{v}_n)$$

which says that $\phi_{\mathcal{B}}$ *preserves* the vector space operations. Functions from one vector space to another that preserve the vector space operations are called *linear transformations* and form the objects of study of the next chapter.

EXERCISES

1. Show that the sum of any collection of subspaces of V is a subspace of V, and that if $S,T \in \mathcal{S}(V)$, then $\text{lub}\{S,T\} = S + T$.
2. Find a vector space V and a subset S of V that is a vector space, using operations that differ from those of V.
3. Referring to Example 1.1, what are the subset relationships, if any, between $\mathit{Seq}(\mathbb{C})$, c_0, ℓ^∞ and ℓ^1?
4. Prove that if $S,T \in \mathcal{S}(V)$, then $S \cup T \in \mathcal{S}(V)$ if and only if $S \subset T$ or $T \subset S$.
5. Let S, T and U be subspaces of V. Show that if $U \subset S$, then

$$S \cap (T+U) = (S \cap T) + U$$

This is called the **modular law**, for the lattice $\mathcal{S}(V)$.

6. Show that the set *Sym* of all symmetric matrices of size $n \times n$ is a subspace of $\mathcal{M}_n$, as is the set *SkewSym* of all skew-symmetric matrices of size $n \times n$.
7. Prove the the first two statements in Theorem 1.7 are equivalent.
8. Show that any subspace of a vector space is a direct summand.
9. Let $dim(V) < \infty$, and suppose that $V = U \oplus S_1$ and $V = U \oplus S_2$. What can you say about the relationship between S_1 and S_2?
10. Show that if S is a subspace of a vector space V, then $dim(S) \leq dim(V)$. Furthermore, if $dim(S) = dim(V) < \infty$, then $S = V$. Give an example to show that the finiteness is required in the second statement. *Hint:* think about the vector space of polynomials $F[x]$.
11. What is the relationship between $S \oplus T$ and $T \oplus S$? Is the direct sum operation commutative? Formulate and prove a similar statement concerning associativity. Is there an "identity" for direct sum? What about "negatives"?
12. Prove that the vector space $\mathcal{F}$ of all functions from $\mathbb{R}$ to $\mathbb{R}$ is infinite dimensional.
13. Prove that the vector space $\mathcal{C}$ of all *continuous* functions from $\mathbb{R}$ to $\mathbb{R}$ is infinite dimensional.
14. Let F be a field, and let V be an infinite dimensional vector space over F. What is the cardinality of V?
15. If $dim(V) = n$ does V necessarily contain a subspace of any dimension r satisfying $0 \leq r \leq n$?
16. Show that Theorem 1.2 does not hold if the base field F is finite.
17. Let S be a subspace of V. The set $\mathbf{v} + S = \{\mathbf{v} + \mathbf{s} \mid \mathbf{s} \in S\}$ is called an **affine subspace** of V.
    a) Under what conditions is an affine subspace of V a subspace of V?
    b) Show that any two affine subspaces of the form $\mathbf{v} + S$ and $\mathbf{w} + S$ are either equal or disjoint.
18. If V and W are vector spaces over F for which $|V| = |W|$, then must it be true that $dim(V) = dim(W)$?
19. Let V be an n-dimensional vector space over a finite field F, with $|F| = q$. What is the cardinality of V?
20. Let F be a field. A **subfield** of F is a subset K of F that is a field in its own right, using the same operations as defined on F.
    a) Show that F is a vector space over a subfield K of F.
    b) Suppose that F is an m-dimensional subspace over a subfield K of F. If V is an n-dimensional vector space over F, show that V is also a vector space over K. What is the dimension of V as a vector space over K?

CHAPTER 2

Linear Transformations

Contents: *Linear Transformations. The Kernel and Image of a Linear Transformation. Isomorphisms. The Rank Plus Nullity Theorem. Linear Transformations from F^n to F^m. Change of Basis Matrices. The Matrix of a Linear Transformation. Change of Bases for Linear Transformations. Equivalence of Matrices. Similarity of Matrices. Invariant Subspaces and Reducing Pairs. Exercises.*

Linear Transformations

Loosely speaking, a linear transformation is a function from one vector space to another that *preserves* the vector space operations. Let us be more precise.

Definition Let V and W be vector spaces over the same field F. A function $\tau:V \rightarrow W$ is said to be a **linear transformation** if

$$\tau(r\mathbf{u}+s\mathbf{v}) = r\tau(\mathbf{u})+s\tau(\mathbf{v})$$

for all scalars $r,s \in F$ and vectors $\mathbf{u},\mathbf{v} \in V$. A linear transformation $\tau:V \rightarrow V$ is called a **linear operator** on V. The set of all linear transformations from V to W is denoted by $\mathcal{L}(V,W)$, and the set of all linear operators on V is denoted by $\mathcal{L}(V)$. □

We should mention that some authors use the term linear operator for any linear transformation from V to W.

Definition The following terms are also employed:
1) **homomorphism** for linear transformation
2) **endomorphism** for linear operator
3) **monomorphism** for injective linear transformation
4) **epimorphism** for surjective linear transformation
5) **isomorphism** for bijective linear transformation. ▯

Example 2.1
1) The derivative $D:\mathfrak{D}\to\mathfrak{D}$ is a linear operator on the vector space of all infinitely differentiable functions on $\mathbb{R}$.
2) The integral operator $\tau:F[x]\to F[x]$ defined by
$$\tau(f) = \int_0^x f(t)\,dt$$
is a linear operator on $F[x]$.
3) Let A be an $m\times n$ matrix over F. The function $\tau_A:F^n\to F^m$ defined by $\tau_A(\mathbf{v}) = A\mathbf{v}$, where all vectors are written as column vectors, is a linear transformation from F^n to F^m. ▯

We next show that $\mathcal{L}(V,W)$ is a vector space in its own right. Moreover, the set $\mathcal{L}(V)$ has the structure of an *algebra*, as given by the following definition.

Definition Let F be a field. An **algebra** $\mathcal{A}$ over F is a nonempty set $\mathcal{A}$, together with three operations, called *addition* (denoted by $+$), *multiplication* (denoted by juxtaposition), and *scalar multiplication* (also denoted by juxtaposition), for which the following properties hold.
1) $\mathcal{A}$ is a vector space under addition and scalar multiplication.
2) $\mathcal{A}$ is a ring under addition and multiplication.
3) If $r\in F$ and $\mathbf{a},\mathbf{b}\in\mathcal{A}$ then
$$r(\mathbf{ab}) = (r\mathbf{a})\mathbf{b} = \mathbf{a}(r\mathbf{b})$$
▯

Thus, an algebra is a vector space in which we can take the product of vectors, or a ring in which we can multiply each element by a scalar (subject, of course, to additional requirements, as given in the definition). We now return to linear transformations.

Theorem 2.1
1) The set $\mathcal{L}(V,W)$ is a vector space, under ordinary addition of functions and scalar multiplication of functions by elements of F.
2) If $\sigma\in\mathcal{L}(U,V)$ and $\tau\in\mathcal{L}(V,W)$, then the composition $\tau\sigma$ is in $\mathcal{L}(U,W)$.
3) If $\tau\in\mathcal{L}(V,W)$ is bijective, then $\tau^{-1}\in\mathcal{L}(W,V)$.

4) The vector space $\mathcal{L}(V)$ is an algebra, where multiplication is composition of functions. The identity map $\iota \in \mathcal{L}(V)$ is the multiplicative identity, and the zero map $0 \in \mathcal{L}(V)$ is the additive identity.

Proof. We prove only part 3. Let $\tau: V \to W$ be a bijective linear transformation. Then $\tau^{-1}: W \to V$ is a well-defined function, and since any two vectors $\mathbf{w}_1$ and $\mathbf{w}_2$ in W have the form $\mathbf{w}_1 = \tau(\mathbf{v}_1)$ and $\mathbf{w}_2 = \tau(\mathbf{v}_2)$, we have

$$\begin{aligned}\tau^{-1}(r\mathbf{w}_1 + s\mathbf{w}_2) &= \tau^{-1}(r\tau(\mathbf{v}_1) + s\tau(\mathbf{v}_2)) \\ &= \tau^{-1}(\tau(r\mathbf{v}_1 + s\mathbf{v}_2)) \\ &= r\mathbf{v}_1 + s\mathbf{v}_2 \\ &= r\tau^{-1}(\mathbf{w}_1) + s\tau^{-1}(\mathbf{w}_2)\end{aligned}$$

which shows that τ^{-1} is linear. ∎

One of the easiest ways to define a linear transformation is to give its values on a basis. The following theorem says that we may assign these values arbitrarily and thereby obtain a *unique* linear transformation.

Theorem 2.2 Let $\mathcal{B}$ be a basis for V, and let W be a vector space. Then we can define a linear transformation $\tau \in \mathcal{L}(V,W)$ by specifying the values of $\tau(\mathbf{b}) \in W$ arbitrarily, for all $\mathbf{b} \in \mathcal{B}$, and extending the domain of τ to all of V by linearity. Moreover, if $\tau, \sigma \in \mathcal{L}(V,W)$ have the property that $\tau(\mathbf{b}) = \sigma(\mathbf{b})$ for all $\mathbf{b} \in \mathcal{B}$, then $\tau = \sigma$.

Proof. Once τ is defined on the basis vectors in $\mathcal{B}$, we extend the definition of τ by letting

$$\tau(r_1\mathbf{b}_1 + \cdots + r_n\mathbf{b}_n) = r_1\tau(\mathbf{b}_1) + \cdots + r_n\tau(\mathbf{b}_n)$$

The crucial point is that this is well-defined, since each vector in V has a *unique* representation as a linear combination of a *finite* number of vectors in $\mathcal{B}$. We leave proof of the linearity of τ, and the uniqueness, to the reader. ∎

If $\tau \in \mathcal{L}(V,W)$, and if S is a subspace of V, then we may restrict the domain of τ to S. The resulting map, denoted by $\tau|_S$, is a linear transformation from S to W, and is called the **restriction** of τ to S. We will have many occasions to use this concept in the sequel.

The Kernel and Image of a Linear Transformation

There are two very important vector spaces associated with a linear transformation τ from V to W.

Definition Let $\tau \in \mathcal{L}(V,W)$. The set

$$ker(\tau) = \{\mathbf{v} \in V \mid \tau(\mathbf{v}) = \mathbf{0}\}$$

is called the **kernel** of τ, and the set

$$im(\tau) = \{\tau(\mathbf{v}) \mid \mathbf{v} \in V\}$$

is called the **image** of τ. The dimension of $ker(\tau)$ is called the **nullity** of τ, and is denoted by $null(\tau)$. The dimension of $im(\tau)$ is called the **rank** of τ, and is denoted by $rk(\tau)$. □

It is routine to show that $ker(\tau)$ is a subspace of V and $im(\tau)$ is a subspace of W. Moreover, we have the following.

Theorem 2.3 Let $\tau \in \mathcal{L}(V,W)$. Then
1) τ is surjective if and only if $im(\tau) = W$
2) τ is injective if and only if $ker(\tau) = \{\mathbf{0}\}$

Proof. The first statement is merely a restatement of the definition of surjectivity. To see the validity of the second statement, observe that

$$(2.1) \qquad \tau(\mathbf{u}) = \tau(\mathbf{v}) \Leftrightarrow \tau(\mathbf{u}-\mathbf{v}) = \mathbf{0} \Leftrightarrow \mathbf{u}-\mathbf{v} \in ker(\tau)$$

Hence, if $ker(\tau) = \{\mathbf{0}\}$, then

$$\tau(\mathbf{u}) = \tau(\mathbf{v}) \Leftrightarrow \mathbf{u} = \mathbf{v}$$

which shows that τ is injective. Conversely, if τ is injective, then (2.1) implies that

$$\mathbf{u}-\mathbf{v} = \mathbf{0} \Leftrightarrow \mathbf{u} = \mathbf{v} \Leftrightarrow \mathbf{u}-\mathbf{v} \in ker(\tau)$$

and so, letting $\mathbf{w} = \mathbf{u}-\mathbf{v}$, we get $\mathbf{w} = \mathbf{0}$ if and only if $\mathbf{w} \in ker(\tau)$, that is, $ker(\tau) = \{\mathbf{0}\}$. ∎

Isomorphisms

Definition A *bijective* linear transformation $\tau:V \to W$ is called an **isomorphism** from V to W. When an isomorphism from V to W exists, we say that V and W are **isomorphic**, and write $V \approx W$. □

Example 2.2 Let $dim(V) = n$. For any ordered basis $\mathcal{B}$ of V, the map $\phi_{\mathcal{B}}:V \to F^n$ that sends each vector $\mathbf{v}$ to its coordinate matrix

$[\mathbf{v}]_{\mathfrak{B}}$ is an isomorphism. Hence, *any n-dimensional vector space over* F *is isomorphic to* F^n. We will refer to the map $\phi_{\mathfrak{B}}$ many times in the sequel. □

When two vector spaces are isomorphic, they behave in essentially the same way with respect to the concepts that depend only on linear operations. The following result provides support for this view.

Theorem 2.4 Suppose that $\tau \in \mathcal{L}(V,W)$ is an isomorphism. Let S be a *set* of vectors in V, and let

$$\tau(S) = \{\tau(\mathbf{s}) \mid \mathbf{s} \in S\}$$

be the set of images of the vectors in S. Then
1) S spans V if and only if $\tau(S)$ spans W.
2) S is linearly independent in V if and only if $\tau(S)$ is linearly independent in W.
3) S is a basis for V if and only if $\tau(S)$ is a basis for W. ∎

An isomorphism can be characterized as a linear transformation $\tau:V\to W$ that maps a basis for V to a basis for W.

Theorem 2.5 Let $\tau \in \mathcal{L}(V,W)$. If $\mathfrak{B}$ is a basis for V and if

$$\tau(\mathfrak{B}) = \{\tau(\mathbf{b}) \mid \mathbf{b} \in \mathfrak{B}\}$$

is a basis for W, then τ is an isomorphism from V onto W. ∎

The following theorem says that, up to isomorphism, there is only one vector space of any given dimension.

Theorem 2.6 Let V and W be vector spaces over F. Then $V \approx W$ if and only if $\dim(V) = \dim(W)$. ∎

In Example 2.2, we saw that any n-dimensional vector space is isomorphic to F^n. To examine the infinite dimensional counterpart of this result, let us observe that each n-tuple $\mathbf{v} = (a_1,\ldots,a_n) \in F^n$ can be thought of as a *function* $f:\{1,\ldots,n\}\to F$, where the value of $f(i)$ is simply the ith coordinate a_i of $\mathbf{v}$. Hence, we may think of F^n as the set F^B of all functions from the set $B = \{1,\ldots,n\}$ to the set F.

Now let us generalize this to include the infinite case. Suppose that B is any set (finite or infinite), called an **index set.** Recall that the vector space of all functions from B to the field F is denoted by F^B, and that the set of all functions in F^B that have finite support is denoted by $(F^B)_0$.

Since any linear combination of functions with finite support also has finite support, $(F^B)_0$ is a subspace of F^B. We leave it to the reader to show that the functions $\delta_b \in (F^B)_0$ defined for all $b \in B$, by

$$\delta_b(x) = \begin{cases} 1 & \text{if } x = b \\ 0 & \text{if } x \neq b \end{cases}$$

form a basis for $(F^B)_0$, called the **standard basis.** Hence, $dim((F^B)_0) = |B|$.

Theorem 2.7 If n is a natural number, then any n-dimensional vector space over F is isomorphic to F^n. If κ is any cardinal number, and if B is a set of cardinality κ, then any κ-dimensional vector space over F is isomorphic to the vector space $(F^B)_0$ of all functions from B to F with finite support. ∎

The Rank Plus Nullity Theorem

Let $\tau \in \mathcal{L}(V,W)$. We know that $ker(\tau)$, being a subspace of V, has a complement $ker(\tau)^c$, that is,

$$V = ker(\tau) \oplus ker(\tau)^c \tag{2.2}$$

Let $\mathcal{K}$ be a basis for $ker(\tau)$, and let $\mathcal{C}$ be a basis for $ker(\tau)^c$. Since $\mathcal{K} \cap \mathcal{C} = \emptyset$ and $\mathcal{K} \cup \mathcal{C}$ is a basis for V, we have

$$dim(V) = dim(ker(\tau)) + dim(ker(\tau)^c)$$

Now, the restriction of τ to $ker(\tau)^c$, which we denote by

$$\tau^c: ker(\tau)^c \to im(\tau)$$

is easily seen to be an isomorphism. In fact, τ^c is injective, since if $\mathbf{v} \in ker(\tau)^c$ (which is the domain of τ^c) and $\tau^c(\mathbf{v}) = \mathbf{0}$, then $\tau(\mathbf{v}) = \mathbf{0}$, and so $\mathbf{v} \in ker(\tau)^c \cap ker(\tau) = \{\mathbf{0}\}$, which implies that $\mathbf{v} = \mathbf{0}$.

To see that τ^c is surjective, suppose that $\tau(\mathbf{v}) \in im(\tau)$. Then (2.2) implies that $\mathbf{v} = \mathbf{u} + \mathbf{w}$, where $\mathbf{u} \in ker(\tau)$ and $\mathbf{w} \in ker(\tau)^c$. Therefore,

$$\tau(\mathbf{v}) = \tau(\mathbf{u}) + \tau(\mathbf{w}) = \tau(\mathbf{w}) = \tau^c(\mathbf{w})$$

which shows that $\tau(\mathbf{v}) \in \mathrm{im}(\tau^c)$. Hence, $im(\tau) \subset im(\tau^c)$, and since the reverse inclusion is obvious, we have $im(\tau^c) = im(\tau)$. Hence, τ^c is onto $im(\tau)$.

Thus, τ^c is an isomorphism, and

$$ker(\tau)^c \approx im(\tau)$$

From this, we deduce several useful facts, which are given in the following theorem.

Theorem 2.8 Let $\tau \in \mathcal{L}(V,W)$. Then we have the following:
1) $ker(\tau)^c \approx im(\tau)$
2) **(The rank plus nullity theorem)**

$$dim(ker(\tau)) + dim(im(\tau)) = dim(V)$$

or, in other notation,

$$rk(\tau) + null(\tau) = dim(V)$$

3) If S is a subspace of V, then all complements of S are isomorphic.
4) If S is a subspace of V, then

$$dim(S) + dim(S^c) = dim(V)$$

Proof. Part (1) has been proven. Part (2) follows from the fact that because $\mathcal{K}$ and $\mathcal{C}$ are disjoint and $|\mathcal{C}| = |\tau(\mathcal{C})| = dim(im(\tau))$,

$$dim(V) = |\mathcal{K} \cup \mathcal{C}| = |\mathcal{K}| + |\mathcal{C}| = dim(ker(\tau)) + dim(im(\tau))$$

To prove parts (3) and (4), let S be a subspace of V, and let T be a particular complement of S. Thus $V = S \oplus T$. Define a map $\rho: V \to T$ as follows. Any $\mathbf{v} \in V$ has the form $\mathbf{v} = \mathbf{u} + \mathbf{w}$, where $\mathbf{u} \in S$ and $\mathbf{w} \in T$. Let $\rho(\mathbf{v}) = \mathbf{w}$. We leave it to the reader to show that $\rho \in \mathcal{L}(V)$, and that $ker(\rho) = S$ and $im(\rho) = T$. Now, part (1), applied to the map ρ, says that $S^c \approx T$, for any complement S^c of S. Hence, any two complements of S are isomorphic to T, and therefore to each other. Part (4) follows from part (2), applied to the map ρ. ∎

Theorem 2.8 has an important corollary.

Corollary 2.9 Let $\tau \in \mathcal{L}(V,W)$, where $dim(V) = dim(W) < \infty$. Then τ is injective if and only if it is surjective. ∎

Linear Transformations from F^n to F^m

For any $m \times n$ matrix A over F, we can define the map "multiplication by A"

$$\tau_A(\mathbf{v}) = A\mathbf{v}$$

where $\mathbf{v} \in F^n$ is written in column form. Note that τ_A is a function from F^n to F^m. It is easy to see that τ_A is a linear transformation.

As it happens, *all* linear transformations $\tau \in \mathcal{L}(F^n, F^m)$ have the form τ_A for some $m \times n$ matrix A. To see this, recall that τ is uniquely defined by its values on a basis for F^n, in particular, by its values on the standard basis

$$\tau(\mathbf{e}_1),\ldots,\tau(\mathbf{e}_n)$$

Now, if A is an $m \times n$ matrix, then

$$A\mathbf{e}_i = \text{ith column of } A$$

Hence, if we let A be the $m \times n$ matrix with columns $\tau(\mathbf{e}_1),\ldots,\tau(\mathbf{e}_n)$, then

$$A\mathbf{e}_i = \tau(\mathbf{e}_i)$$

and so τ and τ_A agree on the standard basis vectors, which implies that $\tau = \tau_A$. Let us summarize.

Theorem 2.10
1) Let A be an $m \times n$ matrix over F. The map τ_A defined by $\tau_A(\mathbf{v}) = A\mathbf{v}$ is in $\mathcal{L}(F^n,F^m)$.
2) Conversely, let $\tau \in \mathcal{L}(F^n,F^m)$. Then there exists a unique $m \times n$ matrix A over F for which $\tau = \tau_A$. This matrix is called the **standard matrix** for τ. The ith column of A is $\tau(\mathbf{e}_i)$. ∎

Example 2.3 Consider the linear transformation $\tau: F^3 \to F^3$ defined by

$$\tau(x,y,z) = (x - 2y, z, x + y + z)$$

Then we have, in column form

$$\tau\begin{bmatrix} x \\ y \\ z \end{bmatrix} = \begin{bmatrix} x-2y \\ z \\ x+y+z \end{bmatrix} = \begin{bmatrix} 1 & -2 & 0 \\ 0 & 0 & 1 \\ 1 & 1 & 1 \end{bmatrix}\begin{bmatrix} x \\ y \\ z \end{bmatrix}$$

and so $\tau = \tau_A$, where

$$A = \begin{bmatrix} 1 & -2 & 0 \\ 0 & 0 & 1 \\ 1 & 1 & 1 \end{bmatrix}$$ □

Since the image of τ_A is the column space of A, we have

$$\mathit{dim}(\mathit{ker}(\tau_A)) + \mathit{rk}(A) = \mathit{dim}(V)$$

This gives the following useful result.

Theorem 2.11 Let A be an $m \times n$ matrix over F.
1) τ_A is injective if and only if $\mathit{rk}(A) = n$.
2) τ_A is surjective if and only if $\mathit{rk}(A) = m$. ∎

Change of Basis Matrices

Let V be a vector space, and suppose that $\mathcal{B} = (\mathbf{b}_1, \ldots, \mathbf{b}_n)$ and $\mathcal{C} = (\mathbf{c}_1, \ldots, \mathbf{c}_n)$ are ordered bases for V. It is a natural question to ask how the coordinate matrices $[\mathbf{v}]_{\mathcal{B}}$ and $[\mathbf{v}]_{\mathcal{C}}$ are related. Figure 2.1 illustrates the situation.

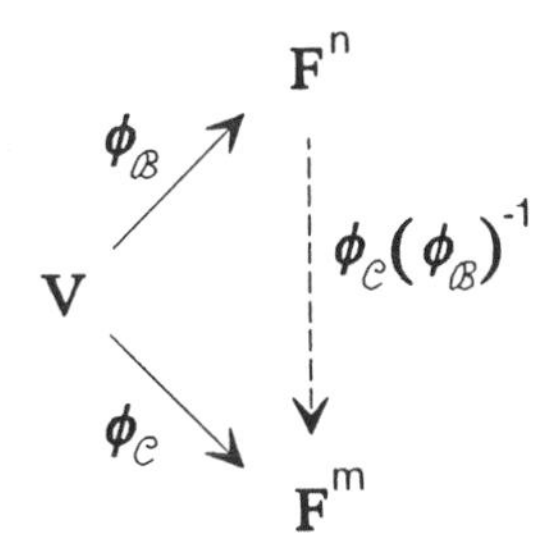

Figure 2.1

Our interest lies in finding an expression for the linear transformation $\theta = \phi_{\mathcal{C}}(\phi_{\mathcal{B}})^{-1}$, since

$$\theta[\mathbf{v}]_{\mathcal{B}} = \phi_{\mathcal{C}}(\phi_{\mathcal{B}})^{-1}[\mathbf{v}]_{\mathcal{B}} = \phi_{\mathcal{C}}(\mathbf{v}) = [\mathbf{v}]_{\mathcal{C}}$$

is the map that describes the relationship between $[\mathbf{v}]_{\mathcal{B}}$ and $[\mathbf{v}]_{\mathcal{C}}$.

Since $\theta \in \mathcal{L}(F^n, F^m)$, we know that $\theta = \tau_A$, for some $m \times n$ matrix A, and furthermore

$$\text{ith column of } A = A\mathbf{e}_i = \tau_A(\mathbf{e}_i) = \theta(\mathbf{e}_i) = \theta([\mathbf{b}_i]_{\mathcal{B}}) = [\mathbf{b}_i]_{\mathcal{C}}$$

Thus, A is just the matrix whose ith column is the coordinate matrix $[\mathbf{b}_i]_{\mathcal{C}}$ of the ith "old" basis vector with respect to the "new" basis $\mathcal{C}$. This matrix is called the **change of basis matrix** from $\mathcal{B}$ to $\mathcal{C}$, which we will denote by $M_{\mathcal{B},\mathcal{C}}$.

Theorem 2.12 Let $\mathcal{B} = (\mathbf{b}_1, \ldots, \mathbf{b}_n)$ and $\mathcal{C}$ be ordered bases for a vector space V. Then

$$[\mathbf{v}]_{\mathcal{C}} = M_{\mathcal{B},\mathcal{C}}[\mathbf{v}]_{\mathcal{B}}$$

where the change of basis matrix $M_{\mathcal{B},\mathcal{C}}$ is the matrix whose ith column is $[\mathbf{b}_i]_{\mathcal{C}}$. ∎

It is worth remarking that given *any* invertible matrix A and *any* ordered basis $\mathcal{B}$ for V, we can find an ordered basis $\mathcal{C}$ for which $A = M_{\mathcal{B},\mathcal{C}}$. To see this, we look again at Figure 2.1, which shows that

$$\phi_{\mathcal{C}}(\mathbf{v}) = \tau_A \phi_{\mathcal{B}}(\mathbf{v}) = A[\mathbf{v}]_{\mathcal{B}}$$

But $\phi_{\mathcal{C}}(\mathbf{v}) = \mathbf{e}_i$ if and only if $\mathbf{v} = \mathbf{c}_i$ is the ith basis vector of $\mathcal{C}$, and so we seek to solve the equation

$$A[\mathbf{v}]_{\mathcal{B}} = \mathbf{e}_i$$

for $\mathbf{v} = \mathbf{c}_i$. This is done simply by multiplying both sides by A^{-1}, to get

$$[\mathbf{v}]_{\mathcal{B}} = A^{-1}\mathbf{e}_i = \text{ith column of } A^{-1}$$

Hence, $\mathbf{c}_i$ is the vector whose coordinate matrix with respect to $\mathcal{B}$ is the ith column of A^{-1}.

The Matrix of a Linear Transformation

Figure 2.2 shows a linear transformation $\tau:V\to W$, along with the pair of linear transformations $\phi_{\mathcal{B}}$ and $\phi_{\mathcal{C}}$, used to represent vectors in V and W in terms of coordinate matrices.

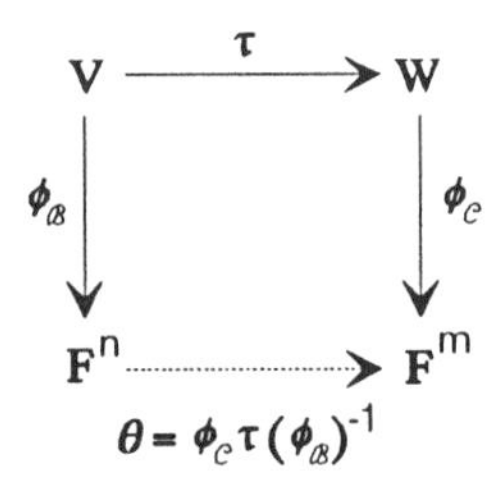

Figure 2.2

Once the ordered bases $\mathcal{B} = (\mathbf{b}_1,\ldots,\mathbf{b}_n)$ and $\mathcal{C} = (\mathbf{c}_1,\ldots,\mathbf{c}_m)$ have been fixed, the linear transformation $\theta = \phi_{\mathcal{C}}\tau(\phi_{\mathcal{B}})^{-1}$ uniquely determines τ, since it determines $\tau(\mathbf{v})$ for all $\mathbf{v} \in V$, by means of

$$[\tau(\mathbf{v})]_{\mathcal{C}} = \phi_{\mathcal{C}}(\tau(\mathbf{v})) = (\phi_{\mathcal{C}}\tau(\phi_{\mathcal{B}})^{-1})([\mathbf{v}]_{\mathcal{B}}) = \theta([\mathbf{v}]_{\mathcal{B}})$$

Moreover, since $\theta:F^n\to F^m$, we know that $\theta = \tau_A$, for some $m\times n$ matrix A. In fact,

$$\text{ith column of } A = A\mathbf{e}_i = \tau_A(\mathbf{e}_i) = \theta(\mathbf{e}_i) = \theta([\mathbf{b}_i]_{\mathcal{B}}) = [\tau(\mathbf{b}_i)]_{\mathcal{C}}$$

This gives the following result.

Theorem 2.13 Let $\tau \in \mathcal{L}(V,W)$, and let $\mathcal{B} = (\mathbf{b}_1,\ldots,\mathbf{b}_n)$ and $\mathcal{C} = (\mathbf{c}_1,\ldots,\mathbf{c}_m)$ be ordered bases for V and W, respectively. Then, referring to Figure 2.2, τ can be represented by a linear transformation $\tau_A \in \mathcal{L}(F^n,F^m)$, that is

$$[\tau(\mathbf{v})]_{\mathcal{C}} = \tau_A([\mathbf{v}]_{\mathcal{B}})$$

where $A = [\tau]_{\mathcal{B},\mathcal{C}}$ is the matrix whose ith column is $[\tau(\mathbf{b}_i)]_{\mathcal{C}}$.

We call $[\tau]_{\mathcal{B},\mathcal{C}}$ the **matrix of** τ **with respect to the bases** $\mathcal{B}$

and $\mathcal{C}$. Thus,

$$[\tau(\mathbf{v})]_{\mathcal{C}} = [\tau]_{\mathcal{B},\mathcal{C}}\,[\mathbf{v}]_{\mathcal{B}}$$

When $V = W$ and $\mathcal{B} = \mathcal{C}$, we denote the matrix $[\tau]_{\mathcal{B},\mathcal{B}}$ of the linear *operator* $\tau \in \mathcal{L}(V)$ with respect to the basis $\mathcal{B}$ by $[\tau]_{\mathcal{B}}$. Thus,

$$[\tau(\mathbf{v})]_{\mathcal{B}} = [\tau]_{\mathcal{B}}\,[\mathbf{v}]_{\mathcal{B}}$$ ∎

Example 2.4 Let $D:\mathcal{P}_2 \to \mathcal{P}_2$ be the derivative operator, defined on the vector space of all polynomials of degree at most 2. Let $\mathcal{B} = \mathcal{C} = (1,x,x^2)$. Then

$$[D(1)]_{\mathcal{C}} = [0]_{\mathcal{C}} = \begin{bmatrix} 0 \\ 0 \\ 0 \end{bmatrix}, \qquad [D(x)]_{\mathcal{C}} = [1]_{\mathcal{C}} = \begin{bmatrix} 1 \\ 0 \\ 0 \end{bmatrix},$$

$$[D(x^2)]_{\mathcal{C}} = [2x]_{\mathcal{C}} = \begin{bmatrix} 0 \\ 2 \\ 0 \end{bmatrix}$$

and so

$$[D]_{\mathcal{B}} = \begin{bmatrix} 0 & 1 & 0 \\ 0 & 0 & 2 \\ 0 & 0 & 0 \end{bmatrix}$$

Hence, for example, if $p(x) = 5 + x + 2x^2$, then

$$[Dp(x)]_{\mathcal{C}} = [D]_{\mathcal{B}}\,[p(x)]_{\mathcal{B}} = \begin{bmatrix} 0 & 1 & 0 \\ 0 & 0 & 2 \\ 0 & 0 & 0 \end{bmatrix} \begin{bmatrix} 5 \\ 1 \\ 2 \end{bmatrix} = \begin{bmatrix} 1 \\ 4 \\ 0 \end{bmatrix}$$

and so $Dp(x) = 1 + 4x$. □

The following result shows that we may work equally well with linear transformations or with the matrices that represent them (with respect to fixed ordered bases $\mathcal{B}$ and $\mathcal{C}$). This applies not only to addition and scalar multiplication, but also to multiplication.

Theorem 2.14 Let V and W be vector spaces over F, with ordered bases $\mathcal{B} = (\mathbf{b}_1,\ldots,\mathbf{b}_n)$ and $\mathcal{C} = (\mathbf{c}_1,\ldots,\mathbf{c}_m)$, respectively.

1) The map

$$\phi:\mathcal{L}(V,W) \to \mathcal{M}_{m,n}(F), \qquad \phi(\tau) = [\tau]_{\mathcal{B},\mathcal{C}}$$

is an isomorphism. Thus

$$\mathcal{L}(V,W) \approx \mathcal{M}_{m,n}(F)$$

2) Furthermore, if $\sigma:U\to V$ and $\tau:V\to W$, and if $\mathfrak{B}$, $\mathfrak{C}$ and $\mathfrak{D}$ are ordered bases for U, V and W, respectively, then
$$[\tau\sigma]_{\mathfrak{B},\mathfrak{D}} = [\tau]_{\mathfrak{C},\mathfrak{D}}[\sigma]_{\mathfrak{B},\mathfrak{C}}$$
In loose terms, the matrix of the product (composition) $\tau\sigma$ is the product of the matrices of τ and σ.

Proof. To see that ϕ is linear, observe that the ith column of the matrix $[s\sigma + t\tau]_{\mathfrak{B},\mathfrak{C}}$ is
$$[(s\sigma + t\tau)(\mathbf{b}_i)]_{\mathfrak{C}} = [s\sigma(\mathbf{b}_i) + t\tau(\mathbf{b}_i)]_{\mathfrak{C}} = s[\sigma(\mathbf{b}_i)]_{\mathfrak{C}} + t[\tau(\mathbf{b}_i)]_{\mathfrak{C}}$$
and so
$$\phi(s\sigma + t\tau) = [s\sigma + t\tau]_{\mathfrak{B},\mathfrak{C}} = s[\sigma]_{\mathfrak{B},\mathfrak{C}} + t[\tau]_{\mathfrak{B},\mathfrak{C}} = s\phi(\sigma) + t\phi(\tau)$$

The map ϕ is surjective, for if A is an $m\times n$ matrix, we simply define $\tau:V\to W$ by the condition
$$\tau_A = \phi_{\mathfrak{C}}\tau(\phi_{\mathfrak{B}})^{-1}$$
that is,
$$\tau = (\phi_{\mathfrak{C}})^{-1}\tau_A\phi_{\mathfrak{B}}$$
Then $\phi(\tau) = A$. Finally, ϕ is injective, since
$$[\tau]_{\mathfrak{B},\mathfrak{C}} = 0 \Rightarrow [\tau(\mathbf{b}_i)]_{\mathfrak{C}} = 0 \text{ for all } i \Rightarrow \tau(\mathbf{b}_i) = 0 \text{ for all } i \Rightarrow \tau = 0$$
Thus, ϕ is an isomorphism.

To prove part (2), observe that
$$[\sigma(\mathbf{v})]_{\mathfrak{C}} = [\sigma]_{\mathfrak{B},\mathfrak{C}}[\mathbf{v}]_{\mathfrak{B}} \quad \text{and} \quad [\tau(\mathbf{w})]_{\mathfrak{D}} = [\tau]_{\mathfrak{C},\mathfrak{D}}[\mathbf{w}]_{\mathfrak{C}}$$
Therefore,
$$\begin{aligned}[\tau]_{\mathfrak{C},\mathfrak{D}}[\sigma]_{\mathfrak{B},\mathfrak{C}}[\mathbf{v}]_{\mathfrak{B}} &= [\tau]_{\mathfrak{C},\mathfrak{D}}[\sigma(\mathbf{v})]_{\mathfrak{C}} \\ &= [\tau(\sigma(\mathbf{v}))]_{\mathfrak{D}} \\ &= [\tau\sigma]_{\mathfrak{B},\mathfrak{D}}[\mathbf{v}]_{\mathfrak{B}}\end{aligned}$$
from which part (2) follows. ∎

Change of Bases for Linear Transformations

Since the matrix $[\tau]_{\mathfrak{B},\mathfrak{C}}$ depends on the ordered bases $\mathfrak{B}$ and $\mathfrak{C}$, it is natural to wonder how to choose these bases in order to make this matrix as simple as possible. For instance, can we always choose the bases so that τ is represented by a diagonal matrix?

As we will see in Chapter 7, the answer to this question is no. In that chapter, we will take up the general question of how best to represent a linear operator by a matrix. For now, let us take the first step and describe the relationship between the matrices of τ with respect to two different pairs $(\mathcal{B},\mathcal{C})$ and $(\mathcal{B}',\mathcal{C}')$ of ordered bases. Figure 2.3 describes the situation.

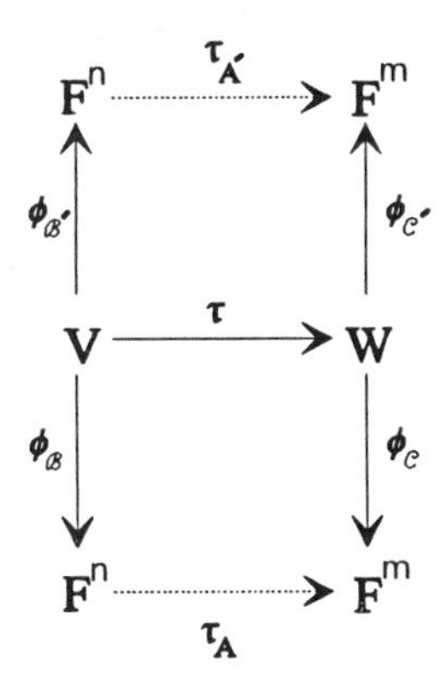

Figure 2.3

As we can see from this figure, τ can be written in two ways

$$\tau = (\phi_{\mathcal{C}})^{-1}\tau_A\phi_{\mathcal{B}} \quad \text{and} \quad \tau = (\phi_{\mathcal{C}'})^{-1}\tau_{A'}\phi_{\mathcal{B}'}$$

Equating these two expressions gives

$$(\phi_{\mathcal{C}})^{-1}\tau_A\phi_{\mathcal{B}} = (\phi_{\mathcal{C}'})^{-1}\tau_{A'}\phi_{\mathcal{B}'}$$

or

$$\tau_{A'} = (\phi_{\mathcal{C}'})(\phi_{\mathcal{C}})^{-1}\tau_A\phi_{\mathcal{B}}(\phi_{\mathcal{B}'})^{-1}$$

or, in matrix terms

$$[\tau]_{\mathcal{B}',\mathcal{C}'} = M_{\mathcal{C},\mathcal{C}'}\,[\tau]_{\mathcal{B},\mathcal{C}}\,M_{\mathcal{B}',\mathcal{B}}$$

This gives the following.

Theorem 2.15 Let $\tau \in \mathcal{L}(V,W)$, and let $(\mathcal{B},\mathcal{C})$ and $(\mathcal{B}',\mathcal{C}')$ be pairs of ordered bases of V and W, respectively. Then the matrix of τ with respect to the ordered bases $(\mathcal{B}',\mathcal{C}')$ can be expressed in terms of the matrix of τ with respect to the ordered bases $(\mathcal{B},\mathcal{C})$ as follows

$$[\tau]_{\mathcal{B}',\mathcal{C}'} = M_{\mathcal{C},\mathcal{C}'}\,[\tau]_{\mathcal{B},\mathcal{C}}\,M_{\mathcal{B}',\mathcal{B}} \tag{2.3}$$ ∎

When $\tau \in \mathcal{L}(V)$ is a linear operator on V, it is customary to represent τ by matrices of the form $[\tau]_{\mathcal{B}}$, where the ordered bases used to represent vectors in the domain and image are the same. We leave it to the reader to show that $M_{\mathcal{B},\mathcal{B}'}$ is invertible and that

$$M_{\mathcal{B},\mathcal{B}'} = (M_{\mathcal{B}',\mathcal{B}})^{-1}$$

Hence, when $\mathcal{B} = \mathcal{C}$, Theorem 2.15 takes the following important form.

Theorem 2.16 Let $\tau \in \mathcal{L}(V)$, and let $\mathcal{B}$ and $\mathcal{B}'$ be ordered bases for V. Then the matrix of τ with respect to $\mathcal{B}'$ can be expressed in terms of the matrix of τ with respect to $\mathcal{B}$ as follows

(2.4) $$[\tau]_{\mathcal{B}'} = M_{\mathcal{B},\mathcal{B}'} [\tau]_{\mathcal{B}} (M_{\mathcal{B},\mathcal{B}'})^{-1}$$ ∎

Equivalence of Matrices

Since change of basis matrices are invertible, (2.3) has the form

$$[\tau]_{\mathcal{B}',\mathcal{C}'} = P[\tau]_{\mathcal{B},\mathcal{C}}Q^{-1}$$

where P and Q are invertible matrices. This leads to the following definition.

Definition Two matrices A and B are **equivalent** if there exist invertible matrices P and Q for which

$$B = PAQ^{-1}$$ □

We remarked in Chapter 0 that B is equivalent to A if and only if B can be obtained from A by a series of elementary row and column operations. Performing the row operations is equivalent to multiplying the matrix A on the left by P, and performing the column operations is equivalent to multiplying A on the right by Q^{-1}.

In terms of (2.3), we see that performing row operations (premultiplying by P) is equivalent to changing the basis used to represent vectors in the image, and performing column operations (postmultiplying by Q^{-1}) is equivalent to changing the basis used to represent vectors in the domain.

According to Theorem 2.15, if A and B are matrices that represent τ with respect to possibly different ordered bases, then A and B are equivalent. The converse of this also holds.

Theorem 2.17 The following statements are equivalent for matrices A and B.

1) If A represents a linear transformation $\tau: V \rightarrow W$, with respect to ordered bases $\mathcal{B}$ and $\mathcal{C}$, then B also represents τ, but perhaps with respect to different ordered bases. That is, if

$$A = [\tau]_{\mathcal{B},\mathcal{C}}$$

then there exist ordered bases $\mathcal{B}'$ and $\mathcal{C}'$ for which

$$B = [\tau]_{\mathcal{B}',\mathcal{C}'}$$

2) A and B are equivalent.

Proof. Suppose that (1) holds. The matrix A represents the linear transformation τ_A, with respect to the standard bases. Hence, B must also represent τ_A, but with respect to possibly different ordered bases, which means that (2.3) holds, and so A and B are equivalent.

Conversely, suppose that A and B are equivalent, and so

$$B = PAQ^{-1}$$

where P and Q are invertible. Let $\tau \in \mathcal{L}(V,W)$, let $\mathcal{B}$ and $\mathcal{C}$ be ordered bases for V and W, respectively, and suppose that

$$A = [\tau]_{\mathcal{B},\mathcal{C}}$$

We have seen earlier (after Theorem 2.12) that there exist ordered bases $\mathcal{B}'$ and $\mathcal{C}'$ for which $P = M_{\mathcal{C},\mathcal{C}'}$ and $Q^{-1} = M_{\mathcal{B}',\mathcal{B}}$. Hence,

$$B = M_{\mathcal{C},\mathcal{C}'} [\tau]_{\mathcal{B},\mathcal{C}} M_{\mathcal{B}',\mathcal{B}}$$

But then, according to Theorem 2.15, $B = [\tau]_{\mathcal{B}',\mathcal{C}'}$. Hence (1) holds. ∎

Similarity of Matrices

As we mentioned earlier, when $\tau \in \mathcal{L}(V)$ is a linear operator on V, it is customary to represent τ by matrices of the form $[\tau]_{\mathcal{B}}$, where the ordered bases used to represent vectors in the domain and image are the same. In this case, (2.4) has the form

$$[\tau]_{\mathcal{B}'} = P\, [\tau]_{\mathcal{B}}\, P^{-1}$$

where P is an invertible matrix. This prompts the following definition.

Definition Two matrices A and B are **similar** if there exists an invertible matrix P for which

$$B = PAP^{-1}$$

The equivalence classes associated with similarity are called **similarity classes.** □

The analog of Theorem 2.17 in this case is the following.

Theorem 2.18 The following statements are equivalent for matrices A and B.

1) If A represents a linear operator $\tau:V\to V$ with respect to an ordered basis $\mathcal{B}$, then B also represents τ, but perhaps with respect to a different ordered basis. That is, if

$$A = [\tau]_{\mathcal{B}}$$

then there exists an ordered basis $\mathcal{B}'$ for which

$$B = [\tau]_{\mathcal{B}'}$$

2) A and B are similar. ∎

Theorem 2.18 can be paraphrased by saying that two matrices represent the same linear operators on V if and only if they are similar. We will devote much effort in Chapter 7 to finding a canonical form for similarity.

Invariant Subspaces and Reducing Pairs

Let τ be a linear operator on V. If S is a subspace of V, there is no guarantee that, for a given $\mathbf{s} \in S$, the vector $\tau(\mathbf{s})$ will also be in S. This prompts us to make the following definition.

Definition Let τ be a linear operator on V. A subspace S of V is said to be **invariant** under τ if $\tau(S) \subset S$, that is, if $\tau(\mathbf{s}) \in S$ for all $\mathbf{s} \in S$. Put another way, S is invariant under τ if the restriction $\tau|_S$, which a priori, maps S to V, is actually a linear *operator* on S. □

If S is a subspace of V and if S^c is a complementary subspace to S, then

$$V = S \oplus S^c$$

However, this does *not* imply that S^c is also invariant under τ. (The reader may wish to supply a simple example with $V = \mathbb{R}^2$.) This leads us to make the following definition.

Definition Let τ be a linear operator on V. If $V = S \oplus T$ and if both S and T are invariant under τ, we say that the pair (S,T) **reduces** τ. Put another way, (S,T) reduces τ if the restrictions $\tau|_S$ and $\tau|_T$ are linear *operators* on S and T, respectively. □

Definition Let ρ be a linear operator on V. Then we write $\rho = \sigma \oplus \tau$, and call ρ the **direct sum** of σ and τ, if there exist subspaces S and T of V for which (S,T) reduces ρ and

$$\sigma = \rho|_S \quad \text{and} \quad \tau = \rho|_T$$

□

The concept of the direct sum of linear operators will play a key role in the study of the structure of a linear operator. For, in a sense we will make precise later, if $\rho = \sigma \oplus \tau$, then we have a *decomposition* of ρ into *simpler* linear operators σ and τ.

EXERCISES

1. Can you think of any other examples of algebras besides $\mathcal{L}(V)$?
2. Prove Corollary 2.9, and find an example to show that it does not hold without the finiteness condition.
3. Let $\tau \in \mathcal{L}(V,W)$. Prove that if $\mathfrak{B}$ is a basis for V and if $\tau(\mathfrak{B}) = \{\tau(\mathbf{b}) \mid \mathbf{b} \in \mathfrak{B}\}$ is a basis for W, then τ is an isomorphism from V onto W.
4. Let V and W be vector spaces over F. Show that $V \approx W$ if and only if $dim(V) = dim(W)$.
5. Let $\tau \in \mathcal{L}(V,W)$. Prove that τ is injective if and only if whenever $\mathbf{v}_1,\ldots,\mathbf{v}_n$ are linearly independent in V, then $\tau\mathbf{v}_1,\ldots,\tau\mathbf{v}_n$ are linearly independent in W.
6. Let $\tau \in \mathcal{L}(V,W)$. Prove that τ is an isomorphism if and only if it carries a basis for V to a basis for W.
7. If $\tau \in \mathcal{L}(V_1,W_1)$ and $\sigma \in \mathcal{L}(V_2,W_2)$, we define the **external direct sum** $\tau \boxplus \sigma \in \mathcal{L}(V_1 \oplus V_2, W_1 \oplus W_2)$ by
$$(\tau \boxplus \sigma)((\mathbf{v}_1,\mathbf{v}_2)) = (\tau(\mathbf{v}_1),\sigma(\mathbf{v}_2))$$
Show that $\tau \boxplus \sigma$ is a linear transformation.
8. Prove that the kernel and image of a linear transformation $\tau{:}V \to W$ are subspaces of V and W, respectively.
9. Let $V = S \oplus T$. Prove that $S \oplus T \approx S \boxplus T$, where $\boxplus$ stands for the external direct sum. Thus, up to isomorphism, internal and external direct sums are the same.
10. Let A be an $m \times n$ matrix. Show that $im(\tau_A)$ is the column space of A. Show that $rk(\tau_A) = rk(A)$.
11. Let $\tau \in \mathcal{L}(V)$, where $dim(V) < \infty$. If $rk(\tau^2) = rk(\tau)$ show that $im(\tau) \cap ker(\tau) = \{\mathbf{0}\}$.
12. Let $\tau \in \mathcal{L}(U,V)$ and $\sigma \in \mathcal{L}(V,W)$. Show that
$$rk(\tau\sigma) \le \min\{rk(\tau),\ rk(\sigma)\}$$
13. Let $\tau \in \mathcal{L}(U,V)$ and $\sigma \in \mathcal{L}(V,W)$. Show that
$$null(\tau\sigma) \le null(\tau) + null(\sigma)$$
14. Let $\tau,\sigma \in \mathcal{L}(V)$, where τ is nonsingular. Show that $rk(\tau\sigma) = rk(\sigma\tau) = rk(\sigma)$.
15. Let $\tau,\sigma \in \mathcal{L}(V,W)$. Show that $rk(\tau + \sigma) \le rk(\tau) + rk(\sigma)$.
16. Let S be a subspace of V. Show that there is a $\tau \in \mathcal{L}(V)$ for

which $ker(\tau) = W$. Show also that there exists a $\sigma \in \mathcal{L}(V)$ for which $im(\sigma) = W$.

17. Prove that any change of basis matrix $M_{\mathfrak{B},\mathfrak{C}}$ is nonsingular and that
$$M_{\mathfrak{C},\mathfrak{B}} = (M_{\mathfrak{B},\mathfrak{C}})^{-1}$$
18. Describe the counterpart of Theorem 2.11 for a linear transformation $\tau:V \rightarrow W$.
19. Let $V = S_1 \oplus S_2$. Define linear operators ρ_i on V by $\rho_i(s_1 + s_2) = s_i$, for $i = 1, 2$. These are referred to as **projection operators.** Show that
    1) $\rho_i^2 = \rho_i$
    2) $\rho_1 + \rho_2 = I$, where I is the identity map on V.
    3) $\rho_i\rho_j = 0$, for $i \neq j$, where 0 is the zero map.
    4) $V = im(\rho_1) \oplus im(\rho_2)$
20. Suppose that $T \in \mathcal{L}(V)$ has the property that $T^2 = T \circ T = 0$. Show that $2rk(T) \leq dim(V)$.
21. Let A be an $m \times n$ matrix over F. What is the relationship between the linear transformation $\tau_A:F^n \rightarrow F^m$ and the system of equations $AX = B$? Use your knowledge of linear transformations to state and prove various results concerning the system $AX = B$, especially when $B = 0$.
22. Draw a figure similar in spirit to Figure 2.3 to show the situation where a single matrix M represents two different linear transformations $\tau_1:V \rightarrow W$ and $\tau_2:V \rightarrow W$. What is the connection between τ_1 and τ_2?
23. Find an example of a vector space V, and a *proper* subspace S of V, for which $V \approx S$.
24. Let $dim(V) < \infty$. If $\tau, \sigma \in \mathcal{L}(V)$, prove that $\sigma\tau = \iota$ implies that τ and σ are invertible, and that $\sigma = p(\tau)$ for some polynomial $p(x) \in F[x]$.
25. Let $\tau \in \mathcal{L}(V)$, where $dim(V) < \infty$. If $\tau\sigma = \sigma\tau$ for all $\sigma \in \mathcal{L}(V)$, show that $\tau = r\iota$, for some $r \in F$. (ι is the identity map.)

CHAPTER 3
The Isomorphism Theorems

Contents: *Quotient Spaces. The First Isomorphism Theorem. The Dimension of a Quotient Space. Additional Isomorphism Theorems. Linear Functionals. Dual Bases. Reflexivity. Annihilators. Operator Adjoints. Exercises.*

Quotient Spaces

Let S be a subspace of a vector space V, and let $\equiv_S$ be the binary relation on V defined by

$$\mathbf{u} \equiv_S \mathbf{v} \Leftrightarrow \mathbf{u} - \mathbf{v} \in S$$

It is easy to see that $\equiv_S$ is an equivalence relation. When $\mathbf{u} \equiv_S \mathbf{v}$, we say that $\mathbf{u}$ and $\mathbf{v}$ are **congruent modulo** S. The term *mod* is used as a colloquialism for *modulo*, and $\mathbf{u} \equiv_S \mathbf{v}$ is often written

$$\mathbf{u} \equiv \mathbf{v} \bmod S$$

When the subspace in question is clear, we will simply write $\mathbf{u} \equiv \mathbf{v}$.

To see what the equivalence classes look like, observe that

$$\begin{aligned}
[\mathbf{v}] &= \{\mathbf{u} \in V \mid \mathbf{u} \equiv \mathbf{v}\} \\
&= \{\mathbf{u} \in V \mid \mathbf{u} - \mathbf{v} \in S\} \\
&= \{\mathbf{u} \in V \mid \mathbf{u} = \mathbf{v} + \mathbf{s} \text{ for some } \mathbf{s} \in S\} \\
&= \{\mathbf{v} + \mathbf{s} \mid \mathbf{s} \in S\} \\
&= \mathbf{v} + S
\end{aligned}$$

The set $\mathcal{S} = \mathbf{v}+S = \{\mathbf{v}+\mathbf{s} \mid \mathbf{s} \in S\}$ is called a **coset** of S in V.

Thus, the equivalence classes for congruence mod S are the cosets $\mathbf{v}+S$ of S in V. The set of all cosets is denoted by

$$\frac{V}{S} = \{\mathbf{v}+S \mid \mathbf{v} \in V\}$$

This is read "V mod S" and is called the **quotient space of V modulo S**. Of course, the term *space* is a hint that we intend to define vector space operations on V/S.

Before doing so, however, observe that Theorem 0.8 implies

(3.1) $$\mathbf{u}+S = \mathbf{v}+S \Leftrightarrow \mathbf{v} \in \mathbf{u}+S \Leftrightarrow \mathbf{u} \in \mathbf{v}+S$$

and

$$\mathbf{u},\mathbf{v} \in V \Rightarrow \mathbf{u}+S = \mathbf{v}+S \text{ or } (\mathbf{u}+S) \cap (\mathbf{v}+S) = \emptyset$$

Thus, a coset may be written in the form $\mathbf{v}+S$ for many different vectors $\mathbf{v}$. In fact, (3.1) implies that

$$\mathbf{u}+S = \mathbf{v}+S \Leftrightarrow \mathbf{u}-\mathbf{v} \in S$$

When a coset $\mathcal{S}$ is written $\mathbf{v}+S$, the vector $\mathbf{v}$ is called the **coset representative** for $\mathcal{S}$. Clearly, any vector in the coset can be a coset representative.

Observe also that

$$\begin{aligned} \mathbf{u} \equiv \mathbf{v} &\Rightarrow \mathbf{u}-\mathbf{v} \in S \\ &\Rightarrow r(\mathbf{u}-\mathbf{v}) \in S \text{ for all } r \in F \\ &\Rightarrow r\mathbf{u}-r\mathbf{v} \in S \text{ for all } r \in F \\ &\Rightarrow r\mathbf{u} \equiv r\mathbf{v} \end{aligned}$$

and so

(3.2) $$\mathbf{u} \equiv \mathbf{v} \Rightarrow r\mathbf{u} \equiv r\mathbf{v} \text{ for all } r \in F$$

In addition,

$$\begin{aligned} \mathbf{u}_1 \equiv \mathbf{v}_1 \text{ and } \mathbf{u}_2 \equiv \mathbf{v}_2 &\Rightarrow \mathbf{u}_1-\mathbf{v}_1 \in S \text{ and } \mathbf{u}_2-\mathbf{v}_2 \in S \\ &\Rightarrow (\mathbf{u}_1-\mathbf{v}_1)+(\mathbf{u}_2-\mathbf{v}_2) \in S \\ &\Rightarrow (\mathbf{u}_1+\mathbf{u}_2)-(\mathbf{v}_1+\mathbf{v}_2) \in S \\ &\Rightarrow (\mathbf{u}_1+\mathbf{u}_2) \equiv (\mathbf{v}_1+\mathbf{v}_2) \end{aligned}$$

and so

(3.3) $$\mathbf{u}_1 \equiv \mathbf{v}_1 \text{ and } \mathbf{u}_2 \equiv \mathbf{v}_2 \Rightarrow (\mathbf{u}_1+\mathbf{u}_2) \equiv (\mathbf{v}_1+\mathbf{v}_2)$$

Properties (3.2) and (3.3) imply that congruence mod S *preserves* the vector space operations on V.

A natural choice for vector space operations on V/S is

$$(\mathbf{u}+S)+(\mathbf{v}+S) = (\mathbf{u}+\mathbf{v})+S$$

and

$$r(\mathbf{u}+S) = r\mathbf{u}+S$$

However, a coset generally has many different coset representatives, and these definitions seem to depend on which representative is chosen. In order to show that they are well-defined, it is necessary to show that they do not depend on the choice of coset representatives, that is,

$$\mathbf{u}_1+S = \mathbf{u}_2+S \text{ and } \mathbf{v}_1+S = \mathbf{v}_2+S \Rightarrow (\mathbf{u}_1+\mathbf{v}_1)+S = (\mathbf{u}_2+\mathbf{v}_2)+S$$

and

$$\mathbf{u}_1+S = \mathbf{u}_2+S \Rightarrow r(\mathbf{u}_1+S) = r(\mathbf{u}_2+S)$$

The straightforward details of this are left to the reader. Let us summarize.

Theorem 3.1 Let S be a subspace of V. The binary relation

$$\mathbf{u} \equiv \mathbf{v} \Leftrightarrow \mathbf{u}-\mathbf{v} \in S$$

is an equivalence relation on V, whose equivalence classes are the *cosets*

$$\mathcal{S} = \mathbf{v}+S = \{\mathbf{v}+\mathbf{s} \mid \mathbf{s} \in S\}$$

of S in V. The set V/S of all cosets of S in V, called the quotient space of V modulo S, is a vector space under the well-defined operations

$$(\mathbf{u}+S)+(\mathbf{v}+S) = (\mathbf{u}+\mathbf{v})+S \quad\text{and}\quad r(\mathbf{u}+S) = r\mathbf{u}+S$$

The zero vector in V/S is the coset $\mathbf{0}+S = S$. ∎

Let S be a subspace of V, and define a map $\pi_S: V \to V/S$ by

$$\pi_S(\mathbf{v}) = \mathbf{v}+S$$

for all $\mathbf{v} \in S$. This map is called the **canonical projection**, or **natural projection**, of V onto S, or simply **projection modulo** S. It is easily seen to be linear, for we have (writing π for π_S)

$$\pi(r\mathbf{u}+s\mathbf{v}) = (r\mathbf{u}+s\mathbf{v})+S = r(\mathbf{u}+S)+s(\mathbf{v}+S) = r\pi(\mathbf{u})+s\pi(\mathbf{v})$$

The canonical projection is surjective, since $\mathbf{v}+S = \pi(\mathbf{v})$ for any coset $\mathbf{v}+S$. To determine the kernel of π, note that

$$\mathbf{v} \in \mathit{ker}(\pi) \Leftrightarrow \pi(\mathbf{v}) = \mathbf{0} \Leftrightarrow \mathbf{v}+S = S \Leftrightarrow \mathbf{v} \in S$$

and so

$$\mathit{ker}(\pi) = S$$

Theorem 3.2 The canonical projection $\pi_S: V \to V/S$ defined by

$$\pi_S(\mathbf{v}) = \mathbf{v}+S$$

is a surjective linear transformation, with $\mathit{ker}(\pi_S) = S$. ∎

The First Isomorphism Theorem

Let S be a subspace of V. Figure 3.1 shows a linear transformation $\tau \in \mathcal{L}(V,W)$, along with the canonical projection π_S from V to the quotient space V/S.

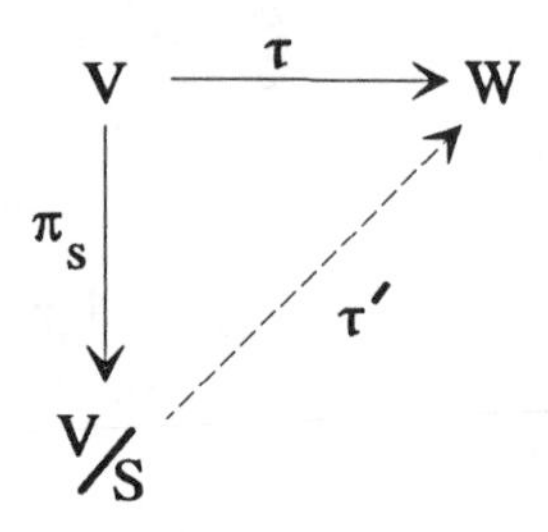

Figure 3.1

This figure suggests the existence of a map τ' from the quotient space V/S to W with the property that

$$\tau' \circ \pi_S = \tau \tag{3.4}$$

that is,

$$\tau(\mathbf{v}) = (\tau' \circ \pi_S)(\mathbf{v}) = \tau'(\mathbf{v}+S)$$

So let us define a *function* τ' from V/S to W by

$$\tau'(\mathbf{v}+S) = \tau(\mathbf{v})$$

This function is well-defined if and only if

$$\mathbf{v}+S = \mathbf{u}+S \ \Rightarrow \ \tau'(\mathbf{v}+S) = \tau'(\mathbf{u}+S)$$

or, equivalently,

$$\mathbf{v}+S = \mathbf{u}+S \ \Rightarrow \ \tau(\mathbf{v}) = \tau(\mathbf{u})$$

But this is equivalent to

$$\mathbf{v}-\mathbf{u} \in S \ \Rightarrow \ \tau(\mathbf{v}-\mathbf{u}) = \mathbf{0}$$

or, replacing $\mathbf{v}-\mathbf{u}$ by $\mathbf{x}$,

$$\mathbf{x} \in S \ \Rightarrow \ \tau(\mathbf{x}) = \mathbf{0}$$

Thus, τ' is well-defined if and only if $S \subset ker(\tau)$.

Let us suppose that $S \subset ker(\tau)$, and hence that τ' is well-defined. Then $\tau': V/S \to W$ is a linear transformation, with image

$$im(\tau') = \{\tau'(\mathbf{v}+S) \mid \mathbf{v}+S \in V/S\} = \{\tau(\mathbf{v}) \mid \mathbf{v} \in V\} = im(\tau)$$

and kernel

$$\begin{aligned} ker(\tau') &= \{\mathbf{v}+S \mid \tau'(\mathbf{v}+S) = \mathbf{0}\} \\ &= \{\mathbf{v}+S \mid \tau(\mathbf{v}) = \mathbf{0}\} \end{aligned}$$

$$= \{\mathbf{v}+S \mid \mathbf{v} \in ker(\tau)\}$$

Also, τ' is unique in the sense that there is only one map $\tau'{:}V/S \to W$ with the property that $\tau' \circ \pi_S = \tau$.

Theorem 3.3 Let $\tau \in \mathcal{L}(V,W)$ and let $S \subset ker(\tau)$ be a subspace of V. Then, as pictured in Figure 3.1, there is a unique linear transformation $\tau'{:}V/S \to W$ with the property that

$$\tau' \circ \pi_S = \tau$$

Moreover, $ker(\tau') = \{\mathbf{v}+S \mid \mathbf{v} \in ker(\tau)\}$ and $im(\tau') = im(\tau)$. ∎

The situation illustrated in Figure 3.1 is often described by saying that any linear transformation $\tau{:}V \to W$ can be *factored through* the projection map π_S for $S \subset ker(\tau)$.

Theorem 3.3 has a very important corollary, which is often called the *first isomorphism theorem*, and is obtained by taking $S = ker(\tau)$.

Theorem 3.4 (The first isomorphism theorem) Let $\tau{:}V \to W$ be a linear transformation. Then the linear transformation $\tau'{:}V/ker(\tau) \to W$ defined by

$$\tau'(\mathbf{v}+ker(\tau)) = \tau(\mathbf{v})$$

is injective, and so

$$\frac{V}{ker(\tau)} \approx im(\tau)$$ ∎

According to Theorem 3.4, the image of any linear transformation with domain V is isomorphic to a quotient space of V. Thus, by identifying isomorphic spaces as being essentially the same, we can say that the images of linear transformations on V are just the quotient spaces of V. Conversely, any quotient space V/S of V is the image of a linear transformation on V, in particular, V/S is the image of the surjective canonical projection map $\pi_S{:}V \to V/S$. Thus, *up to isomorphism*, images of linear transformations on V are the same as quotient spaces of V.

The Dimension of a Quotient Space

The first isomorphism theorem gives further insight into quotient spaces. Recall that any subspace S of V has a complement S^c, for which

$$V = S \oplus S^c$$

Since every vector $\mathbf{v} \in V$ has the form $\mathbf{v} = \mathbf{s} + \mathbf{s}^c$, for unique vectors

$\mathbf{s} \in S$ and $\mathbf{s}^c \in S^c$, we can define a linear operator $\rho:V \to V$ by setting

$$\rho(\mathbf{s}+\mathbf{s}^c) = \mathbf{s}^c$$

Because $\mathbf{s}$ and $\mathbf{s}^c$ are unique, ρ is well-defined. It is called **projection onto** S^c. (Note the word *onto*, rather than *modulo*.) It is clear that

$$im(\rho) = S^c$$

and

$$ker(\rho) = \{\mathbf{s}+\mathbf{s}^c \in V \mid \mathbf{s}^c = \mathbf{0}\} = S$$

Hence, the first isomorphism theorem implies that

$$\frac{V}{S} \approx S^c$$

In other words, we have the following.

Theorem 3.5 Let S be a subspace of V. Then any complement of S in V is isomorphic to the quotient space V/S. ▮

Corollary 3.6 Let S be a subspace of a vector space V. Then

$$dim(V) = dim(S) + dim(V/S)$$ □

The dimension of the quotient space V/S is often called the **codimension** of S in V.

Additional Isomorphism Theorems

There are several other isomorphism theorems that are consequences of the first isomorphism theorem.

Theorem 3.7 **(The second isomorphism theorem)** Let V be a vector space, and let S and T be subspaces of V. Then

$$\frac{S+T}{T} \approx \frac{S}{S \cap T}$$

Proof. Let $\tau:(S+T) \to S/(S \cap T)$ be defined by

$$\tau(\mathbf{s}+\mathbf{t}) = \mathbf{s} + (S \cap T)$$

We leave it to the reader to show that τ is a well-defined surjective linear transformation, with kernel T. An application of the first isomorphism theorem then completes the proof. ▮

Theorem 3.8 **(The third isomorphism theorem)** Let V be a vector space, and suppose that $S \subset T \subset V$ are subspaces of V. Then

$$\frac{V/S}{T/S} \approx \frac{V}{T}$$

Proof. Let $\tau: V/S \to V/T$ be defined by $\tau(\mathbf{v}+S) = \mathbf{v}+T$. We leave it to the reader to show that τ is a well-defined surjective linear transformation whose kernel is T/S. The rest follows from the first isomorphism theorem. ∎

Theorem 3.9 Let V be a vector space, and let S be a subspace of V. Suppose that $V = V_1 \oplus V_2$ and $S = S_1 \oplus S_2$. Then

$$\frac{V}{S} = \frac{V_1 \oplus V_2}{S_1 \oplus S_2} \approx \frac{V_1}{S_1} \boxplus \frac{V_2}{S_2}$$

(Recall that $\boxplus$ stands for the external direct sum.)

Proof. Let $\tau: V \to (V_1/S_1) \boxplus (V_2/S_2)$ be defined by

$$\tau(\mathbf{v}_1 + \mathbf{v}_2) = (\mathbf{v}_1 + S_1, \mathbf{v}_2 + S_2)$$

This map is well-defined, since the sum $V = V_1 \oplus V_2$ is direct. We leave it to the reader to show that τ is a surjective linear transformation, whose kernel is $S_1 \oplus S_2$. The rest follows from the first isomorphism theorem. ∎

Linear Functionals

Linear transformations from V to the base field F (thought of as a vector space over itself) are extremely important.

Definition Let V be a vector space over F. A linear transformation $f \in \mathcal{L}(V,F)$, whose values lie in the base field F is called a **linear functional** (or simply **functional**) on V. The set of all linear functionals on V is denoted by V^* and is called the **algebraic dual space** of V. □

The adjective *algebraic* is needed here, since there is another type of dual space that is defined on *normed* vector spaces, where continuity of linear transformations makes sense. We will discuss the *continuous* dual space briefly in Chapter 13.

To help distinguish linear functionals from other types of linear transformations, we will usually denote linear functionals by lower-case Roman letters, such as f, g, and h.

Note that, according to Theorem 2.1, the dual space V^* is a vector space.

Example 3.1 The map $f: F[x] \to F$, defined by $f(p(x)) = p(0)$, is a linear functional, known as **evaluation at** 0. □

Example 3.2 Let $\mathcal{C}[a,b]$ denote the vector space of all continuous functions on $[a,b] \subset \mathbb{R}$. Let $f:\mathcal{C}[a,b] \to \mathbb{R}$ be defined by

$$f(\alpha(x)) = \int_a^b \alpha(x)\,dx$$

Then $f \in \mathcal{C}[a,b]^*$. □

According to Theorem 2.8, for any $f \in V^*$,

$$dim(ker(f)) + dim(im(f)) = dim(V)$$

But, since $im(f) \subset F$, we have either $im(f) = \{\mathbf{0}\}$, in which case f is the zero linear functional, or $im(f) = F$, in which case f is surjective. In other words, a nonzero linear functional is surjective. Moreover, if $dim(V) < \infty$, then

$$dim(ker(f)) = dim(V) - 1$$

Thus, in loose terms, the kernel of a linear functional is a relatively "large" subspace of the domain V. Even if V is infinite dimensional, we can say that $ker(f)$ has *codimension* 0 or 1.

The following theorem will prove very useful.

Theorem 3.10

1) For any nonzero vector $\mathbf{v} \in V$, there exists a linear functional $f \in V^*$ for which $f(\mathbf{v}) \neq 0$.
2) A vector $\mathbf{v} \in V$ is zero if and only if $f(\mathbf{v}) = 0$ for all $f \in V^*$. ∎

Dual Bases

Suppose that V is finite dimensional, and let $\mathcal{B} = \{\mathbf{v}_1, \ldots, \mathbf{v}_n\}$ be a basis for V. For each $1 \leq i \leq n$, we can define a linear functional $\nu_i \in V^*$, by the *orthogonality condition*

$$\nu_i(\mathbf{v}_j) = \delta_{i,j} \quad \text{for } j = 1, \ldots, n$$

where $\delta_{i,j}$, known as the **Kronecker delta function**, is defined by

$$\delta_{i,j} = \begin{cases} 1 & \text{if } i = j \\ 0 & \text{if } i \neq j \end{cases}$$

Theorem 3.11 Let $\mathcal{B} = \{\mathbf{v}_1, \ldots, \mathbf{v}_n\}$ be a basis for V. Then the linear functionals $\nu_1, \ldots, \nu_n$ defined by

$$\nu_i(\mathbf{v}_j) = \delta_{i,j} \quad \text{for } j = 1, \ldots, n$$

form a basis for the dual space V^*. This basis $\mathcal{B}^* = \{\nu_1, \ldots, \nu_n\}$ is called the **dual basis** for $\mathcal{B}$.

Proof. If

$$0 = r_1\nu_1 + \cdots + r_n\nu_n$$

where 0 represents the zero linear functional, then we may apply both sides of this to the basis vector $\mathbf{v}_i$, to get

$$\mathbf{0} = 0(\mathbf{v}_i) = \sum_j r_j\nu_j(\mathbf{v}_i) = \sum_j r_j\delta_{i,j} = r_i$$

and so $r_i = 0$, for all i. Hence, $\mathcal{B}^*$ is linearly independent.

Any $f \in V^*$ is uniquely determined by its values on the basis vectors $\mathbf{v}_i$, say

$$f(\mathbf{v}_i) = a_i$$

If we let g be the linear functional

$$g = a_1\nu_1 + \cdots + a_n\nu_n$$

then

$$g(\mathbf{v}_j) = a_j = f(\mathbf{v}_j)$$

and so $f = g \in span\{\nu_1, \ldots, \nu_n\}$, which proves that $\mathcal{B}^*$ spans V^*. Hence, $\mathcal{B}^*$ is a basis for V^*. ▮

Corollary 3.12 If $dim(V) < \infty$, then $dim(V^*) = dim(V)$. ▮

The next example shows that Corollary 3.12 does not hold without the finiteness condition.

Example 3.3 Let V be an *infinite dimensional* vector space over the field $F = \mathbb{Z}_2 = \{0,1\}$, with basis $\mathcal{B}$. Since the only coefficients in F are 0 and 1, a finite linear combination over F is just a finite sum. Hence, V is the set of all *finite sums* of vectors in $\mathcal{B}$, and so according to Theorem 0.11,

$$|V| = |\mathcal{P}_0(\mathcal{B})| = |\mathcal{B}|$$

(The finite sums in $\mathcal{B}$ are in one-to-one correspondence with the finite subsets of $\mathcal{B}$.)

On the other hand, each linear functional $f \in V^*$ is uniquely defined by specifying its values on the basis $\mathcal{B}$. Since these values must be either 0 or 1, specifying a linear functional is equivalent to specifying the subset of $\mathcal{B}$ on which f takes the value 1. In other words, there is a one-to-one correspondence between linear functionals on V and *all* subsets of $\mathcal{B}$. Hence,

$$|V^*| = |\mathcal{P}(\mathcal{B})| > |\mathcal{B}| = |V|$$

This shows that V^* cannot be isomorphic to V, nor to any proper subset of V. Hence, $dim(V^*) > dim(V)$. □

Reflexivity

If V is a vector space, then so is the dual space V^*. Hence, we may form the **double dual space** V^{**}, which consists of all linear functionals $\Sigma:V^* \to F$. In other words, an element Σ of V^{**} is a linear map that assigns a scalar to each *linear functional* on V.

With this firmly in mind, there is one rather obvious way to obtain an element of V^{**}. Namely, if $\mathbf{v} \in V$, consider the map $\overline{\mathbf{v}}:V^* \to F$ defined by

$$\overline{\mathbf{v}}(f) = f(\mathbf{v})$$

which sends the linear functional f to the scalar $f(\mathbf{v})$. For obvious reasons, this map is called **evaluation at** $\mathbf{v}$. To see that $\overline{\mathbf{v}}$ is in V^{**}, we must show that it is linear. But if $f,g \in V^*$, then

$$\overline{\mathbf{v}}(rf+sg) = (rf+sg)(\mathbf{v}) = rf(\mathbf{v})+sg(\mathbf{v}) = r\overline{\mathbf{v}}(f)+s\overline{\mathbf{v}}(g)$$

and so $\overline{\mathbf{v}}$ is indeed linear.

Since evaluation at $\mathbf{v}$ is in V^{**} for all $\mathbf{v} \in V$, we can define a map $\tau:V \to V^{**}$ by

$$\tau(\mathbf{v}) = \overline{\mathbf{v}}$$

This is called the **canonical map** (or the **natural map**) from V to V^{**}. It is injective and, in the finite dimensional case, it is also surjective.

Theorem 3.13 The canonical map $\tau:V \to V^{**}$ defined by letting $\tau(\mathbf{v})$ be evaluation at $\mathbf{v}$, is a monomorphism. Furthermore, if V is finite dimensional, then τ is an isomorphism.

Proof. To see that τ is linear, we observe that

$$\tau(r\mathbf{u}+s\mathbf{v}) = \overline{r\mathbf{u}+s\mathbf{v}}$$

is evaluation at $r\mathbf{u}+s\mathbf{v}$. But

$$\overline{r\mathbf{u}+s\mathbf{v}}(f) = f(r\mathbf{u}+s\mathbf{v}) = rf(\mathbf{u})+sf(\mathbf{v}) = (r\overline{\mathbf{u}}+s\overline{\mathbf{v}})(f)$$

for all $f \in V^*$, and so

$$\tau(r\mathbf{u}+s\mathbf{v}) = \overline{r\mathbf{u}+s\mathbf{v}} = r\overline{\mathbf{u}}+s\overline{\mathbf{v}} = r\tau(\mathbf{u})+s\tau(\mathbf{v})$$

which shows that τ is linear.

To determine the kernel of τ, we observe that

$$\begin{aligned}\tau(\mathbf{v}) = 0 \;&\Rightarrow\; \overline{\mathbf{v}} = 0\\ &\Rightarrow\; \overline{\mathbf{v}}(f) = 0 \text{ for all } f \in V^*\\ &\Rightarrow\; f(\mathbf{v}) = 0 \text{ for all } f \in V^*\\ &\Rightarrow\; \mathbf{v} = \mathbf{0}\end{aligned}$$

by Theorem 3.10, and so $\mathit{ker}(\tau) = \{\mathbf{0}\}$, that is, τ is injective.

In the finite dimensional case, we have $dim(V^{**}) = dim(V^*) = dim(V)$, and so τ is also surjective. Thus, τ is an isomorphism. ∎

Note that if $dim(V) < \infty$, then since the dimensions of V and V^{**} are the same, we deduce immediately that $V \approx V^{**}$. This is not the point of Theorem 3.13. The point is that the *natural map* $\mathbf{v} \to \overline{\mathbf{v}}$ is an isomorphism. Because of this, V is said to be **algebraically reflexive.** Thus, Theorem 3.13 implies that all finite dimensional vector spaces are algebraically reflexive.

If V is finite dimensional, it is customary to identify the double dual space V^{**} with V, and to think of the elements of V^{**} simply as vectors in V.

Let us consider an example of a vector space that is not algebraically reflexive.

Example 3.4 Let V be a vector space over $F = \{0,1\}$, with a *countably infinite ordered basis* $\mathfrak{B} = (\mathbf{b}_1, \mathbf{b}_2, \ldots)$. Then any vector $\mathbf{v} \in V$ can be identified with its *coordinate sequence*

$$\mathbf{v} = (a_1, a_2, \ldots)$$

where $a_i \in \{0,1\}$ and only a *finite* number of the a_i are equal to 1.

On the other hand, any $f \in V^*$ is uniquely determined by its values on the vectors in $\mathfrak{B}$, and since these values can be arbitrarily chosen, f can be identified with a binary sequence

$$f = (\alpha_1, \alpha_2, \ldots)$$

with no restriction on the number of 1s in the sequence.

Now, we define the **support** of a binary sequence $x = (x_1, x_2, \ldots)$, denoted by $supp(x)$, to be the set of coordinate positions where $x_i = 1$. Thus, a vector in V is a binary sequence with *finite* support, whereas a linear functional on V is any binary sequence.

A moments reflection on the representation of f will reveal that

$$\overline{\mathbf{v}}(f) = f(\mathbf{v}) = |\, supp(\mathbf{v}) \cap supp(f) \,|$$

We can show that the canonical map $\tau: \mathbf{v} \to \overline{\mathbf{v}}$ is not surjective by finding a linear functional $\phi \in V^{**}$ that does not have the form $\overline{\mathbf{v}}$, for any $\mathbf{v} \in V$. To this end, define linear functionals $e_k \in V^*$ by

$$e_k = (0, \ldots, 0, 1, 0, \ldots)$$

where the 1 appears in the kth position. Then $supp(e_k) = \{k\}$, and so

$$\overline{\mathbf{v}}(e_k) = |\, supp(\mathbf{v}) \cap supp(e_k) \,| = |\, supp(\mathbf{v}) \cap \{k\} \,|$$

Hence, for any $\mathbf{v} \in V$, the map $\overline{\mathbf{v}}$ has the property that

(3.5) $$k > \max\{supp(\overline{\mathbf{v}})\} \Rightarrow \overline{\mathbf{v}}(\mathbf{e}_k) = 0$$

Now, since the linear functionals $\mathbf{e}_k$ are linearly independent, we may extend the set $\{\mathbf{e}_k\}$ to a basis $\mathfrak{B}$ for V^*. Let us define a map $\phi \in V^{**}$ by setting $\phi(\mathbf{e}_k) = 1$ for all k, and then extending ϕ to all vectors in the basis $\mathfrak{B}$ arbitrarily. (In other words, we don't care how ϕ is defined on the other elements of $\mathfrak{B}$.) Then ϕ defines a linear functional in V^{**}, with the property that

$$\phi(\mathbf{e}_k) = 1$$

for all k. But this shows, in conjunction with (3.5), that ϕ cannot have the form $\overline{\mathbf{v}}$, for any $\mathbf{v} \in V$. Hence, the canonical map is not surjective, and V is not algebraically reflexive. ▯

Annihilators

If $f \in V^*$, then f is defined on vectors in V. However, we may also define f on subsets M of V by letting

$$f(M) = \{f(\mathbf{v}) \mid \mathbf{v} \in M\}$$

Definition Let M be a nonempty *subset* of a vector space V. The **annihilator** M^0 of M is

$$M^0 = \{f \in V^* \mid f(M) = 0\}$$ ▯

The term *annihilator* is quite descriptive, since M^0 consists of all linear functionals that annihilate (send to 0) every vector in M.

It is not hard to see that M^0 is a *subspace* of V^*, even when M is not. Subject to this, we prove the following.

Theorem 3.14 If S is a subspace of a finite dimensional vector space V, then

$$dim(S^0) = dim(V) - dim(S)$$

Proof. Let $\{\mathbf{u}_1, \ldots, \mathbf{u}_k\}$ be a basis for S, and extend it to a basis

$$\mathfrak{B} = \{\mathbf{u}_1, \ldots, \mathbf{u}_k, \mathbf{v}_1, \ldots, \mathbf{v}_{n-k}\}$$

for V. Let

$$\mathfrak{B}^* = \{\mu_1, \ldots, \mu_k, \nu_1, \ldots, \nu_{n-k}\}$$

be the dual basis to $\mathfrak{B}$. We show that $\{\nu_1, \ldots, \nu_{n-k}\}$ is a basis for S^0. Certainly, this set is linearly independent, so we need only show that it spans S^0. But if $f \in S^0$, then since $f \in V^*$, we have

$$f = r_1\mu_1 + \cdots + r_k\mu_k + s_1\nu_1 + \cdots + s_{n-k}\nu_{n-k}$$

and since $f(\mathbf{v}) = 0$ for all $\mathbf{v} \in S$, we have for $i = 1, \ldots, k$,

$$0 = f(\mathbf{u}_i) = r_i$$

and so

$$f = s_1\nu_1 + \cdots + s_{n-k}\nu_{n-k}$$

which shows that $\{\nu_1,\ldots,\nu_{n-k}\}$ spans S^0. ∎

Example 3.5 To see what can happen with regard to Theorem 3.14, in the infinite dimensional case, let us continue Example 3.4. Thus, V is a vector space over $F = \{0,1\}$, with a *countably infinite ordered basis* $\mathfrak{B} = (\mathbf{b}_1,\mathbf{b}_2,\ldots)$. Let S be the subspace of V with ordered basis $\mathfrak{C} = (\mathbf{b}_1)$, and let T be the subspace of V with ordered basis $\mathfrak{D} = (\mathbf{b}_2,\mathbf{b}_3,\ldots)$. Since $|\mathfrak{D}| = |\mathfrak{B}|$, we have $T \approx V$.

Now consider the annihilator S^0. Any linear $f \in T^*$ can be extended to a linear functional $\hat{f} \in V^*$ by setting $\hat{f}(\mathbf{b}_1) = 0$. Moreover, any linear functional in S^0 has the form $\hat{f}$, for some $f \in T^*$, since $f \in S^0$ implies that $f(\mathbf{b}_1) = 0$. Hence, there is a one-to-one correspondence between S^0 and T^*, and so $|S^0| = |T^*|$. But $T \approx V$ implies that $T^* \approx V^*$, and so $|T^*| = |V^*|$, which implies that

$$|S^0| = |V^*|$$

But, we have seen in Example 3.4 that $|V^*| > |V|$, and so $|S^0| > |V|$, which implies that S^0 cannot be isomorphic to V, or any subspace of V. Hence, $dim(S^0) > dim(V)$. □

The basic properties of annihilators are contained in the following theorem.

Theorem 3.15

1) For any subsets M and N of V,

$$M \subset N \Rightarrow N^0 \subset M^0$$

2) If $dim(V) < \infty$ then, identifying V^{**} with V under the natural map, we have

$$M^{00} = span(M)$$

In particular, if S is a subspace of V, then $S^{00} = S$.

3) If $dim(V) < \infty$ and S and T are subspaces of V, then

$$(S \cap T)^0 = S^0 + T^0 \quad \text{and} \quad (S + T)^0 = S^0 \cap T^0$$ ∎

The annihilator provides a way to describe the dual space of a direct sum.

Theorem 3.16 Let $V = S \oplus T$. Then

1) $S^* \approx T^0$ and $T^* \approx S^0$
2) $(S \oplus T)^* = S^0 \oplus T^0$

Proof. First, let us prove that $S^* \approx T^0$. Roughly speaking, this says that any linear functional on V that annihilates the direct summand T is nothing more than a linear functional on its complement S, which certainly seems reasonable. The proof consists of making this precise. (Note that, in the finite dimensional case, a dimension argument establishes the isomorphism.)

For this purpose, let $f \in T^0 \subset V^*$. Thus, $f(T) = 0$. The map

$$\tau: f \to f|_S$$

that takes a functional $f \in V^*$ to its restriction $f|_S$, which is in S^*, is linear. Moreover, if $f|_S = 0$, then $f(S) = 0$, and since $f(T) = 0$, we have $f = 0$. Hence, τ is injective. Finally, we must show that τ is surjective. That is, for $g \in S^*$, we must find an $f \in T^0$ for which

$$f|_S(\mathbf{s}) = g(\mathbf{s})$$

for all $\mathbf{s} \in S$. In other words, we want to "extend" g to all of V, in such a way that the extension is in T^0. But that is easy – we just define the extension to be 0 on T. In particular, let $f \in V^*$ be defined by

$$f(\mathbf{s} + \mathbf{t}) = g(\mathbf{s})$$

Then f is well-defined and linear. Moreover, $f \in T^0$, since $f(\mathbf{t}) = f(\mathbf{0} + \mathbf{t}) = g(\mathbf{0}) = 0$, for all $\mathbf{t} \in T$. Finally, $f|_S$ is indeed g, and so τ is an isomorphism, which proves that $T^0 \approx S^*$. By symmetry, we also have $S^0 \approx T^*$.

To prove part 2, let $f \in S^0 \cap T^0$. Then $f(S) = 0 = f(T)$, which implies that $f = 0$. Hence, $S^0 \cap T^0 = \{0\}$. Since S^0 and T^0 are subspaces of V^*, we have $(S \oplus T)^* \supset S^0 \oplus T^0$.

On the other hand, if $f \in (S \oplus T)^*$, then we define g, $h \in (S \oplus T)^*$ by

$$g(\mathbf{s} + \mathbf{t}) = f(\mathbf{t}) \quad \text{and} \quad h(\mathbf{s} + \mathbf{t}) = f(\mathbf{s})$$

It is easy to see that these maps are well-defined and linear. Moreover,

$$g(S) = 0 \quad \text{and} \quad h(T) = 0$$

and so $g \in S^0$ and $h \in T^0$. Finally,

$$f(\mathbf{s} + \mathbf{t}) = f(\mathbf{t}) + f(\mathbf{s}) = g(\mathbf{s} + \mathbf{t}) + h(\mathbf{s} + \mathbf{t}) = (g + h)(\mathbf{s} + \mathbf{t})$$

and so $f = g + h \in S^0 \oplus T^0$. Hence, $(S \oplus T)^* \subset S^0 \oplus T^0$, which completes the proof. ∎

Operator Adjoints

If $\tau \in \mathcal{L}(V,W)$, then we may define a map $\tau^{\times}:W^* \to V^*$ by

$$\tau^{\times}(f) = f \circ \tau = f\tau$$

for $f \in W^*$. This makes sense, since $\tau:V \to W$ and $f:W \to F$, and so the composition $f\tau:V \to F$ is in V^*. Thus

$$\tau^{\times}(f)(\mathbf{v}) = f(\tau(\mathbf{v}))$$

for any $\mathbf{v} \in V$. The map $\tau^{\times}$ is called the **operator adjoint** of τ.

Let us establish the basic properties of the operator adjoint.

Theorem 3.17

1) $(\tau + \sigma)^{\times} = \tau^{\times} + \sigma^{\times}$ for $\tau,\sigma \in \mathcal{L}(V,W)$
2) $(r\tau)^{\times} = r\tau^{\times}$ for any $r \in F$ and $\tau \in \mathcal{L}(V,W)$
3) $(\tau\sigma)^{\times} = \sigma^{\times}\tau^{\times}$ for $\tau \in \mathcal{L}(V,W)$ and $\sigma \in \mathcal{L}(W,U)$
4) $(\tau^{-1})^{\times} = (\tau^{\times})^{-1}$ for any invertible $\tau \in \mathcal{L}(V)$

Proof. We prove parts 3 and 4. Part 3 follows from the fact that

$$(\tau\sigma)^{\times}(f) = f\tau\sigma = \sigma^{\times}(f\tau) = \tau^{\times}(\sigma^{\times}(f)) = (\tau^{\times}\sigma^{\times})(f)$$

for all $f \in U^*$. Part 4 follows from

$$\tau^{\times}(\tau^{-1})^{\times} = (\tau^{-1}\tau)^{\times} = \iota^{\times} = \iota$$

and, in the same way, $(\tau^{-1})^{\times}\tau^{\times} = \iota$. ∎

If $\tau \in \mathcal{L}(V,W)$ then $\tau^{\times} \in \mathcal{L}(W^*,V^*)$, and we may form $\tau^{\times\times} \in \mathcal{L}(V^{**},W^{**})$. Of course, $\tau^{\times\times}$ is not equal to τ. However, in the finite dimensional case, if we use the natural maps to identify V^{**} with V and W^{**} with W, then we can think of $\tau^{\times\times}$ as being in $\mathcal{L}(V,W)$. With this in mind, $\tau^{\times\times}$ is equal to τ.

Theorem 3.18 Let V be finite dimensional, and let $\tau \in \mathcal{L}(V,W)$. If we identify V^{**} with V and W^{**} with W, using the natural maps, then $\tau^{\times\times} = \tau$.

Proof. Before making any identifications, we have $\tau^{\times\times}:V^{**} \to W^{**}$, and

$$\tau^{\times\times}(\overline{\mathbf{v}})(f) = \overline{\mathbf{v}}\tau^{\times}(f) = \overline{\mathbf{v}}(f\tau) = f\tau(\mathbf{v}) = \overline{\tau(\mathbf{v})}(f)$$

for all $f \in W^*$, and so

$$\tau^{\times\times}(\overline{\mathbf{v}}) = \overline{\tau(\mathbf{v})}$$

Therefore, with the appropriate identifications,

$$\tau^{\times\times}(\mathbf{v}) = \tau(\mathbf{v})$$

for all $\mathbf{v} \in V$, and so $\tau^{\times\times} = \tau$. ∎

The next result describes the kernel and image of the operator adjoint.

Theorem 3.19 Let $\tau \in \mathcal{L}(V,W)$. Then

1) $ker(\tau^{\times}) = im(\tau)^0$
2) $im(\tau^{\times})^0 = ker(\tau)$, under the natural identification
3) $im(\tau^{\times}) \subset ker(\tau)^0$
4) if $dim(V) < \infty$ and $dim(W) < \infty$ then $im(\tau^{\times}) = ker(\tau)^0$

Proof. For reference, note that $\tau: V \to W$ and $\tau^{\times}: W^* \to V^*$. To prove (1), observe that

$$\begin{aligned} f \in ker(\tau^{\times}) &\Leftrightarrow \tau^{\times}(f) = 0 \\ &\Leftrightarrow f\tau = 0 \\ &\Leftrightarrow f(\tau(\mathbf{v})) = 0 \text{ for all } \mathbf{v} \in V \\ &\Leftrightarrow f(im(\tau)) = 0 \\ &\Leftrightarrow f \in im(\tau)^0 \end{aligned}$$

To prove part 2, we have

$$\begin{aligned} \mathbf{v} \in ker(\tau) &\Leftrightarrow \tau(\mathbf{v}) = \mathbf{0} \\ &\Leftrightarrow f(\tau(\mathbf{v})) = 0 \text{ for all } f \in W^* \\ &\Leftrightarrow \tau^{\times}(f)(\mathbf{v}) = 0 \text{ for all } f \in W^* \\ &\Leftrightarrow \overline{\mathbf{v}}(\tau^{\times}(f)) = 0 \text{ for all } f \in W^* \\ &\Leftrightarrow \overline{\mathbf{v}} \in im(\tau^{\times})^0 \end{aligned}$$

Part 3 is proved as follows. For all $\mathbf{v} \in ker(\tau)$ and $f \in W^*$,

$$\tau^{\times}(f)(\mathbf{v}) = f(\tau(\mathbf{v})) = 0$$

and so $\tau^{\times}(f)(ker(\tau)) = 0$, that is, $\tau^{\times}(f) \in ker(\tau)^0$. Since this holds for all $f \in W^*$, we have

$$im(\tau^{\times}) \subset ker(\tau)^0$$

As for part 4, when the vector spaces are finite dimensional, part 2 gives

$$im(\tau^{\times}) \approx im(\tau^{\times})^{00} \approx ker(\tau)^0$$

But, according to part 3, $im(\tau^{\times}) \subset ker(\tau)^0$, and so these spaces must be equal. ∎

Corollary 3.20 Let $\tau \in \mathcal{L}(V,W)$, where V and W are finite dimensional. Then $rk(\tau) = rk(\tau^{\times})$. ∎

In the finite dimensional case, $\tau \in \mathcal{L}(V,W)$ and $\tau^{\times} \in \mathcal{L}(W^*,V^*)$ can both be represented by matrices. To explore the connection between these matrices, suppose that

$$\mathcal{B} = (\mathbf{b}_1,\ldots,\mathbf{b}_n) \quad \text{and} \quad \mathcal{C} = (\mathbf{c}_1,\ldots,\mathbf{c}_m)$$

are ordered bases for V and W, respectively, and that

$$\mathcal{B}^* = (\mathbf{b}_1^*,\ldots,\mathbf{b}_n^*) \quad \text{and} \quad \mathcal{C}^* = (\mathbf{c}_1^*,\ldots,\mathbf{c}_m^*)$$

are the corresponding dual bases. If we let

$$[\tau]_{\mathcal{B},\mathcal{C}} = (a_{i,j})$$

then $a_{i,j}$ is the coordinate of $\mathbf{c}_i$ in $\tau(\mathbf{b}_j)$.

On the other hand, if

$$[\tau^{\times}]_{\mathcal{C}^*,\mathcal{B}^*} = (\alpha_{i,j})$$

then $\alpha_{i,j}$ is the coordinate of $\mathbf{b}_i^*$ in $\tau^{\times}(\mathbf{c}_j^*)$. But this coordinate is

$$\tau^{\times}(\mathbf{c}_j^*)(\mathbf{b}_i) = \mathbf{c}_j^*(\tau(\mathbf{b}_i))$$

which is the coordinate of $\mathbf{c}_j$ in $\tau(\mathbf{b}_i)$, and this in turn is $a_{j,i}$. In short, $\alpha_{i,j} = a_{j,i}$, and so

$$[\tau^{\times}]_{\mathcal{C}^*,\mathcal{B}^*} = ([\tau]_{\mathcal{B},\mathcal{C}})^{\mathsf{T}}$$

We have established the following.

Theorem 3.21 Let $\tau \in \mathcal{L}(V,W)$, where V and W are finite dimensional. If $\mathcal{B}$ is an ordered basis for V, $\mathcal{C}$ is an ordered basis for W, and $\mathcal{B}^*$ and $\mathcal{C}^*$ are the corresponding dual bases, then

$$[\tau^{\times}]_{\mathcal{C}^*,\mathcal{B}^*} = ([\tau]_{\mathcal{B},\mathcal{C}})^{\mathsf{T}}$$

In words, the matrix of the adjoint $\tau^{\times}$ is the transpose of the matrix of τ. ∎

EXERCISES

1. If S is a subspace of V, show that $\mathbf{u} \equiv \mathbf{v} \Leftrightarrow \mathbf{u} - \mathbf{v} \in S$ is an equivalence relation on V.
2. Prove that the operations of coset addition and scalar multiplication are well-defined.
3. Prove that there is only one map $U: V/S \to W$ with the property described in Figure 3.1.
4. Prove the first isomorphism theorem.
5. Let S be a subspace of a vector space V. Show that

$$dim(V) = dim(S) + dim(V/S)$$

6. Complete the proof of Theorem 3.9.
7. Let S be a subspace of V. Can you describe a relationship between the set of all subspaces S' of V for which $S \subset S' \subset V$ and the set of all subspaces of the quotient space V/S?
8. Let S be a subspace of V. Starting with a basis $\{s_1,\ldots,s_k\}$ for S, how would you find a basis for V/S?
9. Use the First Isomorphism Theorem to prove that if $\tau:V\to W$, then

$$dim(ker(\tau)) + dim(im(\tau)) = dim(V)$$

10. Let $\tau \in \mathcal{L}(V)$, and suppose that S is a subspace of V. Define a map by

$$\tau':V/S\to V/S \qquad \tau'(\mathbf{v}+S) = \tau(\mathbf{v})+S$$

When is τ' well-defined? If τ' is well-defined, is it a linear transformation? What are $im(\tau')$ and $ker(\tau')$?
11. Show that, for any nonzero vector $\mathbf{v} \in V$, there exists a linear functional $f \in V^*$ for which $f(\mathbf{v}) \neq 0$.
12. Show that a vector $\mathbf{v} \in V$ is zero if and only if $f(\mathbf{v}) = 0$ for all $f \in V^*$.
13. Let S be a proper subspace of a finite dimensional vector space V, and let $\mathbf{v} \in V - S$. Show that there is a linear functional $f \in V^*$ for which $f(\mathbf{v}) = 1$ and $f(\mathbf{s}) = 0$ for all $\mathbf{s} \in S$.
14. Let S be an (n−1)-dimensional subspace of an n-dimensional vector space V. Show that there is a linear functional $f \in V^*$ whose kernel is S. If f and g are two such functionals, must there be any relationship between them?
15. Let $\mathcal{B}$ be a basis for an *infinite dimensional* vector space V, and define, for all $\mathbf{b} \in \mathcal{B}$, the map $\mathbf{b}' \in V^*$ by $\mathbf{b}'(\mathbf{c}) = 1$ if $\mathbf{c} = \mathbf{b}$, and 0 otherwise. Does $\{\mathbf{b}' \mid \mathbf{b} \in \mathcal{B}\}$ form a basis for V^*? What do you conclude about the concept of a dual basis?
16. Show that M^0 is a subspace of V^*, for any nonempty subset M of V.
17. Prove that $(S \oplus T)^* \approx S^* \oplus T^*$.
18. Prove that $0^\times = 0$, and that $\iota^\times = \iota$, where 0 is the zero linear operator and ι is the identity.
19. Let S be a subspace of V. Prove that $(V/S)^* \approx S^0$.
20. Verify that
    (a) $(\tau + \sigma)^\times = \tau^\times + \sigma^\times$ for $\tau,\sigma \in \mathcal{L}(V,W)$.
    (b) $(r\tau)^\times = r\tau^\times$ for any $r \in F$ and $\tau \in \mathcal{L}(V,W)$
21. Let $\tau \in \mathcal{L}(V,W)$, where V and W are finite dimensional. Prove that $rk(\tau) = rk(\tau^\times)$.

The Number of Subspaces of a Vector Space over a Finite Field

22. Let F be a finite field of size q, and let V be an n-dimensional vector space over F. The purpose of this exercise is to show that there are

$$\binom{n}{k}_q = \frac{(q^n - 1)\cdots(q-1)}{(q^k - 1)\cdots(q-1)(q^{n-k} - 1)\cdots(q-1)}$$

subspaces of V of dimension k. The expressions $\binom{n}{k}_q$ are called **Gaussian coefficients**, and have properties similar to those of the binomial coefficients.

a) Let $S(n,k)$ be the number of k-dimensional subspaces of V. Let $N(n,k)$ be the number of k-tuples of linearly independent vectors $(v_1,\ldots,v_k)$ in V. Show that

$$N(n,k) = (q^n - 1)(q^n - q)\cdots(q^n - q^{n-k+1})$$

b) Now, each of the k-tuples in (a) can be obtained by first choosing a subspace of V of dimension k, and then selecting the vectors from this subspace. Show that, for any k-dimensional subspace of V, the number of k-tuples of independent vectors in this subspace is

$$(q^k - 1)(q^k - q)\cdots(q^k - q^{k-1})$$

c) Show that

$$N(n,k) = S(n,k)(q^k - 1)(q^k - q)\cdots(q^k - q^{k-1})$$

How does this complete the proof?

CHAPTER 4

Modules I

Contents: *Motivation. Modules. Submodules. Direct Sums. Spanning Sets. Linear Independence. Homomorphisms. Free Modules. Summary. Exercises.*

Motivation

Let V be a vector space over a field F, and let $\tau \in \mathcal{L}(V)$. Then for any polynomial $p(x) \in F[x]$, the operator $p(\tau)$ is well-defined. For instance, if $p(x) = 1 + 2x + x^3$, then

$$p(\tau) = \iota + 2\tau + \tau^3$$

where ι is the identity operator, and τ^3 is the threefold composition $\tau \circ \tau \circ \tau$.

We can now define the product of a polynomial $p(x) \in F[x]$ and a vector $\mathbf{v} \in V$ by

$$p(x)\mathbf{v} = p(\tau)(\mathbf{v}) \tag{4.1}$$

This product satisfies the usual properties of scalar multiplication, namely, for all $r(x), s(x) \in F[x]$ and $\mathbf{u},\mathbf{v} \in V$,

$$\begin{aligned} r(x)(\mathbf{u}+\mathbf{v}) &= r(x)\mathbf{u} + r(x)\mathbf{v} \\ (r(x)+s(x))\mathbf{u} &= r(x)\mathbf{u} + s(x)\mathbf{u} \\ [r(x)s(x)]\mathbf{u} &= r(x)[s(x)\mathbf{u} \\ 1\mathbf{u} &= \mathbf{u} \end{aligned}$$

Thus, for a fixed $\tau \in \mathcal{L}(V)$, V is an algebraic structure under the operations of addition and scalar multiplication by polynomials in $F[x]$.

Note, however, that since the ring $F[x]$ is not a field, these two operations do not make V into a vector space. Nevertheless, this important situation, which we will study extensively in the sequel, motivates the following definition.

Modules

Definition Let R be a *commutative ring with identity*, whose elements are called **scalars**. An **R-module** (or a **module over R**) is a nonempty set M, together with two operations. The first operation, called *addition* and denoted by $+$, assigns to each pair $(u,v) \in M \times M$, an element $u+v \in M$. The second operation, denoted by juxtaposition, assigns to each pair $(r,u) \in R \times M$, an element $rv \in M$. Furthermore, the following properties must hold.

1) M is an abelian group under addition.
2) For all $r, s \in R$ we have

$$\begin{aligned} r(u+v) &= ru + rv \\ (r+s)u &= ru + su \\ (rs)u &= r(su) \\ 1u &= u \end{aligned}$$

for all $u,v \in M$. ▯

The definition of a module requires that the ring R of scalars be commutative. This requirement is sometimes omitted, but modules over noncommutative rings can behave quite differently than modules over commutative rings. For instance, it is possible for a module over a noncommutative ring to have bases of different sizes. Since such modules will not be needed for the sequel, we require commutativity.

Even with the requirement of commutativity, modules behave quite differently than vector spaces. For example, there are modules that do not have *any* linearly independent elements. Of course, such a module cannot have a basis.

The connection between modules and vector spaces is very simple: *a vector space is a module over a field.*

Example 4.1

1) If R is a ring, the set R^n of all ordered n-tuples, whose components lie in R, is an R-module, with addition and scalar multiplication defined componentwise (just as in F^n),

$$(a_1,\ldots,a_n) + (b_1,\ldots,b_n) = (a_1+b_1,\ldots,a_n+b_n)$$

and

$$r(a_1,\ldots,a_n) = (ra_1,\ldots,ra_n)$$

for a_i, b_i, $r \in R$. For example, $\mathbb{Z}^n$ is the Z-module of all ordered n-tuples of integers.

2) If R is a ring, the set $\mathcal{M}_{m,n}(R)$ of all matrices of size $m \times n$, is an R-module, under the usual operations of matrix addition and scalar multiplication over R. Since R is a ring, we can also take the product of matrices in $\mathcal{M}_{m,n}(R)$. One important example is when $R = F[x]$, whence $\mathcal{M}_{m,n}(F[x])$ is the F[x]-module of all $m \times n$ matrices whose entries are polynomials.

3) Any commutative ring R with identity is a module over itself, that is, R is an R-module. In this case, scalar multiplication is just multiplication by elements of R, that is, scalar multiplication is the ring multiplication. The defining properties of a ring imply that the defining properties of an R-module are satisfied. We shall use this example many times in the sequel. □

When we turn in a later chapter to the study of the structure of a linear transformation $\tau \in \mathcal{L}(V)$, we will think of V as having the structure of a vector space over F, *as well as* a module, over $F[x]$. Put another way, V is an abelian group under addition, with *two* scalar multiplications – one whose scalars are elements of F and one whose scalars are *polynomials* over F. This viewpoint will be of tremendous benefit for the study of τ. For now, we concentrate only on modules.

Many of the basic concepts that we defined for vector spaces can also be defined for modules, although their properties are often quite different.

Submodules

The definition of submodule parallels that of subspace.

Definition A **submodule** of an R-module M is a subset S of M that is an R-module in its own right, under the operations obtained by restricting the operations of M to S. □

Theorem 4.1 A nonempty subset S of an R-module M is a submodule if and only if

$$r,s \in R,\ u,v \in S \Rightarrow ru + sv \in S$$ ∎

Theorem 4.2 If S and T are submodules of M, then $S \cap T$ and

$$S + T = \{u + v \mid u \in S,\ v \in T\}$$

are also submodules of M. □

Recall that a commutative ring R with identity is a module over itself. Thus, in a sense, R plays two roles – it is a ring and it is an R-module. Now suppose that S is a submodule of R. According to Theorem 4.1, if $a,b \in S$ and $r \in R$, then $a-b \in S$ and $ra \in S$. Hence, S is an *ideal* of the *ring* R. Conversely, if $\mathfrak{I}$ is an ideal of the ring R, then $\mathfrak{I}$ is also a submodule of the module R. In other words, *the subrings of the R-module R are precisely the ideals of the ring R.*

Direct Sums

The definition of direct sum is the same for modules as for vector spaces. We will confine our attention to the direct sum of a finite number of modules.

Definition Let M be an R-module. We say that M is the **direct sum** of the submodules $S_1,\ldots,S_n$ if every $v \in M$ can be written, in a *unique* way (except for order), as a sum of elements from the submodules S_i. More specifically, M is the direct sum of $S_1,\ldots,S_n$ if, for all $v \in M$, we have

$$v = u_1 + \cdots + u_n$$

for some $u_i \in S_i$, and furthermore, if

$$v = w_1 + \cdots + w_n$$

where $w_i \in S_i$, then $w_i = u_i$ for all $i = 1,\ldots,n$.

If M is the direct sum of $S_1,\ldots,S_n$, we write

$$M = S_1 \oplus \cdots \oplus S_n$$

and refer to each S_i as a **direct summand** of M. If $M = S \oplus S^c$, we refer to S^c as a **complement** of S in M. ▯

In the case of vector spaces, every subspace has a complement. However, as the next example shows, this is *not* true for modules.

Example 4.2 The set $\mathbb{Z}$ of integers is a $\mathbb{Z}$-module, that is, $\mathbb{Z}$ is a module over itself. Let us examine the nature of the submodules of $\mathbb{Z}$. Since the submodules of the $\mathbb{Z}$-module $\mathbb{Z}$ are precisely the ideals of the ring $\mathbb{Z}$, and since $\mathbb{Z}$ is a principal ideal domain (see Chapter 0), the submodules of $\mathbb{Z}$ are precisely the sets

$$\langle n \rangle = \mathbb{Z}n = \{zn \mid z \in \mathbb{Z}\}$$

Thus, all *nonzero* submodules of $\mathbb{Z}$ are of the form $\mathbb{Z}u$, for some *positive* $u \in \mathbb{Z}$. As a result, we see that any two nonzero submodules of $\mathbb{Z}$ have nonzero intersection. For if $u,v > 0$, then $0 \neq uv \in \mathbb{Z}u \cap \mathbb{Z}v$.

Hence, none of the submodules $\mathbb{Z}u$, for $u \neq 0$ or 1, have complements. □

As with vector spaces, we have the following useful characterization of direct sums.

Theorem 4.3 A module M is the direct sum of submodules $S_1, \ldots, S_n$ if and only if

1) $M = S_1 + \cdots + S_n$
2) For each $i = 1, \ldots, n$

$$S_i \cap \Big(\sum_{j \neq i} S_j\Big) = \{0\}$$

■

Spanning Sets

The concept of spanning set carries over to modules as well.

Definition The submodule **spanned** (or **generated**) by a subset S of a module M is the set of all linear combinations of elements of S

$$\langle S \rangle = span(S) = \{r_1 v_1 + \cdots + r_n v_n \mid r_i \in R,\ v_i \in M\}$$

A subset $S \subset M$ is said to **span** M, or **generate** M, if

$$M = span(S)$$

that is, if every $v \in M$ can be written in the form

$$v = r_1 v_1 + \cdots + r_n v_n$$

for some $r_1, \ldots, r_n \in R$, and $v_1, \ldots, v_n \in M$. □

Observe that $\langle v \rangle = Rv = \{rv \mid r \in R\}$ is just the set of all scalar multiples of v. Since modules of this type are extremely important, they have a special name.

Definition Let M be an R-module. A submodule of the form $\langle v \rangle = Rv = \{rv \mid r \in R\}$, for $v \in M$, is called the **cyclic submodule** generated by v. □

For reasons that will become clear soon, we need the following definition.

Definition An R-module M is said to be **finitely generated** if it contains a *finite* set that generates M. □

Of course, a vector space is finitely generated if and only if it has a finite basis, that is, if and only if it is finite dimensional. However, for modules, things are not quite as simple. The following is an example of a finitely generated module that has a submodule that is *not* finitely generated.

Example 4.3 Let R be the ring $F[x_1,x_2,\ldots]$ of all polynomials in infinitely many variables over a field F. It will be convenient to use the boldface letter $\mathbf{x}$ to denote $x_1,x_2,\ldots$, and write a polynomial in R in the form $p(\mathbf{x})$. (Each polynomial in R, being a finite sum, involves only finitely many variables, however.) Then R is an R-module, and as such, is finitely generated by the identity element $p(\mathbf{x}) = 1$.

Now, consider the submodule S of all polynomials with zero constant term. This module is generated by the variables themselves,

$$S = \langle x_1,x_2,\ldots\rangle$$

However, S is not generated by any finite set of polynomials. For suppose that $\{p_1,\ldots,p_n\}$ is a finite generating set for S. Then, for each k, there exist polynomials $a_{k,1}(\mathbf{x}),\ldots,a_{k,n}(\mathbf{x})$ for which

$$x_k = \sum_{i=1}^{n} a_{k,i}(\mathbf{x})p_i(\mathbf{x}) \tag{4.2}$$

Note that since $p_i(\mathbf{x}) \in S$, it has zero constant term.

Since there are only a finite number of variables involved in all of the $p_i(\mathbf{x})$'s, we can choose an index k for which $p_1(\mathbf{x}),\ldots,p_n(\mathbf{x})$ do not involve x_k. For each $a_{k,j}(\mathbf{x})$, let us collect all terms involving x_k, and all terms not involving x_k,

$$a_{k,j}(\mathbf{x}) = x_k q_j(\mathbf{x}) + r_j(\mathbf{x}) \tag{4.3}$$

where $q_j(\mathbf{x})$ is any polynomial in R, and $r_j(\mathbf{x})$ does not involve x_k. Now (4.2) and (4.3) give

$$x_k = \sum_{i=1}^{n} [x_k q_j(\mathbf{x}) + r_j(\mathbf{x})]p_i(\mathbf{x})$$

$$= x_k \sum_{i=1}^{n} q_j(\mathbf{x})p_i(\mathbf{x}) + \sum_{i=1}^{n} r_j(\mathbf{x})p_i(\mathbf{x})$$

The last sum does not involve x_k and so it must equal 0. Hence, the first sum must equal 1, but this is not possible, since the $p_i(\mathbf{x})$'s have no constant terms. Hence, S has no finite generating set. $\square$

Linear Independence

The concept of linear independence also carries over to modules.

Definition A nonempty subset S of a module M is **linearly independent** if for any $v_1,\ldots,v_n \in M$,

$$r_1v_1 + \cdots + r_nv_n = 0 \Rightarrow r_1 = \cdots = r_n = 0$$

If a set S is not linearly independent, we say that it is **linearly dependent.** ▯

It is clear from the definition that any nonempty subset of a linearly independent set is linearly independent.

In a vector space, the set $S = \{v\}$, consisting of a single *nonzero* vector $\mathbf{v}$, is linearly independent. However, in a module, this need not be the case.

Example 4.4 The abelian group $\mathbb{Z}_n = \{0,1,\ldots,n-1\}$ is a $\mathbb{Z}$-module, with scalar multiplication defined by $za = (z \cdot a) \bmod n$, for all $n \in \mathbb{Z}$ and $a \in \mathbb{Z}_n$. However, since $na = 0$ for all $a \in \mathbb{Z}_n$, we see that no singleton set $\{a\}$ is linearly independent. ▯

Recall that, in a vector space, a set S of vectors is linearly *dependent* if and only if some vector in S is a linear combination of the other vectors in S. For arbitrary modules, this is not true.

Example 4.5 Consider the $\mathbb{Z}$-module $\mathbb{Z}^2$, consisting of all ordered pairs of integers. Then the ordered pairs $(2,0)$ and $(3,0)$ are linearly dependent, since

$$3(2,0) - 2(3,0) = (0,0)$$

but neither one of these ordered pairs is a linear combination (i.e., scalar multiple) of the other! ▯

The problem in the previous example is that

$$r_1v_1 + \cdots + r_nv_n = 0$$

and (say) $r_1 \neq 0$ together imply that

$$r_1v_1 = -r_2v_2 - \cdots - r_nv_n$$

but, in general, we *cannot* divide both sides by r_1, since it may not have a multiplicative inverse in the ring R.

We can now define the concept of a basis for a module.

Definition Let M be an R-module. A subset $\mathcal{B}$ of M is a **basis** if $\mathcal{B}$ is linearly independent and spans M. □

Theorem 4.4 A subset $\mathcal{B}$ of a module M is a basis if and only if, for every $v \in M$, there is a *unique* set of scalars $r_1, \ldots, r_n$ for which

$$v = r_1 v_1 + \cdots + r_n v_n$$ ∎

In a vector space, a set of vectors is a basis if and only if it is a minimal spanning set, or equivalently, a maximal linearly independent set. For modules, the following is the best we can do, in general.

Theorem 4.5 Let $\mathcal{B}$ be a basis for an R-module M. Then
1) $\mathcal{B}$ is a minimal spanning set.
2) $\mathcal{B}$ is a maximal linearly independent set. ∎

The $\mathbb{Z}$-module of Example 4.4 is an example of a module that has no basis, since it has no linearly independent sets. But since the entire module is a spanning set, we deduce that a minimal spanning set need not be a basis. In the exercises, the reader is asked to give an example of a module M that has a finite basis, but with the property that not every spanning set in M contains a basis, and not every linearly independent set in M is contained in a basis.

We will continue our discussion of bases for modules in a moment, but first let us discuss the module counterpart of linear transformations.

Homomorphisms

The term linear transformation is special to vector spaces. However, the *concept* applies to most algebraic structures.

Definition Let M and N be R-modules. A function $\tau: M \to N$ is said to be a **homomorphism** if

$$\tau(ru + sv) = r\tau(u) + s\tau(v)$$

for all scalars $r,s \in R$ and $u,v \in M$. The set of all homomorphisms from M to N is denoted by $Hom(M,N)$. Moreover, we have the following definitions.
1) An **endomorphism** is a homomorphism from M to M.
2) A **monomorphism** is an injective homomorphism.
3) An **epimorphism** is a surjective homomorphism.
4) An **isomorphism** is a bijective homomorphism. □

Theorem 4.6 Let $\tau \in Hom(M,N)$. The kernel and image of τ, defined as for linear transformations, by

$$ker(\tau) = \{v \in M \mid \tau(v) = 0\}$$

and

$$im(\tau) = \{\tau(v) \mid v \in M\}$$

are submodules of M and N, respectively. ∎

Free Modules

The fact that not all modules have a basis leads us to make the following definition.

Definition An R-module M is said to be **free** if it has a basis. If $\mathfrak{B}$ is a basis for M, we say that M is **free on** $\mathfrak{B}$. □

The next example shows that even free modules are not very much like vector spaces. It is an example of a free module that has a submodule that is not free!

Example 4.6 The set $\mathbb{Z} \times \mathbb{Z}$ is a free module over itself, with basis $\{(1,1)\}$. To see this, observe that $(1,1)$ is linearly independent, since

$$(n,m)(1,1) = (0,0) \Rightarrow (n,m) = (0,0)$$

Also, $(1,1)$ spans $\mathbb{Z} \times \mathbb{Z}$, since $(n,m) = (n,m)(1,1)$.

But the submodule $S = \mathbb{Z} \times \{0\}$ is *not* free, since it has no linearly independent elements, and hence no basis. This follows from the fact that, if $(n,0) \neq (0,0)$, then, for instance $(0,1)(n,0) = (0,0)$, and so $\{(n,0)\}$ is not linearly independent. □

Since all bases for a vector space V have the same cardinality, the concept of vector space dimension is well-defined. We now turn to the same issue for modules. The next example shows what can happen if the ring R is not commutative – it is an example of a module over a noncommutative ring that has a basis of size n for any natural number n!

Example 4.7 Let V be a vector space over F, with a countably infinite basis $\mathfrak{B} = \{\mathbf{b}_1, \mathbf{b}_2, \ldots\}$. Let $R = \mathcal{L}(V)$ be the ring of linear operators on V. Observe that R is not commutative, since composition of functions is not commutative.

The ring R is an R-module, and as such, the identity map ι forms a basis for R. However, we can also construct a basis for R of any desired finite size n. We begin by partitioning $\mathfrak{B}$ into n blocks.

For each $s = 0,\ldots,n-1$, let

$$\mathfrak{B}_s = \{\mathbf{b}_i \mid i \equiv s \bmod n\} = \{\mathbf{b}_i \mid i = kn+s \text{ for some } k\}$$

Now we define elements $\beta_s \in R = \mathcal{L}(V)$ by their action on the basis vectors in $\mathfrak{B}$ as follows. The intention is that β_s is zero on all basis vectors *not* in $\mathfrak{B}_s$, and β_s takes $\mathbf{b}_{kn+s} \in \mathfrak{B}_s$ to $\mathbf{b}_k$.

Since any nonnegative integer i has the form $kn+t$ for unique k and t satisfying $0 \le t < n$, we can define β_s by

$$\beta_s(\mathbf{b}_{kn+t}) = \begin{cases} \mathbf{b}_k & \text{if } t = s \\ \mathbf{0} & \text{if } t \neq s \end{cases}$$

Now $\mathfrak{C}_n = \{\beta_0,\ldots,\beta_{n-1}\}$ is linearly independent. For if $\alpha_s \in \mathcal{L}(V)$, and

$$0 = \alpha_0\beta_0 + \cdots + \alpha_{n-1}\beta_{n-1}$$

then, applying this to $\mathbf{b}_{kn+t}$ gives

$$\mathbf{0} = \alpha_t\beta_t(\mathbf{b}_{kn+t}) = \alpha_t(\mathbf{b}_k)$$

for all k. Hence, $\alpha_t = 0$.

Also, $\mathfrak{C}_n$ spans $R = \mathcal{L}(V)$. For if $\tau \in \mathcal{L}(V)$, we define $\alpha_s \in \mathcal{L}(V)$ by

$$\alpha_s(\mathbf{b}_k) = \tau(\mathbf{b}_{kn+s})$$

Then

$$(\alpha_0\beta_0 + \cdots + \alpha_{n-1}\beta_{n-1})(\mathbf{b}_{kn+t}) = \alpha_t\beta_t(\mathbf{b}_{kn+t}) = \alpha_t(\mathbf{b}_k) = \tau(\mathbf{b}_{kn+t})$$

and so

$$\tau = \alpha_0\beta_0 + \cdots + \alpha_{n-1}\beta_{n-1}$$

which shows that $\tau \in span\{\beta_0,\ldots,\beta_{n-1}\}$. Thus, $\mathfrak{C}_n = \{\beta_0,\ldots,\beta_{n-1}\}$ is a basis for $\mathcal{L}(V)$, and we have shown that $\mathcal{L}(V)$ has a basis of any finite size n. □

Example 4.7 shows that modules over noncommutative rings can behave very poorly when it comes to bases. Fortunately, when the ring of scalars is commutative, things are much nicer. We will postpone the proof of the following theorem to the next chapter.

Theorem 4.7 Let M be a *free* R-module. (By our definition, R is a commutative ring with identity.) Then any two bases of M have the same cardinality. ∎

Theorem 4.7 allows us to define the rank of a free module. (In the case of modules, it is customary to use the term rank, rather than dimension.)

Definition Let M be a free R-module. We define the **rank** $rk(M)$ of M to be the cardinality of any basis for M. □

Recall that if B is a basis for a vector space V over F, then V is isomorphic to the vector space $(F^B)_0$ of all functions from B to F that have finite support. A similar result holds for *free* R-modules. We begin by establishing that $(R^B)_0$ is a free R-module.

Theorem 4.8 Let B be any set, and let R be a ring. The set $(R^B)_0$ of all functions from B to R that have finite support is a free R-module, with basis $\mathfrak{B} = \{\delta_b\}$ defined by

$$\delta_b(x) = \begin{cases} 1 & \text{if } x = b \\ 0 & \text{if } x \neq b \end{cases}$$

and rank $|B|$. This basis is referred to as the **standard basis** for $(R^B)_0$. ∎

Theorem 4.9 Let M be an R-module. If B is a basis for M, then M is isomorphic to $(R^B)_0$.

Proof. Since B is a basis for M, any $v \in M$ has a unique representation (up to order) as a linear combination of elements of B. If

$$v = r_1 b_1 + \cdots + r_n b_n$$

then we let $\overline{v} \in (R^B)_0$ be the function defined by

$$\overline{v}(b) = \begin{cases} r_i & \text{if } b = b_i \text{ for some } i \\ 0 & \text{if } b \neq b_i \text{ for any } i \end{cases}$$

In words, $\overline{v}$ is the function that assigns to each basis element $b \in B$, the coefficient of b in the expression of v as a linear combination of basis elements. This defines a map $\tau: M \to (R^B)_0$, by $\tau(v) = \overline{v}$.

It is easy to see that τ is a module homomorphism from M to $(R^B)_0$. Furthermore, τ is injective, since $\tau(v) = \overline{v} = 0$ implies that the coordinates of v with respect to *all* basis elements are 0, and so v must be 0. Also, τ is surjective, since if $f \in (R^B)_0$, then we may define $v \in M$ by

$$v = \sum_{b \in B} f(b)b$$

Since f has finite support, this is a finite linear combination of basis elements. Moreover, $\tau(v) = \overline{v}$ has the property that $\overline{v}(b) = f(b)$ for all $b \in B$, and so $\tau(v) = \overline{v} = f$. Thus, τ is an isomorphism from M to $(R^B)_0$. ∎

Corollary 4.10 Two free R-modules are isomorphic if and only if they have the same rank.

Proof. If $M \approx N$, then any isomorphism τ from M to N maps a basis for M to a basis for N. Since τ is a bijection, we have $rk(M) = rk(N)$. Conversely, suppose that $rk(M) = rk(N)$. Let $\mathfrak{B}$ be a basis for M and let $\mathfrak{C}$ be a basis for N. Since $|\mathfrak{B}| = |\mathfrak{C}|$, there is a bijective map $\tau:\mathfrak{B}\to\mathfrak{C}$. This map can be extended by linearity to an isomorphism of M onto N, and so $M \approx N$. ∎

Summary

Here is a list of some of the properties of modules that emphasize the differences between modules and vector spaces.

1) A submodule of a module need not have a complement.
2) A submodule of a finitely generated module need not be finitely generated.
3) There exist modules with no linearly independent elements, and hence with no basis.
4) In a module, there may exist a set S of linearly dependent elements for which no element in S is a linear combination of the other elements in S.
5) In a module, a minimal spanning set is not necessarily a basis.
6) In a module, a maximal linearly independent set is not necessarily a basis. In fact, maximal linearly independent sets need not even exist.
7) A module over a *noncommutative* ring may have bases of different sizes. However, all bases for a free module over a commutative ring with identity have the same size.
8) There exist free modules with linearly independent sets that are not contained in a basis, and spanning sets that do not contain a basis.

EXERCISES

1. Give the details to show that any commutative ring with identity is a module over itself.
2. Let M be an R-module, and let I be an ideal in R. Let IM be the set of all finite sums of the form

$$r_1v_1 + \cdots + r_nv_n$$

where $r_i \in I$ and $v_i \in M$. Is IM a submodule of M?

3. Show that if S and T are submodules of M, then (with respect to set inclusion)
$$S \cap T = \text{glb}\{S,T\} \quad \text{and} \quad S + T = \text{lub}\{S,T\}$$
4. Let $S_1 \subset S_2 \subset \cdots$ be an ascending sequence of submodules of an R-module M. Prove that the union $\bigcup S_i$ is a submodule of M.
5. Is it true that a subset S of a module M is linearly independent if and only if every element of *span*(S) can be expressed as a *unique* linear combination of elements of S? Explain.
6. Consider the $\mathbb{Z}$-module $\mathbb{Z}_n = \{0,\ldots,n-1\}$, with scalar multiplication defined by
$$zu = (z \cdot u) \bmod n$$
for $z \in \mathbb{Z}$ and $u \in \mathbb{Z}_n$. Which subsets of $\mathbb{Z}_n$ (if any) are linearly independent?
7. Give an example of a module M that has a finite basis, but with the property that not every spanning set in M contains a basis, and not every linearly independent set in M is contained in a basis.
8. Let $\tau \in \mathit{Hom}(M,N)$ be an isomorphism. If $\mathfrak{B}$ is a basis for M, prove that $\tau(\mathfrak{B}) = \{\tau(b) \mid b \in \mathfrak{B}\}$ is a basis for N.
9. Consider the ring $R = F[x,y]$ of polynomials in two variables. Show that the set M consisting of all polynomials in R that have zero constant term, is an R-module. Show that M is not a free R-module.
10. Referring to Example 4.7, where $R = \mathcal{L}(V)$, show that R^n is isomorphic to R^m for all n and m. (By R^n, we mean the set of all ordered n-tuples of elements of R.)
11. How does the proof of Corollary 4.10 use the fact that R is a commutative ring?
12. Prove that if a ring R has the property that every finitely generated R-module is free, then either R is the zero ring or R is a field.
13. Let I be an ideal in R. Prove that I is a free R-module if and only if I is a principle ideal, generated by an element in R that is *not* a zero divisor.
14. Let M be an R-module. An element $v \in M$ is called a **torsion element** if there exists a nonzero $r \in R$ for which $rv = 0$.
    a) Prove that if R is an integral domain, then the set *Tor*(M) of torsion elements in M forms a submodule of M.
    b) Find an example of a ring R with the property that, thinking of R as an R-module, the set *Tor*(R) is not a submodule of R.

CHAPTER 5

Modules II

Contents: *Quotient Modules. Quotient Rings and Maximal Ideals. Noetherian Modules. The Hilbert Basis Theorem. Exercises.*

Quotient Modules

The procedure for defining quotient modules is the same as that for defining quotient spaces. We summarize in the following theorem.

Theorem 5.1 Let S be a submodule of an R-module M. The binary relation

$$\mathbf{u} \equiv \mathbf{v} \;\Leftrightarrow\; \mathbf{u} - \mathbf{v} \in S$$

is an equivalence relation on M, whose equivalence classes are the **cosets**

$$\mathcal{S} = \mathbf{v}{+}S = \{\mathbf{v} + \mathbf{s} \mid \mathbf{s} \in S\}$$

of S in M. The set M/S of all cosets of S in M, called the **quotient module** of M modulo S, is an R-module under the well-defined operations

$$(\mathbf{u}{+}S) + (\mathbf{v}{+}S) = (\mathbf{u} + \mathbf{v}){+}S \quad \text{and} \quad r(\mathbf{u}{+}S) = r\mathbf{u}{+}S$$

The zero element in M/S is the coset $\mathbf{0} + S = S$. ∎

It is left to the reader to formulate and prove precise statements of the three isomorphism theorems for modules that correspond to the isomorphism theorems of Chapter 3.

One question that immediately comes to mind is whether or not a quotient space of a free module need be free. As the next example shows, the answer is no.

Example 5.1 As a module over itself, the $\mathbb{Z}$-module $\mathbb{Z}$ is free on the set $\{1\}$. The set $\mathbb{Z}n = \{zn \mid z \in \mathbb{Z}\}$ is a (cyclic) submodule of $\mathbb{Z}$, but the quotient $\mathbb{Z}$-module $\mathbb{Z}/\mathbb{Z}n$ is isomorphic to $\mathbb{Z}_n$, via the map

$$\tau(u+\mathbb{Z}n) = u \bmod n$$

and since $\mathbb{Z}_n$ is not free as a $\mathbb{Z}$-module, neither is $\mathbb{Z}/\mathbb{Z}n$. □

Quotient Rings and Maximal Ideals

In order to prove Theorem 4.7, we need a few more facts about rings. The construction of quotient spaces and quotient modules works equally well for other algebraic structures. For rings, it proceeds as follows. Let S be a subring of a commutative ring R with identity. Then the set of all cosets

$$R/S = \{r+S \mid r \in R\}$$

is easily seen to be an abelian group under coset addition

$$(a+S) + (b+S) = (a+b)+S$$

In order for the product

$$(a+S)(b+S) = ab+S$$

to be well-defined, we must have

$$b+S = b'+S \;\Rightarrow\; ab+S = ab'+S$$

or, equivalently,

$$b - b' \in S \;\Rightarrow\; a(b-b') \in S$$

But $b - b'$ may be any element of S, and a may be any element of R, and so this condition implies that S must be an ideal. Conversely, if S is an ideal, then coset multiplication is well-defined.

Theorem 5.2 Let R be a commutative ring with identity. If $\mathfrak{I}$ is any ideal of R, then the set $R/\mathfrak{I}$ of all cosets of $\mathfrak{I}$ in R is a ring, called the **quotient ring** of R modulo $\mathfrak{I}$, where addition and multiplication are defined by

$$(a+S) + (b+S) = (a+b)+S$$

$$(a+S)(b+S) = ab+S$$

▮

Definition An ideal $\mathfrak{I}$ in a ring R is a **maximal ideal** if $\mathfrak{I} \neq R$, and if whenever $\mathfrak{J}$ is an ideal satisfying $\mathfrak{I} \subset \mathfrak{J} \subset R$, then either $\mathfrak{J} = \mathfrak{I}$ or $\mathfrak{J} = R$. □

Here is one reason why maximal ideals are important.

Theorem 5.3 Let R be a commutative ring with identity. Then the quotient ring $R/\mathfrak{I}$ is a field if and only if $\mathfrak{I}$ is a maximal ideal.

Proof. Suppose first that $R/\mathfrak{I}$ is a field. Assume that $\mathfrak{I}$ is not maximal, and so there exists an ideal $\mathfrak{J}$ with the property that $\mathfrak{I} \subsetneq \mathfrak{J} \subsetneq R$. Let $j \in \mathfrak{J} - \mathfrak{I}$, and consider the ideal

$$\mathfrak{K} = \langle j, \mathfrak{I} \rangle \subset \mathfrak{J}$$

generated by j and $\mathfrak{I}$. Since $j \notin \mathfrak{I}$, we have $j+\mathfrak{I} \neq 0$ and since $R/\mathfrak{I}$ is a field, $j+\mathfrak{I}$ must have an inverse, say $j'+\mathfrak{I}$, for which

$$(j+\mathfrak{I})(j'+\mathfrak{I}) = jj'+\mathfrak{I} = 1+\mathfrak{I}$$

Therefore, $1 - jj' \in \mathfrak{I} \subset \mathfrak{K}$ and since $jj' \in \mathfrak{K}$, we have $1 \in \mathfrak{K}$, which implies that $\mathfrak{K} = R$. But $\mathfrak{K} \subset \mathfrak{J}$ and $\mathfrak{J}$ is a *proper* subset of R. This contradiction implies that $\mathfrak{I}$ is maximal.

Conversely, suppose that $\mathfrak{I}$ is maximal. We want to show that any nonzero $r+\mathfrak{I} \in R/\mathfrak{I}$ has an inverse. But if $0 \neq r+\mathfrak{I}$, then $r \notin \mathfrak{I}$, and so the ideal $\mathfrak{J} = \langle r, \mathfrak{I} \rangle$ is strictly larger than $\mathfrak{I}$. Since $\mathfrak{I}$ is maximal, we must then have $\mathfrak{J} = R$. This implies that $1 \in \mathfrak{J}$, and so there exists $s \in R$ for which $1 = sr+i$, for some $i \in \mathfrak{I}$. Hence,

$$(s+\mathfrak{I})(r+\mathfrak{I}) = sr+\mathfrak{I} = (1-i)+\mathfrak{I} = 1+\mathfrak{I}$$

and so $(r+\mathfrak{I})^{-1} = s+\mathfrak{I}$. Hence, $R/\mathfrak{I}$ is a field. ∎

We need one more fact in order to prove Theorem 4.7.

Theorem 5.4 Any commutative ring R with identity contains a maximal ideal.

Proof. Since R is not the zero ring, it has a proper ideal, namely, $\{0\}$. (By a *proper* ideal, we mean an ideal different from R itself.) Let $\mathcal{S}$ be the collection of all proper ideals of R. Then $\mathcal{S}$ is nonempty. If

$$\mathfrak{I}_1 \subset \mathfrak{I}_2 \subset \cdots$$

is a chain of proper ideals in R, then the union $\mathfrak{J} = \bigcup \mathfrak{I}_j$ is also an ideal. Furthermore, if $\mathfrak{J} = R$, then $1 \in \mathfrak{J}$, and so $1 \in \mathfrak{I}_k$, for some k, which implies that $\mathfrak{I}_k = R$, and this contradicts the fact that $\mathfrak{I}_k$ is proper. Hence, $\mathfrak{J} \in \mathcal{S}$. Thus, any chain in $\mathcal{S}$ has an upper bound, and so Zorn's lemma implies that $\mathcal{S}$ has a maximal element. This shows that R has a maximal ideal. ∎

We are now ready for the proof of Theorem 4.7.

Theorem 5.5 Let M be a *free* R-module. Then any two bases of M have the same cardinality.

Proof. Our plan is quite straightforward. We seek to find a vector space V with the property that, for any basis for M, there is a basis of the same cardinality for V. Then we can appeal to the corresponding result for vector spaces, which we proved in Chapter 1.

Now, according to Theorem 5.4, R has a maximal ideal $\mathfrak{I}$, and according to Theorem 5.3, $R/\mathfrak{I}$ is a field. Let

$$\mathfrak{I}M = \{a_1v_1 + \cdots + a_nv_n \mid a_i \in \mathfrak{I},\ v_i \in M\}$$

Then $\mathfrak{I}M$ is a submodule of M, and so we may form the quotient module $M/\mathfrak{I}M$.

We want to show that $M/\mathfrak{I}M$ is a vector space over $R/\mathfrak{I}$, with scalar multiplication defined by

$$(r+\mathfrak{I})(u+\mathfrak{I}M) = ru+\mathfrak{I}M$$

To see that this is well-defined, suppose that

$$r+\mathfrak{I} = r'+\mathfrak{I} \quad \text{and} \quad u+\mathfrak{I}M = u'+\mathfrak{I}M$$

We must show that

$$ru+\mathfrak{I}M = r'u'+\mathfrak{I}M$$

Equivalently, we must show that

$$r - r' \in \mathfrak{I},\ u - u' \in \mathfrak{I}M \ \Rightarrow\ ru - r'u' \in \mathfrak{I}M$$

But

$$\begin{aligned} r - r' \in \mathfrak{I},\ u - u' \in \mathfrak{I}M \ &\Rightarrow\ (r - r')u' \in \mathfrak{I}M \text{ and } r(u - u') \in \mathfrak{I}M \\ &\Rightarrow\ (r - r')u' + r(u - u') = ru - r'u' \in \mathfrak{I}M \end{aligned}$$

Hence, scalar multiplication is well-defined. We leave it to the reader to show that the necessary properties of scalar multiplication are satisfied, and so $M/\mathfrak{I}M$ is indeed a vector space over $R/\mathfrak{I}$.

Let $\mathfrak{B}$ be a basis for M over R. If b_i and b_j are in $\mathfrak{B}$ then $b_i+\mathfrak{I}M$ and $b_j+\mathfrak{I}M$ are distinct, for if

$$b_i+\mathfrak{I}M = b_j+\mathfrak{I}M$$

then $b_i - b_j \in \mathfrak{I}M$, and so

$$b_i - b_j = a_1v_1 + \cdots + a_nv_n$$

for $a_i \in \mathfrak{I}$, $v_i \in M$. But each v_i is a linear combination of the basis vectors in $\mathfrak{B}$. Let us suppose that the coefficient of b_i in v_k is r_k, for $k = 1,\ldots,n$. Equating coefficients of b_i on both sides gives

$$1 = a_1r_1 + \cdots + a_nr_n$$

But the sum on the right side of this equation is in the ideal $\mathfrak{I}$, and so $1 \in \mathfrak{I}$, which is a contradiction to the fact that $\mathfrak{I}$ is maximal (and hence proper). Thus, the set

$$\mathfrak{B}' = \{b+\mathfrak{I}M \mid b \in \mathfrak{B}\}$$

has the same cardinality as the basis $\mathfrak{B}$ of M. We need only show that $\mathfrak{B}'$ is a basis for the vector space $\mathfrak{I}M$ over $R/\mathfrak{I}$.

It is clear that $\mathfrak{B}'$ generates $M/\mathfrak{I}M$ over $R/\mathfrak{I}$, since $\mathfrak{B}$ generates M. To see that $\mathfrak{B}'$ is linearly independent, observe that

$$\sum_{j \in U} (r_j+\mathfrak{I})(b_j+\mathfrak{I}M) = 0 \Rightarrow \sum_{j \in U} (r_j b_j+\mathfrak{I}M) = 0$$

$$\Rightarrow \sum_{j \in U} r_j b_j \in \mathfrak{I}M \Rightarrow \sum_{j \in U} r_j b_j = \sum_{i \in V} a_i b_i, \text{ for } a_i \in \mathfrak{I}$$

Equating coefficients of b_j on both sides shows that $r_j \in \mathfrak{I}$, and so $r_j+\mathfrak{I} = 0$. This shows that $\mathfrak{B}'$ is linearly independent. Hence $\mathfrak{B}'$ is a basis for $M/\mathfrak{I}M$. Thus, $|\mathfrak{B}| = dim(M/\mathfrak{I}M)$ is independent of the choice of basis $\mathfrak{B}$. ∎

Noetherian Modules

One of the most desirable properties of a finitely generated R-module M is that all of its submodules be finitely generated. Example 4.3 shows that this is not always the case, and leads us to search for conditions on the ring R that will guarantee that any finitely generated R-module has only finitely generated submodules.

Definition An R-module M is said to satisfy the **ascending chain condition on submodules** if, for any ascending sequence of submodules

$$S_1 \subset S_2 \subset S_3 \subset \cdots$$

of M, there exists an index k for which $S_k = S_{k+1} = S_{k+2} = \cdots$. □

Put less formally, an R-module satisfies the ascending chain condition (abbreviated a.c.c.) on submodules if any ascending chain of submodules eventually becomes constant.

Theorem 5.6 The following are equivalent for an R-module M.

1) Every submodule of M is finitely generated.
2) M satisfies the a.c.c. on submodules.

Any module that satisfies either of these conditions is called a **noetherian module** (after *Emmy Noether*, one of the pioneers of module theory).

Proof. Suppose that all submodules of M are finitely generated, and that M contains an infinite ascending sequence

$$S_1 \subset S_2 \subset S_3 \subset \cdots \tag{5.1}$$

of submodules. Then the union

$$S = \bigcup_j S_j$$

is easily seen to be a submodule of M. Hence, S is finitely generated, and $S = \langle u_1, \ldots, u_n \rangle$, for some $u_i \in M$. Since $u_i \in S$, there exists an index k_i such that $u_i \in S_{k_i}$. Therefore, if $k = \max\{k_1, \ldots, k_n\}$, we have

$$u_i \in S_k \quad \text{for all } k = 1, \ldots, n$$

and so

$$S = \langle u_1, \ldots, u_n \rangle \subset S_k \subset S_{k+1} \subset S_{k+2} \subset \cdots \subset S$$

which shows that the submodules in the chain (5.1), from S_k on, are equal.

For the converse, we must show that if M satisfies the a.c.c on submodules, then every submodule of M is finitely generated. To this end, let S be a submodule of M. Pick $u_1 \in S$, and consider the submodule $S_1 = \langle u_1 \rangle \subset S$ generated by u_1. If $S_1 = S$, then S is finitely generated. If $S_1 \neq S$, then there is a $u_2 \in S - S_1$. Now let $S_2 = \langle u_1, u_2 \rangle$. If $S_2 = S$, then S is finitely generated. If $S_2 \neq S$, then pick $u_3 \in S - S_2$, and consider the submodule $S_3 = \langle u_1, u_2, u_3 \rangle$.

Continuing in this way, we get an ascending chain of submodules

$$\langle u_1 \rangle \subset \langle u_1, u_2 \rangle \subset \langle u_1, u_2, u_3 \rangle \subset \cdots \subset S$$

If none of these submodules is equal to S, we would have an infinite ascending chain of submodules, each properly contained in the next, which contradicts the fact that M satisfies the a.c.c. on submodules. Hence, $S = \langle u_1, \ldots, u_n \rangle$, for some n, and so S is finitely generated. ∎

Since a ring R is a module over itself, and since the submodules of the module R are precisely the ideals of the ring R, the preceding may be formulated for rings as follows.

Definition A ring R is said to satisfy the **ascending chain condition on ideals** if, for any ascending sequence of ideals

$$\mathcal{I}_1 \subset \mathcal{I}_2 \subset \mathcal{I}_3 \subset \cdots$$

of R, there exists an index k for which $\mathcal{I}_k = \mathcal{I}_{k+1} = \mathcal{I}_{k+2} = \cdots$. □

Theorem 5.7 The following are equivalent for a ring R.

1) Every ideal of R is finitely generated (as an R-module).
2) R satisfies the a.c.c. on ideals.

Any ring that satisfies either of the conditions is called a **noetherian ring.** ∎

Now we are ready for the main result of this section.

Theorem 5.8 If R is noetherian, then so is any finitely generated R-module.

Proof. Let M be a finitely generated R-module, say $M = \langle u_1, \ldots, u_n \rangle$. Consider the epimorphism $\tau: R^n \to M$ defined by

$$\tau(r_1, \ldots, r_n) = r_1 u_1 + \cdots + r_n u_n$$

Let S be a submodule of M. Then

$$\tau^{-1}(S) = \{u \in R^n \mid \tau(u) \in S\}$$

is a submodule of R^n, and $\tau(\tau^{-1}(S)) = S$. Now suppose that R^n has only finitely generated submodules, and so $\tau^{-1}(S)$ is finitely generated, say, $\tau^{-1}(S) = \langle v_1, \ldots, v_k \rangle$. Then if $w \in S$, we have $w = \tau(v)$ for some $v \in \tau^{-1}(S)$, and since

$$v = r_1 v_1 + \cdots + r_k v_k$$

we get

$$w = \tau(v) = r_1 \tau(v_1) + \cdots + r_k \tau(v_k)$$

which implies that S is finitely generated, by $\{\tau(v_1), \ldots, \tau(v_k)\}$. Therefore, the proof will be complete if we can show that every submodule of R^n is finitely generated.

We do this by induction on n. If $n = 1$, the result is clear. Suppose that R^k has only finitely generated submodules, for all $1 \leq k < n$. Let S be a submodule of R^n, and consider the sets

$$S_1 = \{(s_1, \ldots, s_{n-1}, 0) \mid (s_1, \ldots, s_{n-1}, s_n) \in S \text{ for some } s_n\}$$

and

$$S_2 = \{(0, \ldots, 0, s_n) \mid (s_1, \ldots, s_{n-1}, s_n) \in S \text{ for some } s_n\}$$

It is easy to see that S_1 and S_2 are submodules of R^n. Moreover, S_1 is isomorphic to a submodule of R^{n-1}, obtained by simply dropping the last coordinate,

$$S_1 \approx \{(s_1, \ldots, s_{n-1}) \mid (s_1, \ldots, s_{n-1}, 0) \in S_1\} \subset R^{n-1}$$

and similarly, S_2 is isomorphic to a submodule of R,

$$S_2 \approx \{s_n \mid (0, \ldots, 0, s_n) \in S_2\} \subset R$$

Therefore, the induction hypothesis (and the isomorphisms) imply that S_1 and S_2 are finitely generated, say

$$S_1 = \langle u_1,\ldots,u_s\rangle \text{ and } S_2 = \langle v_1,\ldots,v_t\rangle$$

Hence,

$$S = S_1 \oplus S_2$$

is finitely generated, by $\{u_1,\ldots,u_s,v_1,\ldots,v_t\}$. ∎

The Hilbert Basis Theorem

Theorem 5.8 naturally leads us to ask which familiar rings are noetherian. We leave it to the reader to show that a commutative ring R with identity is a field if and only if its only ideals are $\{0\}$ and R. Hence, a field is a noetherian ring. Also, any principal ideal domain is noetherian. The following theorem describes some additional noetherian rings.

Theorem 5.9 (Hilbert basis theorem) If a ring R is noetherian, then so is the polynomial ring $R[x]$.

Proof. We wish to show that any ideal $\mathfrak{I}$ in $R[x]$ is finitely generated. Let L denote the set of all leading coefficients of polynomials in $\mathfrak{I}$, together with the 0 element of R. Then L is an ideal of R.

To see this, observe that if $\alpha \in L$ is the leading coefficient of $f(x) \in \mathfrak{I}$, and if $r \in R$, then either $r\alpha = 0$ or else $r\alpha$ is the leading coefficient of $rf(x) \in \mathfrak{I}$. In either case, $r\alpha \in L$. Similarly, suppose that $\beta \in L$ is the leading coefficient of $g(x) \in \mathfrak{I}$. We may assume that $\deg f(x) = i$ and $\deg g(x) = j$, with $i \leq j$. Then $h(x) = x^{j-i}f(x)$ is in $\mathfrak{I}$, has leading coefficient α, and has the same degree as $g(x)$. Hence, $\alpha - \beta$ is either 0 or it is the leading coefficient of $h(x) - g(x) \in \mathfrak{I}$. In either case $\alpha - \beta \in L$.

Since L is an ideal of the noetherian ring R, it must be finitely generated, say $L = \langle a_1,\ldots,a_k\rangle$. Since $a_i \in L$, there exist polynomials $f_i(x)$ with leading coefficients a_i. By multiplying each $f_i(x)$ by a suitable power of x, we may assume that $\deg f_i(x) = d$ for all $i = 1,\ldots,k$.

Now, let

$$g(x) = g_0 + g_1x + \cdots + g_nx^n$$

be any polynomial in $\mathfrak{I}$ with $\deg g(x) \geq d$. Since $g_n \in L$, we have

$$g_n = r_1a_1 + \cdots + r_ka_k$$

and so

$$h(x) = g(x) - \sum_i r_if_i(x) \tag{5.2}$$

has coefficient of x^n equal to 0. In other words, $\deg h(x) < \deg g(x)$.

We now have the basis for an induction argument. Any polynomial in $\mathfrak{I}$ of degree less than d is certainly generated by the set

$$S = \{1, x, \ldots, x^{d-1}, f_1(x), \ldots, f_k(x)\}$$

Assume, for the purposes of induction, that any polynomial of degree at most $n-1$ is generated by S. Let $g(x)$ have degree n. Referring to (5.2), we see that $\deg h(x) \leq n-1$, and so $h(x) \in \mathfrak{I}$ is generated by S. But then

$$g(x) = h(x) + \sum_i r_i f_i(x)$$

is also generated by S. ∎

EXERCISES

1. State and prove the first isomorphism theorem for modules.
2. State and prove the second isomorphism theorem for modules.
3. State and prove the third isomorphism theorem for modules.
4. If M is a free R-module, and $\tau: M \to N$ is an epimorphism, must N also be free?
5. Let I be an ideal of R. Prove that if R/I is a free R-module, then I is the zero ideal.
6. Show that the submodules of the R-module R are the same as the ideals of the ring R.
7. Let R be a commutative ring with identity. An ideal $\mathfrak{I}$ in R is called a **prime ideal** if $r, s \in R$ and $rs \in \mathfrak{I}$ implies that $r \in \mathfrak{I}$ or $s \in \mathfrak{I}$. Show that $R/\mathfrak{I}$ is an integral domain if and only if $\mathfrak{I}$ is a prime ideal and $\mathfrak{I} \neq R$.
8. Prove that the union of an *ascending chain* of submodules is a submodule.
9. Prove that a commutative ring R with identity is a field if and only if it has no ideals other than $\{0\}$ and R.
10. Let S be a submodule of an R-module M. Show that if M is finitely generated, so is the quotient module M/S.
11. Let S be a submodule of an R-module. Show that if both S and M/S are finitely generated, so is M.
12. Referring to the proof of Theorem 5.5, show that the necessary properties of scalar multiplication are satisfied and so $M/\mathfrak{I}M$ is indeed a vector space over $R/\mathfrak{I}$.
13. Show that an R-module M satisfies the a.c.c. for submodules if and only if the following condition holds. Every nonempty collection $\mathcal{S}$ of submodules of M has a maximal element. That is, for every nonempty collection $\mathcal{S}$ of submodules of M, there is an $S \in \mathcal{S}$ with the property that $T \in \mathcal{S} \Rightarrow T \subset S$.

14. Let $\tau:M \to N$ be a homomorphism of R-modules.
    a) Show that if M is finitely generated, then so is $im(\tau)$.
    b) Show that if $ker(\tau)$ and $im(\tau)$ are finitely generated, then $M = ker(\tau) + im(\tau)$ is finitely generated.
15. If R is a noetherian ring, show that any proper ideal of R is contained in a maximal ideal.
16. If R is noetherian, and $\mathfrak{I}$ is an ideal of R, show that $R/\mathfrak{I}$ is also noetherian.
17. Prove that if R is noetherian, then so is $R[x_1,\ldots,x_n]$.

CHAPTER 6

Modules over Principal Ideal Domains

Contents: *Free Modules over a Principal Ideal Domain. Torsion Modules. The Primary Decomposition Theorem. The Cyclic Decomposition Theorem for Primary Modules. Uniqueness. The Cyclic Decomposition Theorem. Exercises.*

Free Modules over a Principal Ideal Domain

When a ring R has nice properties (such as being noetherian), then its R-modules tend to have nice properties (such as being noetherian, at least in the finitely generated case). Since principal ideal domains (abbreviated p.i.d.s) have very nice properties, we expect the same for modules over p.i.d.s.

For instance, Example 4.6 showed that a submodule of a free module need not be free. However, if the ring of scalars is a principal ideal domain, this cannot happen.

Theorem 6.1 Let M be a free module over a *principal ideal domain* R. Then any submodule S of M is also free. Moreover, $rk(S) \le rk(M)$.

Proof. We will give the proof only for modules of finite rank, although the theorem is true for all free modules. Thus, since $M \approx R^n$, we may in fact assume that $M = R^n$. Our plan is to proceed by induction on n.

For $n = 1$, we have $M = R$, and any submodule S of R is just an ideal of R. Hence, $S = \langle a \rangle$ is principal. But since R is an integral domain, we have $ra \neq 0$ for all $r \neq 0$, and so the map

$$\tau: R \to S, \qquad \tau(r) = ra$$

is an isomorphism from R to S. Hence, S is free.

Now assume that any submodule of R^k is free, for $1 \leq k \leq n-1$, and let S be a submodule of R^n. Consider the sets

$$S_1 = \{(s_1, \ldots, s_{n-1}, 0) \mid (s_1, \ldots, s_{n-1}, s_n) \in S \text{ for some } s_n\}$$

and

$$S_2 = \{(0, \ldots, 0, s_n) \mid (s_1, \ldots, s_{n-1}, s_n) \in S \text{ for some } s_n\}$$

It is easy to see that S_1 and S_2 are submodules of R^n, and that

$$S = S_1 \oplus S_2$$

Moreover, S_1 is isomorphic to a submodule of R^{n-1}, obtained by simply dropping the last coordinate,

$$S_1 \approx \{(s_1, \ldots, s_{n-1}) \mid (s_1, \ldots, s_{n-1}, 0) \in S_1\} \subset R^{n-1}$$

and S_2 is isomorphic to a submodule of R,

$$S_2 \approx \{s_n \mid (0, \ldots, 0, s_n) \in S_2\} \subset R$$

Therefore, the induction hypothesis (and the isomorphisms) imply that S_1 and S_2 are free. If S_1 is free on $\{u_1, \ldots, u_s\}$, where $s \leq n-1$, and S_2 is free on $\{v_1\}$, then S is free on $\{u_1, \ldots, u_s, v_1\}$, where $s + 1 \leq n$. ∎

Torsion Modules

In a vector space V over a field F, if $r \in F$ and $\mathbf{v} \in V$ are nonzero, then $r\mathbf{v}$ is nonzero. In a module, this need not be the case and leads to the following definition.

Definition Let M be an R-module. If $v \in M$ has the property that $rv = 0$ for some *nonzero* $r \in R$, then v is called a **torsion element** of M. A module that has no nonzero torsion elements is said to be **torsion free**. If all elements of M are torsion elements, then M is a **torsion module**. □

If M is a module, it is not hard to see that the set M_{tor} of all torsion elements is a submodule of M, and that M/M_{tor} is torsion free. Moreover, any free module over a principal ideal domain is torsion free. The following is a partial converse.

Theorem 6.2 Let M be a torsion free, finitely generated module over a principal ideal domain R. Then M is free.

Proof. Since M is finitely generated, we have $M = \langle v_1,\ldots,v_n\rangle$ for some $v_i \in M$. Now let us take a maximal linearly independent subset of these generators, say $S = \{u_1,\ldots,u_k\}$, and renumber to get

$$M = \langle u_1,\ldots,u_k,v_1,\ldots,v_{n-k}\rangle$$

Thus, for each v_i, the set $\{u_1,\ldots,u_k,v_i\}$ is linearly dependent, and so there exists a_i and $r_1,\ldots,r_k$ for which

$$a_i v_i + r_1 u_1 + \cdots + r_k u_k = 0$$

Now, if we let $a = a_1 \cdots a_{n-k}$ be the product of the coefficients of the various v_i's, then $av_i \in span(S)$, for all $i = 1,\ldots,n-k$.

Hence, the module $aM = \{av \mid v \in M\}$ is a submodule of $span(S)$. But $span(S)$ is a free module, with basis S, and so by Theorem 6.1, aM is also free. Finally, $M \approx aM$, since the map

$$\tau(v) = av$$

is an epimorphism, that happens to be injective, because M is torsion free. Thus M, being isomorphic to aM, is also free. ∎

Our goal in this section is to show that any module M over a principal ideal domain is the direct sum

$$M = M_{tor} \oplus M_{free}$$

where M_{free} is a free module. This is the first step in the decomposition of a module over a principal ideal domain.

Since the quotient module M/M_{tor} is torsion free and since M/M_{tor} is finitely generated when M is finitely generated, we deduce from Theorem 6.2 that M/M_{tor} is a free module. Consider the natural projection

$$\pi: M \to M/M_{tor}, \quad \pi(v) = v + M_{tor}$$

It is tempting to infer (as we would for vector spaces) that M is isomorphic to the direct sum of $ker(\pi)$ and $im(\pi)$, and since $ker(\pi) = M_{tor}$, and $im(\pi) = M/M_{tor}$ is free, the desired result would follow. Happily, this is the case for modules as well.

Theorem 6.3 Let M be a finitely generated module over a principal ideal domain R. Then

$$M = M_{tor} \oplus M_{free}$$

where M_{free} is a free R-module.

Proof. Consider the epimorphism $\pi: M \to M/M_{tor}$ from M onto the free module M/M_{tor}. Let $\mathfrak{B}$ be a basis for M/M_{tor}. For each $b \in B$, choose a $b' \in M$ with the property that $\pi(b') = b$. Let $\mathfrak{B}'$ be the set of all such elements of M. We leave it to the reader to show that $\mathcal{B}'$ is linearly independent. Hence, $S = span(\mathcal{B}')$ is a free submodule of M. Moreover,

$$\begin{aligned} v \in M_{tor} \cap S = ker(\pi) \cap S &\Rightarrow \pi(v) = 0 \text{ and } v = \Sigma r_i b_i' \\ &\Rightarrow 0 = \Sigma r_i \pi(b_i') = \Sigma r_i b_i \\ &\Rightarrow r_i = 0 \text{ for all } i \\ &\Rightarrow v = 0 \end{aligned}$$

and so $M_{tor} \cap S = \{0\}$. Furthermore, if $v \in M$, then $\pi(v) = \Sigma s_i b_i$, for some $s_i \in R$. Now let $u = \Sigma s_i b_i' \in S$. Then

$$\pi(v - u) = \pi(v) - \pi(\Sigma s_i b_i') = \Sigma s_i b_i - \Sigma s_i b_i = 0$$

and so $x = v - u \in ker(\pi)$. Hence, $v = x + u \in ker(\pi) + S$. This shows that $M = ker(\pi) \oplus S = M_{tor} \oplus S$. ∎

In view of Theorem 6.3, we can turn our attention to the decomposition of finitely generated *torsion* modules over a principal ideal domain.

The Primary Decomposition Theorem

To show that every finitely generated torsion module over a principal ideal domain is the direct sum of cyclic submodules, we need some definitions.

Definition Let M be an R-module. The **annihilator** of $v \in M$ is

$$ann(v) = \{r \in R \mid rv = 0\}$$

and the **annihilator** of M is

$$ann(M) = \{r \in R \mid rM = \{0\}\}$$

where $rM = \{rv \mid v \in M\}$. □

It is easy to see that $ann(v)$ and $ann(M)$ are ideals of R. Clearly, $v \in M$ is a torsion element if and only if $ann(v) \neq \{0\}$.

If M is a finitely generated torsion module over a principal ideal domain, say $M = \langle u_1, \ldots, u_n \rangle$, then there exists nonzero $a_i \in ann(u_i)$, for $i = 1, \ldots, n$. Hence, the nonzero product $a = a_1 \cdots a_n$ satisfies $av = 0$ for all $v \in M$, and so $a \in ann(M)$. This shows that $ann(M) \neq \{0\}$.

Definition Let M be a finitely generated torsion module over a principal ideal domain. Any generator of the principal ideal $ann(v)$ is called an **order** of v. Any generator of the nonzero principal ideal *ann*(M) is called an **order** of M. ▯

Annihilators are also referred to as **order ideals.** Note that any two orders μ and ν of M (or of $v \in M$) are associates, that is,

$$ann(M) = \langle\mu\rangle = \langle\nu\rangle \;\Rightarrow\; \mu = u\nu \text{ for some unit } u \in R$$

Hence, an order of M is uniquely determined up to multiplicative unit, and so μ and ν have the same factorization into a product of prime elements in R, up to multiplication by a unit.

Definition A module M is said to be **primary** if its annihilator has the form $ann(M) = \langle p^e \rangle$, where p is a prime and e is a positive integer. In other words, M is primary if it has order a positive power of a prime. ▯

Note that a finitely generated torsion module M over a principal ideal domain is primary if and only if every element of M has order a power of a fixed prime p.

Our plan for the decomposition of a torsion module M is to first decompose M as a direct sum of *primary* submodules.

Theorem 6.4 (The primary decomposition theorem) Let M be a nonzero finitely generated torsion module over a principal ideal domain, with order

$$\mu = p_1^{e_1} \cdots p_n^{e_n}$$

where the p_i's are distinct primes. Then M is the direct sum

$$M = M_{p_1} \oplus \cdots \oplus M_{p_n}$$

where

$$M_{p_i} = \{v \in M \mid p_i^{e_i} v = 0\}$$

is a primary submodule, with order $p_i^{e_i}$.

Proof. Let $\mu = pq$, where $\gcd(p,q) = 1$, and consider the sets

$$M_p = \{v \in M \mid pv = 0\} \quad \text{and} \quad M_q = \{v \in M \mid qv = 0\}$$

We wish to show that $M = M_p \oplus M_q$ and that M_p and M_q have annihilators $\langle p \rangle$ and $\langle q \rangle$, respectively.

Since p and q are relatively prime, there exist $a,b \in R$ such that

$$ap + bq = 1 \tag{6.1}$$

(This follows from the fact that the ideal $\langle p,q \rangle$ is generated by $\gcd(p,q) = 1$, and so $1 \in \langle p,q \rangle$.) Now, if $v \in M_p \cap M_q$, then $pv =$

$qv = 0$ and so

$$v = 1v = (ap + bq)v = 0$$

Thus $M_p \cap M_q = \{0\}$. From (6.1), we also get, for any $v \in M$,

$$v = 1v = apv + bqv$$

Moreover, $q(apv) = a(pq)v = a\mu v = 0$ implies that $apv \in M_q$, and similarly, $bqv \in M_p$. Hence, $v \in M_p + M_q$.

Now suppose that $rM_p = 0$. Then, for any $v = v_1 + v_2 \in M_p \oplus M_q = M$, we have

$$rqv = rq(v_1 + v_2) = qrv_1 + rqv_2 = 0$$

and so $rq \in ann(M)$, which implies that $\mu = pq \mid rq$. Hence $p \mid r$, which shows that $ann(M_p) = \langle p \rangle$. Similarly, $ann(M_q) = \langle q \rangle$.

Finally, since μ can be written as a product of primes, say

$$\mu = p_1^{e_1} \cdots p_n^{e_n}$$

we can use the preceding argument to write

$$M = M_{p_1} \oplus N$$

where N is a submodule with annihilator $\langle \mu / p_1^{e_1} \rangle$. Repeating the process gives the desired decomposition. ∎

The Cyclic Decomposition Theorem for Primary Modules

The next step is to decompose primary modules.

Theorem 6.5 **(The cyclic decomposition theorem)** Let M be a nonzero primary finitely generated torsion module over a principal ideal domain R, with order p^e. Then M is the direct sum

$$M = C_1 \oplus \cdots \oplus C_n \tag{6.2}$$

of cyclic submodules, with orders $p^{e_1}, \ldots, p^{e_n}$ satisfying

$$e = e_1 \geq e_2 \geq \cdots \geq e_n \tag{6.3}$$

or, equivalently,

$$p^{e_n} \mid p^{e_{n-1}} \mid \cdots \mid p^{e_1} \tag{6.4}$$

Proof. Once (6.2) is established, (6.4) will follow easily, since

$$p^e \in ann(M) \subset ann(C_i)$$

and so if $ann(C_i) = \langle \alpha_i \rangle$, then $\alpha_i \mid p^e$. Hence, $\alpha_i = p^{e_i}$, for some $e_i \leq e$. Then we may rearrange the order of the summands to get (6.4).

To prove (6.2), we begin by observing that there is an element

$v_1 \in M$ with $ann(v_1) = ann(M) = \langle p^e \rangle$. For if not, then for all $v \in M$, we would have $ann(v) = \langle p^k \rangle$ with $k < e$, and so $p^{e-1} \in ann(M)$. But this implies that $p^e \mid p^{e-1}$, which is impossible.

Our goal now is to show that the cyclic submodule $\langle v_1 \rangle$ is a direct summand of M, that is,

$$M = \langle v_1 \rangle \oplus S \tag{6.5}$$

for some submodule S of M. Then since S is also a finitely generated primary torsion module over R, we can repeat the process, to get

$$M = \langle v_1 \rangle \oplus \langle v_2 \rangle \oplus S_2$$

where $ann(v_2) = \langle p^{e_2} \rangle$ with $e_2 \leq e_1$. Continuing in this way, we get an ascending sequence of submodules

$$\langle v_1 \rangle \subset \langle v_1 \rangle \oplus \langle v_2 \rangle \subset \cdots$$

which must terminate, since M satisfies the ascending chain condition on submodules.

Thus, we need only establish (6.5). Since this is the most involved part of the proof, we will approach it slowly. Since M is finitely generated, we may write $M = \langle v_1, u_1, \ldots, u_k \rangle$. Our argument will be by induction on k. If $k = 0$, then let $S = \{0\}$, and we are done. Assume that the result is true for k, and suppose that

$$M = \langle v_1, u_1, \ldots, u_k, u \rangle$$

By the induction hypothesis,

$$\langle v_1, u_1, \ldots, u_k \rangle = \langle v_1 \rangle \oplus S_0$$

for some submodule S_0.

Notice that we may replace u by any element of the form $u - \alpha v_1$, for $\alpha \in R$, without effecting the span, that is,

$$\langle v_1, u_1, \ldots, u_k, u-\alpha v_1 \rangle = \langle v_1, u_1, \ldots, u_k, u \rangle = M$$

and so we seek an $\alpha \in R$ for which

$$\langle v_1 \rangle \cap \langle u - \alpha v_1, S_0 \rangle = \{0\} \tag{6.6}$$

since then we would have

$$M = \langle v_1 \rangle \oplus \langle u - \alpha v_1, S_0 \rangle = \langle v_1 \rangle \oplus S$$

Since any element of $\langle u - \alpha v_1, S_0 \rangle$ has the form $r(u - \alpha v_1) + s_0$, equation (6.6) is equivalent to

$$r(u - \alpha v_1) + s_0 \in \langle v_1 \rangle \;\Rightarrow\; r(u - \alpha v_1) + s_0 = 0$$

for any $r \in R$ and $s_0 \in S_0$, or equivalently

$$r(u-\alpha v_1) \in \langle v_1 \rangle \oplus S_0 \;\Rightarrow\; r(u-\alpha v_1) \in S_0$$

or, finally,

$$\text{(6.7)} \qquad ru \in \langle v_1 \rangle \oplus S_0 \;\Rightarrow\; r(u-\alpha v_1) \in S_0$$

Now, we observe that the set

$$\mathcal{I} = \{r \in R \mid ru \in \langle v_1 \rangle \oplus S_0\}$$

is an ideal of R, and so it is principal, say $\mathcal{I} = \langle a \rangle$. Note, however, that

$$p^e u = 0 \in \langle v_1 \rangle \oplus S_0$$

and so $p^e \in \langle a \rangle$, which implies that $a \mid p^e$, or that $a = p^f$, for some $f \le e$. Thus, we have

$$\begin{aligned} ru \in \langle v_1 \rangle \oplus S_0 \;&\Rightarrow\; r \in \mathcal{I} \;\Rightarrow\; r = qp^f \quad \text{for some } q \in R \\ &\Rightarrow\; r(u-\alpha v_1) = qp^f(u-\alpha v_1) \end{aligned}$$

and so if we can find an $\alpha \in R$ for which

$$\text{(6.8)} \qquad p^f(u-\alpha v_1) \in S_0$$

then (6.7) will be satisfied.

But $p^f \in \mathcal{I}$ and so $p^f u \in \langle v_1 \rangle \oplus S_0$, say

$$\text{(6.9)} \qquad p^f u = tv_1 + s_0$$

for some $t \in R$ and $s_0 \in S_0$. Equation (6.8) then becomes

$$tv_1 + s_0 - \alpha p^f v_1 \in S_0$$

or

$$(t - \alpha p^f)v_1 \in S_0$$

and this happens if and only if $t - \alpha p^f = 0$, that is, if and only if $p^f \mid t$, which is what we must show.

Equation (6.9) implies that

$$0 = p^{e-f}p^f u = p^{e-f}tv_1 + p^{e-f}s_0$$

and since $\langle v_1 \rangle \cap S_0 = \{0\}$, we deduce that $p^{e-f}tv_1 = 0$. Therefore, since v_1 has order p^e, it follows that $p^e \mid p^{e-f}t$, that is, $p^f \mid t$, as desired. This completes the proof. ∎

Uniqueness

Although the decomposition (6.2) is not unique, we will see that the orders p^{e_i} are unique up to multiplication by a unit. The prime p is certainly unique, since it must divide the order p^e of M. Before proceeding further, we need a few preliminary results, whose proofs are left as exercises.

Lemma 6.6 Let R be a principal ideal domain.

1) If $\langle v\rangle$ is a cyclic R-module, with $ann(v) = \langle a\rangle$, then the map
$$\tau:R\to\langle v\rangle \qquad \tau(r) = rv$$
is an epimorphism between R-modules, with kernel $\langle a\rangle$, and so
$$\langle v\rangle \approx \frac{R}{\langle a\rangle}$$
Moreover, if a is a prime, then $\langle a\rangle$ is a maximal ideal in R, and so $R/\langle a\rangle$ is a field.

2) Let $p \in R$ be a prime. If M is an R-module for which $pM = \{0\}$, then M is a *vector space* over $R/\langle p\rangle$, where scalar multiplication is defined by
$$(r+\langle p\rangle)v = rv$$
for all $v \in M$.

3) Let $p \in R$ be a prime. For any submodule S of an R-module M, the set
$$S^{(p)} = \{v \in S \mid pv = 0\}$$
is a submodule of M. Moreover, if $M = S \oplus T$, then $M^{(p)} = S^{(p)} \oplus T^{(p)}$. ∎

Now we are ready for the uniqueness result.

Theorem 6.7 Let M be a nonzero primary finitely generated torsion module over a principal ideal domain R, with order p^e. Suppose that
$$M = C_1 \oplus \cdots \oplus C_n$$
where C_i are nonzero cyclic submodules with orders p^{e_i}, and $e_1 \geq \cdots \geq e_n$. Then if
$$M = D_1 \oplus \cdots \oplus D_m$$
where D_i are nonzero cyclic submodules with orders p^{f_i}, and $f_1 \geq \cdots \geq f_m$, we must have $n = m$ and
$$e_1 = f_1, \ldots, e_n = f_n$$

Proof. Let us begin by showing that $n = m$. According to part (3) of Lemma 6.6,
$$M^{(p)} = C_1^{(p)} \oplus \cdots \oplus C_n^{(p)}$$
and
$$M^{(p)} = D_1^{(p)} \oplus \cdots \oplus D_m^{(p)}$$
where each summand in both decompositions is nonzero. (Why?) Since $pM^{(p)} = \{0\}$ by definition, Lemma 6.6 implies that $M^{(p)}$ is a

vector space over $R/\langle p\rangle$, and so each of the preceding decompositions expresses $M^{(p)}$ as a direct sum of one-dimensional subspaces. Hence, $m = n$.

Next, we show that the exponents e_i and f_i are equal by using induction on e_1. Assume first that $e_1 = 1$, in which case $e_i = 1$ for all i. Then $pM = \{0\}$, and so $f_i = 1$ for all i, since if $f_1 > 1$, and if $D_1 = \langle w\rangle$, then $pw \neq 0$, which is not the case.

Now suppose the result is true whenever $e_1 \leq k-1$, and let $e_1 = k$. To isolate the exponents that equal 1, suppose that

$$(e_1,\ldots,e_n) = (e_1,\ldots,e_s,1,\ldots,1), \quad e_s > 1$$

and

$$(f_1,\ldots,f_n) = (f_1,\ldots,f_t,1,\ldots,1), \quad f_t > 1$$

Then

$$pM = pC_1 \oplus \cdots \oplus pC_s$$

and

$$pM = pD_1 \oplus \cdots \oplus pD_t$$

But pC_i is a cyclic submodule of M, and $ann(pC_i) = \langle p^{e_i-1}\rangle$. To see this, suppose that $C_i = \langle v_i\rangle$. Then

$$pC_i = \{pc \mid c \in C_i\} = \{prv_i \mid r \in R\} = \{r(pv_i) \mid r \in R\} = \langle pv_i\rangle$$

and pv_i has order p^{e_i-1}. Similarly, pD_i is a cyclic submodules of M, with $ann(pD_i) = \langle p^{f_i-1}\rangle$. In particular, $ann(pC_1) = \langle p^{e_i-1}\rangle$, and so, by the induction hypothesis, we have

$$s = t \quad \text{and} \quad e_1 = f_1,\ldots,e_s = f_s$$

which concludes the proof. ∎

The Cyclic Decomposition Theorem

Let us pause to see where we stand. If M is a finitely generated module over a principal ideal domain then, according to Theorem 6.3,

$$M = M_{tor} \oplus M_{free}$$

where M_{tor} is the submodule of all torsion elements and M_{free} is a free submodule of M. If M_{tor} has order

$$\mu = p_1^{e_1} \cdots p_n^{e_n}$$

where the p_i's are distinct primes, the primary decomposition theorem implies that

$$M_{tor} = M_{p_1} \oplus \cdots \oplus M_{p_n}$$

where M_{p_i} is a primary module, with order $p_i^{e_i}$. Hence,

$$M = M_{p_1} \oplus \cdots \oplus M_{p_n} \oplus M_{free}$$

Finally, the cyclic decomposition theorem for primary modules allows us to write each primary module M_{p_i} as a direct sum of cyclic submodules. Let us put the pieces together in one theorem.

Theorem 6.8 (The cyclic decomposition theorem for finitely generated modules over a principal ideal domain – elementary divisor version) Let M be a nonzero finitely generated module over a principal ideal domain R. Then

$$M = M_{tor} \oplus M_{free}$$

where M_{tor} is the set of all torsion elements in M, and M_{free} is a free module, whose rank is uniquely determined by the module M. If M_{tor} has order

$$\mu = p_1^{e_1} \cdots p_n^{e_n}$$

where the p_i's are distinct primes, then M_{tor} is the direct sum

$$M_{tor} = M_{p_1} \oplus \cdots \oplus M_{p_n}$$

where

$$M_{p_i} = \{v \in M \mid p_i^{e_i} v = 0\}$$

is a primary submodule, with order $p_i^{e_i}$.

Moreover, each M_{p_i} can be further decomposed into a direct sum of cyclic submodules

$$M_{p_i} = C_{i,1} \oplus \cdots \oplus C_{i,k_i}$$

with orders $p_i^{e_{i,j}}$, and where

$$e_i = e_{i,1} \geq e_{i,2} \geq \cdots \geq e_{i,k_i}$$

The orders $p_i^{e_{i,j}}$, called the **elementary divisors** of M, are uniquely determined, up to multiplication by a unit, by the module M.

This yields the decomposition of M into a direct sum of cyclic submodules (and a free summand)

$$(6.10)\quad M = (C_{1,1} \oplus \cdots \oplus C_{1,k_1}) \oplus \cdots\cdots \oplus (C_{n,1} \oplus \cdots \oplus C_{1,k_n}) \oplus M_{free} \quad ∎$$

The decomposition of M can be formulated in a slightly different way by observing that if S and T are cyclic submodules of M, and if $ann(S) = \langle a \rangle$ and $ann(T) = \langle b \rangle$ where $\gcd(a,b) = 1$, then $S \cap T = \{0\}$ and $S \oplus T$ is a cyclic submodule with $ann(S \oplus T) = \langle ab \rangle$.

With this in mind, the summands in (6.10) can be collected as follows. Let D_1 be the direct sum of the first summands in each group of summands in (6.10) (by a *group* of summands, we mean the summands associated with a given prime p_i). Thus,

$$D_1 = C_{1,1} \oplus \cdots \oplus C_{n,1}$$

This cyclic submodule has order

$$q_1 = \prod_i p_i^{e_{i,1}}$$

Similarly, let D_2 be the direct sum of the second summands in each group in (6.10). (If a group does not have a second summand, we simply skip that group.) This gives us the following decomposition.

Theorem 6.9 (The cyclic decomposition theorem for finitely generated modules over a principal ideal domain – invariant factor version) Let M be a finitely generated module over a principal ideal domain R. Then

$$M = D_1 \oplus \cdots \oplus D_m \oplus M_{free}$$

where M_{free} is a free submodule, and D_i is a cyclic submodule of M, with order q_i, where

$$q_m \mid q_{m-1},\ q_{m-1} \mid q_{m-2}, \ldots, q_2 \mid q_1$$

Moreover, the scalars q_i, called the **invariant factors** of M, are uniquely determined, up to multiplication by a unit, by the module M. Also, the rank of M_{free} is uniquely determined by M. ∎

EXERCISES

1. Referring to the proof of Theorem 6.1, why may we assume that $M = R^n$?
2. Provide an example of an R-module M in which, for some $0 \neq r \in R$ and $0 \neq u \in M$, we have $ru = 0$.
3. Show that, for any module M, the set M_{tor} of all torsion elements in M is a submodule of M.
4. Show that, for any module M the quotient module M/M_{tor} is torsion free.
5. Show that any free module over a principal ideal domain is torsion free.
6. Referring to the proof of Theorem 6.3, show that $\mathfrak{B}'$ is linearly independent.
7. Let M be an R-module. Show that $ann(v)$ and $ann(M)$ are ideals of R.
8. Let M be a module over a p.i.d. R. If μ and ν are both orders of M, show that μ and ν are associates.
9. What is the order of the zero element in a module? What is the order of $1 \in M$?

10. Let R be a principal ideal domain. Show that the ideal $\langle p,q \rangle$ generated by p and q is also the ideal generated by $\gcd\{p,q\}$.
11. Let M be an R-module. Prove that $ann(M) \subset ann(v)$, for any $v \in M$. Furthermore, when R is a principal ideal domain, and $ann(v) = \langle \nu \rangle$ and $ann(M) = \langle \mu \rangle$, then $\nu \mid \mu$.
12. Prove Lemma 6.6.
13. Show that if S and T are cyclic submodules of M, and if $ann(S) = \langle a \rangle$ and $ann(T) = \langle b \rangle$ with $\gcd(a,b) = 1$, then $S \cap T = \{0\}$ and $S \oplus T$ is a cyclic submodule with $ann(S \oplus T) = \langle ab \rangle$. *Hint*: use the fact that there exists $p,q \in R$ such that $pa + qb = 1$.
14. Show that $ann(M) \subset ann(v)$, for any $v \in M$.
15. Show that, when R is a principal ideal domain, and $ann(v) = \langle \nu \rangle$ and $ann(M) = \langle \mu \rangle$, then $\nu \mid \mu$. In words, an order of $v \in M$ divides an order of M.

CHAPTER 7

The Structure of a Linear Operator

Contents: *A Brief Review. The Module Associated with a Linear Operator. Submodules and Invariant Subspaces. Orders and The Minimal Polynomial. Cyclic Submodules and Cyclic Subspaces. Summary. The Decomposition of V. The Rational Canonical Form. Exercises.*

In this chapter, we study the structure of a linear operator on a finite dimensional vector space, using the powerful module decomposition theorems of the previous chapter. *Unless otherwise noted, all vector spaces will be assumed to be finite dimensional.*

A Brief Review

We have seen that any linear operator on a finite dimensional vector space can be represented by matrix multiplication. Let us restate Theorem 2.13 for linear operators.

Theorem 7.1 Let $\tau \in \mathcal{L}(V)$, and let $\mathcal{B} = (\mathbf{b}_1, \ldots, \mathbf{b}_n)$ be an ordered basis for V. Then τ can be represented by a linear operator $\tau_A \in \mathcal{L}(F^n)$, that is,

$$[\tau(\mathbf{v})]_{\mathcal{B}} = \tau_A([\mathbf{v}]_{\mathcal{B}})$$

where $A = [\tau]_{\mathcal{B}}$ is the matrix whose ith column is $[\tau(\mathbf{b}_i)]_{\mathcal{B}}$. Thus,

$$[\tau(\mathbf{v})]_{\mathcal{B}} = [\tau]_{\mathcal{B}} [\mathbf{v}]_{\mathcal{B}}$$

∎

Since the matrix $[\tau]_{\mathcal{B}}$ depends on the ordered basis $\mathcal{B}$, it is natural to wonder how to choose this basis in order to make the matrix $[\tau]_{\mathcal{B}}$ as simple as possible, and that is the subject of this chapter.

Let us also restate the relationship between the matrices of τ with respect to different ordered bases.

Theorem 7.2 Let $\tau \in \mathcal{L}(V)$, and let $\mathcal{B}$ and $\mathcal{B}'$ be ordered bases for V. Then the matrix of τ with respect to $\mathcal{B}'$ can be expressed in terms of the matrix of τ with respect to $\mathcal{B}$ as follows

$$[\tau]_{\mathcal{B}'} = M_{\mathcal{B},\mathcal{B}'}\,[\tau]_{\mathcal{B}}\,(M_{\mathcal{B},\mathcal{B}'})^{-1}$$

where $M_{\mathcal{B},\mathcal{B}'}$ is the change of basis matrix, whose ith column is $[\mathbf{b}_i]_{\mathcal{B}'}$, where $\mathcal{B} = (\mathbf{b}_1,\ldots,\mathbf{b}_n)$. ∎

Finally, we recall the definition of similarity, and its relevance to the current discussion.

Definition Two matrices A and B are **similar** if there exists an invertible matrix P for which

$$B = PAP^{-1}$$

The equivalence classes associated with similarity are called **similarity classes.** □

Theorem 7.3 The following statements are equivalent for matrices A and B.

1) If A represents a linear operator $\tau: V \to V$ with respect to an ordered basis $\mathcal{B}$, then B also represents τ, but perhaps with respect to a different ordered basis. That is, if

$$A = [\tau]_{\mathcal{B}}$$

then there exists an ordered basis $\mathcal{B}'$ for which

$$B = [\tau]_{\mathcal{B}'}$$

2) A and B are similar. ∎

According to Theorem 7.3, the matrices that represent a given linear operator $\tau \in \mathcal{L}(V)$ are precisely the matrices that lie in a particular similarity class. Hence, in order to best represent τ, we seek a simple representative of that similarity class. More generally, in order to represent all linear operators on V, we would like to find a simple representative of each similarity class, that is, a set of simple *canonical forms* for similarity.

Let us recall the definition of canonical form.

Definition Let $\sim$ be an equivalence relation on S. A subset $C \subset S$ is said to be a set of **canonical forms** for $\sim$ if for every $s \in S$, there is *exactly* one $c \in C$ such that $c \sim s$. □

Now, the simplest type of matrices are probably the diagonal matrices. However, not all linear operators can be represented by diagonal matrices. In other words, the set of diagonal matrices does not form a set of canonical forms for similarity.

This gives rise to two different directions for further study. First, we can search for a characterization of those linear operators that *can* be represented by diagonal matrices. Such operators are called **diagonalizable.** Second, we can search for a different type of "simple" matrix that does provide a set of canonical forms for similarity. We will pursue both of these directions at the same time.

The Module Associated with a Linear Operator

Throughout this chapter, we fix a nonzero linear operator $\tau \in \mathcal{L}(V)$, and think of V not only as a vector space over a field F, but also as a module over $F[x]$ (as described in Chapter 4), with scalar multiplication defined by

$$p(x)\mathbf{v} = p(\tau)(\mathbf{v})$$

Our plan is to translate the language of the previous chapter into the language of V, by relating module concepts and vector spaces concepts.

First, since V is a finite dimensional vector space, the module V is a torsion module. To see this, observe that the vector space $\mathcal{L}(V)$, being isomorphic to $\mathcal{M}_n(F)$, has dimension n^2. Hence, the n^2+1 vectors

$$\iota, \tau, \tau^2, \ldots, \tau^{n^2}$$

are linearly dependent, which implies that $p(\tau) = 0$ for some polynomial $p(x) \in F[x]$. Hence, $p(x)V = \{\mathbf{0}\}$, and so all elements of V are torsion elements.

Also, V is finitely generated as a module. For if $\mathcal{B} = \{\mathbf{v}_1, \ldots, \mathbf{v}_n\}$ is a basis for the vector space V, then every vector $\mathbf{v} \in V$ is a linear combination

$$\mathbf{v} = r_1\mathbf{v}_1 + \cdots + r_n\mathbf{v}_n$$

where $r_i \in F \subset F[x]$, and so $\mathcal{B}$ generates the module V.

Hence, V is a finitely generated torsion module over a principal ideal domain $F[x]$, and so we may apply the decomposition theorems of the previous chapter.

Submodules and Invariant Subspaces

There is a simple connection between the submodules of the module V and the subspaces of the vector space V. Recall that a subspace S of V is *invariant* under τ if $\tau(S) \subset S$.

Theorem 7.4 A subset S of V is a submodule of the F[x]-module V if and only if it is an *invariant* subspace of the vector space V. ∎

Theorem 7.4 raises an issue that we should address. Namely, a submodule S of V can be made into an F[x]-module in two ways – as a submodule of V, and as a module using the restriction $\tau|_S: S \to S$ of τ to S. However, since

$$p(\tau)(\mathbf{s}) = p(\tau|_S)(\mathbf{s})$$

for all $\mathbf{s} \in S$, scalar multiplication is the same in both cases, and so these two modules are identical.

Orders and the Minimal Polynomial

Next, consider the annihilator of V

$$ann(V) = \{p(x) \in F[x] \mid p(x)V = \{0\}\}$$

which is a nonzero principal ideal of F[x]. Since all orders of V (that is, generators of $ann(V)$) are associates, and since the units of F[x] are precisely the nonzero elements of F, there is a unique *monic* order of V. This leads to the following definition.

Definition The unique *monic* order of the module V, that is, the unique monic polynomial that generates $ann(V)$, is called the **minimal polynomial** for τ. We denote this polynomial by $m_\tau(x)$, or $min(\tau)$. Thus,

$$ann(V) = \langle m_\tau(x) \rangle$$

and

$$p(x)V = \{0\} \text{ if and only if } m_\tau(x) \mid p(x)$$

or, equivalently

$$p(\tau) = 0 \text{ if and only if } m_\tau(x) \mid p(x)$$ □

In treatments of linear algebra that do not emphasize the role of the *module* V, the minimal polynomial of a linear operator τ is simply defined as the unique monic polynomial $m_\tau(x)$ of *smallest degree* for which $m_\tau(\tau) = 0$. It is not hard to see that this is equivalent to our definition.

The connection between order and minimal polynomial carries

over to submodules as well.

Theorem 7.5 Let S be a submodule of the module V. Then the monic order of S is the minimal polynomial of the restriction $\tau|_S$.

Proof. This follows from the fact that, if $q(x)$ is the monic order of S, then

$$\langle q(x)\rangle = ann(S) = \{p(x) \mid p(x)S = \{0\}\} = \{p(x) \mid p(\tau|_S)(S) = \{0\}\}$$

and so $q(x)$ is the minimal polynomial of the restriction $\tau|_S$. ∎

The concept of minimal polynomial is also defined for matrices. In particular, if A is a *square* matrix over F, the **minimal polynomial** $m_A(x)$ of A is the unique monic polynomial $p(x) \in F[x]$ of smallest degree for which $p(A) = 0$. We leave it to the reader to verify that this concept is well-defined, and that the following holds.

Theorem 7.6

1) If A and B are similar matrices, then $m_A(x) = m_B(x)$. Thus, the minimal polynomial is an invariant under similarity.
2) The minimal polynomial of $\tau \in \mathcal{L}(V)$ is the same as the minimal polynomial of any matrix that represents τ. ∎

Cyclic Submodules and Cyclic Subspaces

Consider the cyclic submodule

$$\langle \mathbf{v}\rangle = \{p(x)\mathbf{v} \mid p(x) \in F[x]\}$$

and suppose that it has monic order $m(x)$. Thus, $m(x)$ is the minimal polynomial of the restriction $\sigma = \tau|_{\langle \mathbf{v}\rangle}$. If

$$m(x) = a_0 + a_1x + \cdots + a_{n-1}x^{n-1} + x^n$$

then

$$\mathcal{B} = (\mathbf{v}, x\mathbf{v}, \ldots, x^{n-1}\mathbf{v}) = (\mathbf{v}, \sigma(\mathbf{v}), \ldots, \sigma^{n-1}(\mathbf{v}))$$

is an ordered basis for the vector space $\langle \mathbf{v}\rangle$. To see that $\mathcal{B}$ is linearly independent, suppose there exist nonzero scalars for which

$$r_0\mathbf{v} + r_1x\mathbf{v} + \cdots + r_{n-1}x^{n-1}\mathbf{v} = \mathbf{0}$$

that is,

$$(r_0 + r_1x + \cdots + r_{n-1}x^{n-1})\mathbf{v} = \mathbf{0}$$

Then

$$(r_0 + r_1x + \cdots + r_{n-1}x^{n-1})\langle \mathbf{v}\rangle = \{\mathbf{0}\}$$

and so $m(x) \mid r_0 + r_1x + \cdots + r_{n-1}x^{n-1}$, which implies that $r_i = 0$ for all $i = 0,\ldots,n-1$.

To see that $\mathfrak{B}$ spans $\langle \mathbf{v} \rangle$, observe that all elements of $\langle \mathbf{v} \rangle$ have the form $p(x)\mathbf{v}$, for some polynomial $p(x) \in F[x]$. However, dividing $p(x)$ by the minimal polynomial $m(x)$ gives

$$p(x) = q(x)m(x) + r(x)$$

where $\deg r(x) < \deg m(x)$. Since $m(x)\mathbf{v} = 0$, we have

$$p(x)\mathbf{v} = q(x)m(x)\mathbf{v} + r(x)\mathbf{v} = r(x)\mathbf{v}$$

which shows that all elements of $\langle \mathbf{v} \rangle$ have the form $r(x)\mathbf{v}$, where $\deg r(x) < \deg q(x) = n$. In symbols,

$$\langle \mathbf{v} \rangle = \{r(x)\mathbf{v} \mid \deg r(x) < \deg m(x)\}$$

Hence, if

$$r(x) = r_0 + r_1 x + \cdots + r_{n-1}x^{n-1}$$

we have

$$r(x)\mathbf{v} = r_0\mathbf{v} + r_1 x\mathbf{v} + \cdots + r_{n-1}x^{n-1}\mathbf{v} \in \mathit{span}(\mathfrak{B})$$

Thus, $\mathfrak{B}$ is an ordered basis for $\langle \mathbf{v} \rangle$.

To determine the matrix $[\sigma]_{\mathfrak{B}}$ of σ with respect to $\mathfrak{B}$, observe that

$$\sigma(\sigma^i(\mathbf{v})) = \sigma^{i+1}(\mathbf{v})$$

for $i = 0,\ldots,n-2$, and so σ simply "shifts" each basis vector in $\mathfrak{B}$, except the last one, to the next basis vector in $\mathfrak{B}$. Also, $m(\sigma) = 0$ implies that

$$\begin{aligned}\sigma(\sigma^{n-1}(\mathbf{v})) &= \sigma^n(\mathbf{v}) \\ &= -(a_0 + a_1\sigma + \cdots + a_{n-1}\sigma^{n-1})(\mathbf{v}) \\ &= -a_0\mathbf{v} - a_1\sigma(\mathbf{v}) - \cdots - a_{n-1}\sigma^{n-1}(\mathbf{v})\end{aligned}$$

Hence, the matrix of σ, with respect to $\mathfrak{B}$, is

$$C[m(x)] = \begin{bmatrix} 0 & 0 & \cdots & 0 & -a_0 \\ 1 & 0 & \cdots & 0 & -a_1 \\ 0 & 1 & \ddots & & \vdots \\ \vdots & \vdots & \ddots & 0 & -a_{n-2} \\ 0 & 0 & \cdots & 1 & -a_{n-1} \end{bmatrix}$$

This matrix is known as the **companion matrix** for the polynomial

$$m(x) = a_0 + a_1 x + \cdots + a_{n-1}x^{n-1} + x^n$$

Note that companion matrices are defined only for *monic* polynomials.

Let us summarize, beginning with a definition.

Definition Let $\tau \in \mathcal{L}(V)$. A subspace S of V is **τ-cyclic** if there exists a vector $\mathbf{v} \in S$ for which the set

$$\{\mathbf{v},\tau(\mathbf{v}),\ldots,\tau^{m-1}(\mathbf{v})\}$$

is a basis for S, where $m = \mathit{dim}(S)$. ☐

Theorem 7.7

1) A subset $S \subset V$ is a cyclic submodule of V if and only if it is a τ-cyclic subspace of V.
2) Suppose that $\langle\mathbf{v}\rangle$ is a cyclic submodule of V. If the monic order of $\langle\mathbf{v}\rangle$ (that is, the minimal polynomial of $\sigma = \tau\,|_{\langle\mathbf{v}\rangle}$) is

$$m_\sigma(x) = a_0 + a_1x + \cdots + a_{n-1}x^{n-1} + x^n$$

Then

$$\mathcal{B} = (\mathbf{v},x\mathbf{v},\ldots,x^{n-1}\mathbf{v}) = (\mathbf{v},\sigma(\mathbf{v}),\ldots,\sigma^{n-1}(\mathbf{v}))$$

is an ordered basis for $\langle\mathbf{v}\rangle$, and the matrix $[\sigma]_{\mathcal{B}}$ is the companion matrix $C(m_\sigma(x))$ of $m_\sigma(x)$. ∎

Summary

The following table summarizes the connection between the module concepts and the vector space concepts that we have discussed.

Module V	*Vector Space V*
Scalar multiplication: $p(x)\mathbf{v}$	Action of τ: $p(\tau)(\mathbf{v})$
Submodule	Invariant subspace
Annihilator: $\mathit{ann}(v) = \{p(x) \mid p(x)V = \{\mathbf{0}\}\}$	Annihilator: $\mathit{ann}(V) = \{p(x) \mid p(\tau)(V) = \{\mathbf{0}\}\}$
Monic order $m(x)$ of V: $\mathit{ann}(V) = \langle m(x)\rangle$	Minimal polynomial of τ: $m(x)$ is poly. of smallest degree for which $m(\tau) = 0$
Cyclic submodule: $\langle\mathbf{v}\rangle = \{p(x)\mathbf{v} \mid p(x) \in F[x]\}$	τ-cyclic subspace: $\langle\mathbf{v}\rangle = \mathit{span}\{\mathbf{v},\tau(\mathbf{v}),\ldots,\tau^{m-1}(\mathbf{v})\}$

The Decomposition of V

We are now ready to translate the cyclic decomposition theorem (Theorem 6.8) into the language of V.

Theorem 7.8 (The cyclic decomposition theorem for V) Let $\tau \in \mathcal{L}(V)$, where $dim(V) < \infty$. If the minimal polynomial of τ is

$$m_\tau(x) = p_1^{e_1}(x) \cdots p_n^{e_n}(x)$$

where the monic polynomials $p_i(x)$ are distinct and irreducible, then V is the direct sum

$$V = V_{p_1} \oplus \cdots \oplus V_{p_n}$$

where

$$V_{p_i} = \{\mathbf{v} \in V \mid p_i^{e_i}(\tau)(\mathbf{v}) = \mathbf{0}\}$$

is an invariant subspace (submodule) of V, and

$$min(\tau \mid_{V_{p_i}}) = p_i^{e_i}(x)$$

Moreover, each V_{p_i} can be further decomposed into a direct sum of τ-cyclic subspaces (cyclic submodules)

$$V_{p_i} = \langle \mathbf{v}_{i,1} \rangle \oplus \cdots \oplus \langle \mathbf{v}_{i,k_i} \rangle$$

where

$$min(\tau \mid_{\langle \mathbf{v}_{i,j} \rangle}) = p_i^{e_{i,j}}(x)$$

and

$$e_i = e_{i,1} \geq e_{i,2} \geq \cdots \geq e_{i,k_i}$$

The elementary divisors $p_i^{e_{i,j}}(x)$ of V, also known as the **elementary divisors** of τ, are uniquely determined by the operator τ.

This yields the decomposition of V into the direct sum of τ-cyclic subspaces

$$V = (\langle \mathbf{v}_{1,1} \rangle \oplus \cdots \oplus \langle \mathbf{v}_{1,k_1} \rangle) \oplus \cdots \oplus (\langle \mathbf{v}_{n,1} \rangle \oplus \cdots \oplus \langle \mathbf{v}_{n,k_n} \rangle)$$ ∎

The Rational Canonical Form

The cyclic decomposition theorem can be used to determine a set of canonical forms for similarity.

Recall that if $V = S \oplus T$ and if both S and T are invariant under τ, the pair (S,T) is said to *reduce* τ. Put another way, (S,T) reduces τ if the restrictions

$$\tau\mid_S : S \to S \quad \text{and} \quad \tau\mid_T : T \to T$$

are linear *operators* on S and T, respectively. Recall also that we write $\tau = \rho \oplus \sigma$ if there exist subspaces S and T of V for which (S,T) reduces τ and

$$\rho = \tau|_S \text{ and } \sigma = \tau|_T$$

If $\tau = \sigma \oplus \rho$, then any matrix representations of σ and ρ can be used to construct a matrix representation of τ. This is especially relevant to our situation, since according to Theorem 7.8,

$$\tau = \tau|_{\langle \mathbf{v}_{1,1} \rangle} \oplus \cdots \oplus \tau|_{\langle \mathbf{v}_{n,k_n} \rangle} \tag{7.1}$$

Before discussing this further, it will be convenient to introduce the notational device of a *block matrix*. If A is an $n \times n$ matrix, and B is an $m \times m$ matrix, then by the **block matrix**

$$M = \begin{bmatrix} A & 0 \\ 0 & B \end{bmatrix}_{block}$$

we mean the $(n+m) \times (n+m)$ matrix whose upper left $n \times n$ *submatrix* is A, and whose lower left $m \times m$ *submatrix* is B. (Thus, A and B are submatrices of M, and not entries.) All other entries in M are 0. Because of the particular block form of M, we also refer to it as a **block diagonal matrix.** Clearly, this concept can be extended to more than two matrices A and B.

Theorem 7.9 Suppose that $\tau = \tau_1 \oplus \tau_2 \in \mathcal{L}(V)$, with corresponding reducing pair (S,T). Let $\mathcal{C} = (\mathbf{c}_1, \ldots, \mathbf{c}_s)$ be an ordered basis for S, let $\mathcal{D} = (\mathbf{d}_1, \ldots, \mathbf{d}_t)$ be an ordered basis for T, and let

$$\mathcal{B} = (\mathbf{c}_1, \ldots, \mathbf{c}_s, \mathbf{d}_1, \ldots, \mathbf{d}_t)$$

be the corresponding ordered basis for V. Then the matrix $[\tau]_{\mathcal{B}}$ has the block diagonal form

$$[\tau]_{\mathcal{B}} = \begin{bmatrix} [\tau_1]_{\mathcal{C}} & 0 \\ 0 & [\tau_2]_{\mathcal{D}} \end{bmatrix}_{block}$$ ∎

Of course, this theorem may be extended to apply to multiple direct summands. In particular, referring to (7.1), if $\mathcal{B}_{i,j}$ is an ordered basis for the cyclic submodule $\langle \mathbf{v}_{i,j} \rangle$, and if

$$\mathcal{B} = (\mathcal{B}_{1,1}, \ldots, \mathcal{B}_{n,k_n})$$

denotes the ordered basis for V, obtained (as in Theorem 7.9) from these ordered bases, then

$$[\tau]_{\mathfrak{B}} = \begin{bmatrix} [\tau_{1,1}]_{\mathfrak{B}_{1,1}} & & \\ & \ddots & \\ & & [\tau_{n,k_n}]_{\mathfrak{B}_{n,k_n}} \end{bmatrix}_{block}$$

where $\tau_{i,j} = \tau|_{\langle \mathbf{v}_{i,j} \rangle}$.

Now, the cyclic submodule $\langle \mathbf{v}_{i,j} \rangle$ has monic order $p_i^{e_{i,j}}(x)$, that is, the restriction $\tau_{i,j}$ has minimal polynomial $p_i^{e_{i,j}}(x)$. Thus, if

$$\deg p_i^{e_{i,j}}(x) = d_{i,j}$$

then

$$\mathfrak{B}_{i,j} = \left(\mathbf{v}_{i,j}, \tau_{i,j}(\mathbf{v}_{i,j}), \ldots, \tau_{i,j}^{d_{i,j}-1}(\mathbf{v}_{i,j})\right)$$

is an ordered basis for $\langle \mathbf{v}_{i,j} \rangle$. Hence, we arrive at the matrix representation of τ described in the following theorem.

Theorem 7.10 Let $dim(V) < \infty$, and suppose that $\tau \in \mathcal{L}(V)$ has minimal polynomial

$$m_\tau(x) = p_1^{e_1}(x) \cdots p_n^{e_n}(x)$$

where the monic polynomials $p_i(x)$ are distinct and irreducible. Then we can write

$$V = (\langle \mathbf{v}_{1,1} \rangle \oplus \cdots \oplus \langle \mathbf{v}_{1,k_1} \rangle) \oplus \cdots \oplus (\langle \mathbf{v}_{n,1} \rangle \oplus \cdots \oplus \langle \mathbf{v}_{n,k_n} \rangle)$$

where $\langle \mathbf{v}_{i,j} \rangle$ is a τ-cyclic subspace of V. The minimal polynomials for $\tau_{i,j} = \tau|_{\langle \mathbf{v}_{i,j} \rangle}$ are the elementary divisors

$$min(\tau_{i,j}) = p_i^{e_{i,j}}(x)$$

of V, where

$$e_i = e_{i,1} \geq e_{i,2} \geq \cdots \geq e_{i,k_i}$$

These elementary divisors are uniquely determined by τ. Furthermore, if $\deg p_i^{e_{i,j}}(x) = d_{i,j}$, then

$$\mathfrak{B}_{i,j} = (\mathbf{v}_{i,j}, \tau_{i,j}(\mathbf{v}_{i,j}), \ldots, \tau_{i,j}^{d_{i,j}-1}(\mathbf{v}_{i,j}))$$

is an ordered basis for $\langle \mathbf{v}_{i,j} \rangle$, and the matrix of τ with respect to the ordered basis

$$\mathcal{B} = (\mathcal{B}_{1,1}, \ldots, \mathcal{B}_{n,k_n})$$

is the block diagonal matrix

$$(7.2) \qquad [\tau]_{\mathcal{B}} = \begin{bmatrix} C[p_1^{e_{1,1}}(x)] & & & & & \\ & \ddots & & & & \\ & & C[p_1^{e_{1,k_1}}(x)] & & & \\ & & & \ddots & & \\ & & & & C[p_n^{e_{n,n_1}}(x)] & \\ & & & & & \ddots \\ & & & & & & C[p_n^{e_{n,k_n}}(x)] \end{bmatrix}_{block}$$

The matrix on the right is called the **rational canonical form** of τ. ∎

Let us denote the matrix on the right of (7.2) by

$$diag\Big(C[p_1^{e_{1,1}}(x)], \ldots, C[p_n^{e_{n,k_n}}(x)]\Big)$$

Theorem 7.10 implies that, for any $\tau \in \mathcal{L}(V)$, we can find an ordered basis $\mathcal{B}$ for which the matrix $[\tau]_{\mathcal{B}}$ has the rational canonical form (7.2). On the other hand, τ has only one rational canonical form (up to reordering of the blocks on the diagonal). To see this, suppose that, for some ordered basis $\mathcal{C}$ for V, the matrix $[\tau]_{\mathcal{C}}$ has the form

$$[\tau]_{\mathcal{C}} = diag\Big(C[q_1^{f_{1,1}}(x)], \ldots, C[q_m^{f_{m,k_m}}(x)]\Big)$$

Then V can be written as a direct sum of cyclic submodules, with elementary divisors $q_i^{f_{i,j}}(x)$. Hence, the uniqueness of the rational canonical form (up to reordering of the blocks on the diagonal) follows from the uniqueness of the cyclic decomposition of V.

Theorem 7.10 can be reformulated in terms of matrices as follows.

Theorem 7.11 Any square matrix A is similar to a unique (except for the order of the blocks on the diagonal) matrix that is in rational canonical form. ∎

Corollary 7.12 Two matrices over the same field F are similar if and only if they have the same elementary divisors. ∎

We will not go into the details of how best to find the rational canonical form of a matrix, since our main interest in this form is as a theoretical tool. However, for concreteness, here are some examples of rational canonical forms.

Example 7.1 Let τ be a linear operator on the vector space $\mathbb{R}^7$, and suppose that τ has minimal polynomial

$$m_\tau(x) = (x-1)(x^2+1)^2$$

Noting that $x-1$ and $(x^2+1)^2$ are elementary divisors, we have the following possibilities for the list of elementary divisors.

1) $x-1, (x^2+1)^2, x^2+1$

2) $x-1, x-1, x-1, (x^2+1)^2$

These correspond to the following rational canonical forms

1) $\begin{bmatrix} -1 & 0 & 0 & 0 & 0 & 0 & 0 \\ 0 & 0 & 0 & 0 & -1 & 0 & 0 \\ 0 & 1 & 0 & 0 & 0 & 0 & 0 \\ 0 & 0 & 1 & 0 & -2 & 0 & 0 \\ 0 & 0 & 0 & 1 & 0 & 0 & 0 \\ 0 & 0 & 0 & 0 & 0 & 0 & -1 \\ 0 & 0 & 0 & 0 & 0 & 1 & 0 \end{bmatrix}$ 2) $\begin{bmatrix} -1 & 0 & 0 & 0 & 0 & 0 & 0 \\ 0 & -1 & 0 & 0 & 0 & 0 & 0 \\ 0 & 0 & -1 & 0 & 0 & 0 & 0 \\ 0 & 0 & 0 & 0 & 0 & 0 & -1 \\ 0 & 0 & 0 & 1 & 0 & 0 & 0 \\ 0 & 0 & 0 & 0 & 1 & 0 & -2 \\ 0 & 0 & 0 & 0 & 0 & 1 & 0 \end{bmatrix}$ □

EXERCISES

1. Show that a subset S of V is a submodule of the F[x]-module V if and only if it is an *invariant* subspace of the vector space V.
2. Show that the units in F[x] are precisely the nonzero scalars in F.
3. Verify that the concept of the minimal polynomial of a matrix is well-defined. Prove Theorem 7.6.
4. We have seen that any $\tau \in \mathcal{L}(V)$ can be used to make V into an F[x]-module. Does every F[x]-module V come from some $\tau \in \mathcal{L}(V)$? Explain.
5. Formulate an invariant factor version of Theorem 7.10.
6. Referring to the discussion immediately following Theorem 7.10, show that the rational canonical form of τ is unique, up to the order of the block diagonal matrices.

7. Prove that the minimal polynomial of $\tau \in \mathcal{L}(V)$ is the least common multiple of its elementary divisors.
8. Let $\mathbb{Q}$ be the field of rational numbers. Consider the linear operator $\tau \in \mathcal{L}(\mathbb{Q}^2)$ defined by $\tau(\mathbf{e}_1) = \mathbf{e}_2$, $\tau(\mathbf{e}_2) = -\mathbf{e}_1$.
   a) Find the minimal polynomial for τ, and show that the rational canonical form for τ is
   $$R = \begin{bmatrix} 0 & -1 \\ 1 & 0 \end{bmatrix}$$
   What are the elementary divisors of τ?
   b) Now consider the map $\sigma \in \mathcal{L}(\mathbb{C}^2)$ defined by the same rules as τ, namely, $\sigma(\mathbf{e}_1) = \mathbf{e}_2$, $\sigma(\mathbf{e}_2) = -\mathbf{e}_1$. Find the minimal polynomial for σ, and the rational canonical form for σ. What are the elementary divisors of σ?
   c) The invariant factors of τ are defined, using the elementary divisors of τ, in the same way as we did at the end of Chapter 6, for a module M. Describe the invariant factors for the operators in parts (a) and (b).
9. Find all rational canonical forms (up to the order of the blocks on the diagonal) for a linear operator on $\mathbb{R}^6$ having minimal polynomial $(x-1)^2(x+1)^2$.
10. How many possible rational canonical forms (up to order of blocks) are there for linear operators on $\mathbb{R}^6$, with minimal polynomial $(x-1)(x+1)^2$?
11. Prove that if C is the companion matrix of $p(x)$, then $p(C) = 0$, and C has minimal polynomial $p(x)$.
12. Let τ be a linear operator on F^4, with minimal polynomial $m_\tau(x) = (x^2+1)(x^2-2)$. Find the rational canonical form for τ if (a) $F = \mathbb{Q}$, (b) $F = \mathbb{R}$, (c) $F = \mathbb{C}$.

CHAPTER 8

Eigenvalues and Eigenvectors

Contents: *The Characteristic Polynomial of an Operator. Eigenvalues and Eigenvectors. The Cayley-Hamilton Theorem. The Jordan Canonical Form. Geometric and Algebraic Multiplicities. Diagonalizable Operators. Projections. The Algebra of Projections. Resolutions of the Identity. Projections and Diagonalizability. Projections and Invariance. Exercises.*

Unless otherwise noted, we will assume throughout this chapter that all vector spaces are finite dimensional.

The Characteristic Polynomial of an Operator

Let us compute the determinant of the matrix $xI - R$, where R is the rational canonical form of τ. To do this, we need the following result, whose proof is left to the reader.

Lemma 8.1 If a square matrix M has the block diagonal form

$$M = \begin{bmatrix} A & 0 \\ 0 & B \end{bmatrix}_{block}$$

where A and B are square, then $\det(M) = \det(A)\det(B)$. ∎

Now, let $C[p(x)]$ be the companion matrix of the polynomial $p(x) = a_0 + a_1 x + \cdots + a_{n-1}x^{n-1} + x^n$, and let

$$A = xI - C[p(x)] = \begin{bmatrix} x & 0 & \cdots & 0 & a_0 \\ -1 & x & \cdots & 0 & a_1 \\ 0 & -1 & \ddots & & \vdots \\ \vdots & \vdots & \ddots & x & a_{n-2} \\ 0 & 0 & \cdots & -1 & x+a_{n-1} \end{bmatrix}$$

To compute the determinant of this matrix, we indicate its dependence on the coefficients a_i by writing $A = A(x;a_0,\ldots,a_{n-1})$, and then look at some simple cases

$$\det(A(x;a_0,a_1)) = \begin{vmatrix} x & a_0 \\ -1 & x+a_1 \end{vmatrix} = x(x+a_1)+a_0 = a_0+a_1x+x^2 = p(x)$$

$$\det(A(x;a_0,a_1,a_2)) = \begin{vmatrix} x & 0 & a_0 \\ -1 & x & a_1 \\ 0 & -1 & x+a_2 \end{vmatrix} = x\begin{vmatrix} x & a_1 \\ -1 & x+a_2 \end{vmatrix} + a_0\begin{vmatrix} -1 & x \\ 0 & -1 \end{vmatrix}$$

$$= x[x(x+a_2)+a_1]+a_0 = a_0+a_1x+a_2x^2+x^3 = p(x)$$

In general, expanding along the first row gives

$$\begin{aligned} \det(A(x,a_0,\ldots,a_{n-1})) &= x\det(A(x,a_1,\ldots,a_{n-1})) + (-1)^{n+1}(-1)^{n-1}a_0 \\ &= x\det(A(x,a_1,\ldots,a_{n-1})) + a_0 \end{aligned}$$

An induction argument thus leads to the following.

Lemma 8.2 If $C[p(x)]$ is the companion matrix of a polynomial $p(x)$, then

$$\det(xI - C[p(x)]) = p(x)$$ ∎

Combining Lemmas 8.1 and 8.2 gives the following.

Theorem 8.3 If R is the rational canonical form for $\tau \in \mathcal{L}(V)$, then

$$C_\tau(x) = \det(xI - R) = \prod_{i,j} p_i^{e_{i,j}}(x)$$

This determinant is called the **characteristic polynomial** of τ. ∎

The characteristic polynomial is often defined first for matrices and then for linear operators. The **characteristic polynomial** of a square matrix A is defined to be $C_A(x) = \det(xI - A)$.

Theorem 8.4

1) If A is similar to B, then $C_A(x) = C_B(x)$. Thus, the characteristic polynomial is an invariant under similarity.
2) The characteristic polynomial of an operator τ is equal to the characteristic polynomial of any matrix that represents τ.
3) The characteristic polynomial of an operator τ is the product of the elementary divisors of τ. ∎

Even though the characteristic polynomial is an invariant under similarity (as is the minimal polynomial), the matrices

$$A = \begin{bmatrix} \delta & 0 \\ 0 & \delta \end{bmatrix} \quad \text{and} \quad B = \begin{bmatrix} \delta & 0 \\ 1 & \delta \end{bmatrix}$$

which have the same characteristic polynomial but are not similar, show that the characteristic polynomial is not a *complete* invariant.

Eigenvalues and Eigenvectors

Notice that $\lambda \in F$ is a root of the characteristic polynomial $C_\tau(x)$ of a linear operator $\tau \in \mathcal{L}(V)$ if and only if

$$\det(\lambda I - R) = 0 \tag{8.1}$$

that is, if and only if the matrix $\lambda I - R$ is singular. In particular, if $dim(V) = d$, then the rational canonical form R for τ has size $d \times d$, and so (8.1) holds if and only if there exists a *nonzero* vector $\mathbf{x} \in F^d$ for which

$$(\lambda I - R)\mathbf{x} = \mathbf{0}$$

or

$$\tau_R(\mathbf{x}) = \lambda \mathbf{x}$$

If $\mathbf{v} \in V$ is the nonzero vector for which $[\mathbf{v}]_{\mathcal{B}} = \mathbf{x}$, where $\mathcal{B}$ is the ordered basis used to represent τ by R, then this is equivalent to

$$\tau(\mathbf{v}) = \lambda \mathbf{v}$$

This prompts the following definition, which applies to vector spaces of arbitrary dimension.

Definition Let $\tau \in \mathcal{L}(V)$ be a linear operator. A scalar $\lambda \in F$ is an **eigenvalue** (or **characteristic value**) of τ if there exists a *nonzero* vector $\mathbf{v} \in V$ for which

$$\tau(\mathbf{v}) = \lambda\mathbf{v}$$

In this case, $\mathbf{v}$ is an **eigenvector** (or **characteristic vector**) of τ, associated with λ.

If A is a matrix over F, then $\lambda \in F$ is an **eigenvalue** for A if there exists a *nonzero* column vector $\mathbf{x}$ for which

$$A\mathbf{x} = \lambda\mathbf{x}$$

In this case, $\mathbf{x}$ is an **eigenvector** (or **characteristic vector**) for A, associated with λ. ▯

The set of all eigenvectors associated to a given eigenvalue λ, together with the zero vector, forms a subspace of V, called the **eigenspace** of λ. We will denote the eigenspace of an eigenvalue λ by $\mathcal{E}_\lambda$. This applies to both linear operators and matrices.

The following theorems summarize the key facts about eigenvalues and eigenvectors.

Theorem 8.5

1) A scalar $\lambda \in F$ is an eigenvalue of $\tau \in \mathcal{L}(V)$ if and only if it is a root of the characteristic polynomial $C_\tau(x)$ of τ.
2) A scalar $\lambda \in F$ is an eigenvalue of $\tau \in \mathcal{L}(V)$ if and only if it is a root of the minimal polynomial $m_\tau(x)$ of τ.
3) A scalar $\lambda \in F$ is an eigenvalue of τ if and only if it is an eigenvalue of any matrix that represents τ.
4) The eigenvalues of a matrix are invariants under similarity.
5) If λ is an eigenvalue for a matrix A, then the eigenspace $\mathcal{E}_\lambda$ is the solution space to the homogeneous system of equations

$$(\lambda I - A)(\mathbf{x}) = \mathbf{0}$$

Proof. The first part of this theorem has already been established. Part (2) follows from the fact that the prime factors of the characteristic polynomial $C_\tau(x)$ and the minimal polynomial $m_\tau(x)$ are the same.

As for part (3), λ is an eigenvalue of τ if and only if

$$\tau(\mathbf{v}) = \lambda\mathbf{v} \tag{8.2}$$

for some nonzero $\mathbf{v} \in V$. Now, suppose that $dim(V) = d$, let $\mathcal{B}$ be an ordered basis for V, and let $\phi_\mathcal{B}: V \to F^d$ be the isomorphism defined by $\phi_\mathcal{B}(\mathbf{u}) = [\mathbf{u}]_\mathcal{B}$. Then, if $A = [\tau]_\mathcal{B}$, we have (*cf.* Figure 3.2)

$$\tau = (\phi_\mathcal{B})^{-1}\tau_A\phi_\mathcal{B}$$

and so (8.2) is equivalent to

$$(\phi_\mathcal{B})^{-1}\tau_A\phi_\mathcal{B}(\mathbf{v}) = \lambda\mathbf{v} = (\phi_\mathcal{B})^{-1}\lambda\phi_\mathcal{B}(\mathbf{v})$$

or

$$\tau_A \phi_{\mathcal{B}}(\mathbf{v}) = \lambda \phi_{\mathcal{B}}(\mathbf{v})$$

which says that λ is an eigenvalue for A. Hence, λ is an eigenvalue for τ if and only if it is an eigenvalue for A. This proves part (3). Part (4) follows from part (3), and part (5) is evident. ∎

Theorem 8.6 Suppose that $\lambda_1,\ldots,\lambda_k$ are the *distinct* eigenvalues of a linear operator $\tau \in \mathcal{L}(V)$. Then $\mathcal{E}_{\lambda_i} \cap \mathcal{E}_{\lambda_j} = \{\mathbf{0}\}$. Moreover, eigenvectors associated with distinct eigenvalues are linearly independent. That is, if $\mathbf{v}_i \in \mathcal{E}_{\lambda_i}$ for $i = 1,\ldots,k$, then the vectors $\{\mathbf{v}_1,\ldots,\mathbf{v}_k\}$ are linearly independent.

Proof. We leave it to the reader to show that $\mathcal{E}_{\lambda_i} \cap \mathcal{E}_{\lambda_j} = \{\mathbf{0}\}$. Let $\mathbf{v}_i \in \mathcal{E}_{\lambda_i}$, for $i = 1,\ldots,k$, where $\lambda_1,\ldots,\lambda_k$ are distinct eigenvalues of τ. We want to show that the $\mathbf{v}_i$'s are linearly independent. Assuming that the $\mathbf{v}_i$'s are linearly dependent, we may also assume (after renumbering if necessary) that, among all nontrivial linear combinations of these vectors that equal $\mathbf{0}$, the equation

$$r_1\mathbf{v}_1 + \cdots + r_j\mathbf{v}_j = \mathbf{0} \tag{8.3}$$

is the *shortest* such equation (that is, has the fewest terms). Applying τ gives

$$r_1\tau(\mathbf{v}_1) + \cdots + r_j\tau(\mathbf{v}_j) = \mathbf{0}$$

or

$$r_1\lambda_1\mathbf{v}_1 + \cdots + r_j\lambda_j\mathbf{v}_j = \mathbf{0} \tag{8.4}$$

Now we multiply (8.3) by λ_1, and subtract from (8.4), to get

$$r_2(\lambda_2 - \lambda_1)\mathbf{v}_2 + \cdots + r_j(\lambda_j - \lambda_1)\mathbf{v}_j = \mathbf{0}$$

But this is a shorter equation than (8.3), and so all of the coefficients must equal 0, and since the λ_i's are distinct, we deduce that $r_i = 0$ for $i = 2,\ldots,j$, and so $r_1 = 0$ as well. This contradiction implies that the $\mathbf{v}_i$'s are linearly independent. ∎

One way to compute the eigenvalues of a linear operator τ is to first represent τ by a matrix A, and then solve the **characteristic equation**

$$C_A(x) = 0$$

Unfortunately, however, it is quite likely that we cannot solve this polynomial equation when $\deg C_A(x) = \mathit{dim}(V) \geq 3$. As a result, the art of *approximating* the eigenvalues of a matrix is a very important area of applied linear algebra.

The Cayley-Hamilton Theorem

Since the characteristic polynomial $C_\tau(x)$ of a linear operator τ is the product of its elementary divisors, and since the minimal polynomial of τ is the product

$$m_\tau(x) = p_1^{e_1}(x) \cdots p_n^{e_n}(x)$$

we deduce that $m_\tau(x) \mid C_\tau(x)$. This important result is referred to as the *Cayley-Hamilton theorem.*

Theorem 8.7 Let $\tau \in \mathcal{L}(V)$.

1) The minimal polynomial $m_\tau(x)$ and the characteristic polynomial $C_\tau(x)$ have the same prime factors.
2) **(The Cayley-Hamilton theorem)** $m_\tau(x) \mid C_\tau(x)$, or equivalently, τ satisfies its own characteristic polynomial. ∎

The Jordan Canonical Form

One of the virtues of the rational canonical form is that every linear operator τ on a finite dimensional vector space *has* a rational canonical form. That is, the set of all matrices in rational canonical form constitutes a set of canonical forms (at least up to the order of the blocks on the diagonal). Unfortunately, however, the rational canonical form of a matrix may be far from the ideal of simplicity that we had in mind for a set of simple canonical forms.

Fortunately, in certain important cases, we can do better than the rational canonical form. In particular, let us consider the case of a linear operator $\tau \in \mathcal{L}(V)$ whose minimal polynomial factors into a product of *linear* factors

$$m_\tau(x) = (x - \lambda_1)^{e_1} \cdots (x - \lambda_n)^{e_n} \tag{8.5}$$

When a polynomial factors into a product of linear factors over a field F, we say that the polynomial **splits** over F.

To put this in perspective, we note that a field F is said to be **algebraically closed** if every *nonconstant* polynomial over F has a root in F. Thus, the only irreducible polynomials over an algebraically closed field are the linear polynomials, and so any nonconstant polynomial over F splits over F. For example, the complex numbers $\mathbb{C}$ form an algebraically closed field, and so any linear operator over a *complex* vector space has minimal polynomial that splits over $\mathbb{C}$.

In some sense, the "weakness" in the rational canonical form comes in choosing the basis for the cyclic submodules $\langle \mathbf{v}_{i,j} \rangle$, whose monic order is the elementary divisor $p_i^{e_{i,j}}(x)$, of which we know very little in general. Recall that, since $\langle \mathbf{v}_{i,j} \rangle$ is a τ-cyclic subspace of V, we have chosen the ordered basis

$$\mathfrak{B}_{i,j} = (\mathbf{v}_{i,j}, \tau_{i,j}(\mathbf{v}_{i,j}), \ldots, \tau_{i,j}^{d_{i,j}-1}(\mathbf{v}_{i,j}))$$

However, when the minimal polynomial has the form (8.5), then the elementary divisors have the form

$$p_i^{e_{i,j}}(x) = (x - \lambda_i)^{e_{i,j}}$$

In this case, we can make a more judicious choice of ordered basis. Observe that $dim(\langle \mathbf{v}_{i,j} \rangle) = \deg p_i^{e_{i,j}}(x)$, and so it is easy to see that the set

$$\mathfrak{C}_{i,j} = \left(\mathbf{v}_{i,j}, (\tau_{i,j} - \lambda_i)(\mathbf{v}_{i,j}), \ldots, (\tau_{i,j} - \lambda_i)^{e_{i,j}-1}(\mathbf{v}_{i,j})\right)$$

is an ordered basis for $\langle \mathbf{v}_{i,j} \rangle$. Furthermore, denoting the kth basis vector in $\mathfrak{C}_{i,j}$ by $\mathbf{b}_k$, we have for $k = 0, \ldots, e_{i,j}-2$,

$$\begin{aligned}\tau_{i,j}(\mathbf{b}_k) &= \tau_{i,j}[(\tau_{i,j} - \lambda_i)^k(\mathbf{v}_{i,j})] = (\tau_{i,j} - \lambda_i + \lambda_i)[(\tau_{i,j} - \lambda_i)^k(\mathbf{v}_{i,j})] \\ &= (\tau_{i,j} - \lambda_i)^{k+1}(\mathbf{v}_{i,j}) + \lambda_i(\tau_{i,j} - \lambda_i)^k(\mathbf{v}_{i,j}) = \mathbf{b}_{k+1} + \lambda_i \mathbf{b}_k\end{aligned}$$

For $k = e_{i,j} - 1$, a similar computation, using the fact that

$$(\tau_{i,j} - \lambda_i)^{k+1}(\mathbf{v}_{i,j}) = (\tau_{i,j} - \lambda_i)^{e_{i,j}}(\mathbf{v}_{i,j}) = \mathbf{0}$$

gives

$$\tau_{i,j}(\mathbf{b}_{e_{i,j}-1}) = \mathbf{b}_{e_{i,j}-1}$$

Hence, the matrix of $\tau_{i,j} = \tau|_{\langle \mathbf{v}_{i,j} \rangle}$ is the $e_{i,j} \times e_{i,j}$ matrix

$$\mathfrak{J}(\lambda_i, e_{i,j}) = \begin{bmatrix} \lambda_i & 0 & \cdots & \cdots & 0 \\ 1 & \lambda_i & \ddots & & \vdots \\ 0 & 1 & \ddots & \ddots & \vdots \\ \vdots & \ddots & \ddots & \ddots & 0 \\ 0 & \cdots & 0 & 1 & \lambda_i \end{bmatrix}$$

This matrix is referred to as a **Jordan block** associated to the scalar λ_i. Note that a Jordan block has λ_i's on the main diagonal, 1s on the subdiagonal, and 0s elsewhere.

Now we can state the analog of Theorem 7.10 for this new choice of ordered basis.

Theorem 8.8 Suppose that the minimal polynomial of an operator $\tau \in \mathcal{L}(V)$ splits over the base field F, that is,

$$m_\tau(x) = (x - \lambda_1)^{e_1} \cdots (x - \lambda_n)^{e_n}$$

Then we can write

$$V = (\langle \mathbf{v}_{1,1}\rangle \oplus \cdots \oplus \langle \mathbf{v}_{1,k_1}\rangle) \oplus \cdots \oplus (\langle \mathbf{v}_{n,1}\rangle \oplus \cdots \oplus \langle \mathbf{v}_{n,k_n}\rangle)$$

where $\langle \mathbf{v}_{i,j}\rangle$ is a τ-cyclic subspace of V. The minimal polynomials for $\tau_{i,j} = \tau|_{\langle \mathbf{v}_{i,j}\rangle}$ are the elementary divisors

$$min(\tau_{i,j}) = (x - \lambda_i)^{e_{i,j}}$$

of V, where

$$e_i = e_{i,1} \geq e_{i,2} \geq \cdots \geq e_{i,k_i}$$

These elementary divisors are uniquely determined by τ. Furthermore, the set

$$\mathcal{C}_{i,j} = \left(\mathbf{v}_{i,j}, (\tau_{i,j} - \lambda_i)(\mathbf{v}_{i,j}), \ldots, (\tau_{i,j} - \lambda_i)^{e_{i,j}-1}(\mathbf{v}_{i,j})\right) \tag{8.6}$$

is an ordered basis for $\langle \mathbf{v}_{i,j}\rangle$, and the matrix of τ with respect to the ordered basis

$$\mathcal{C} = (\mathcal{C}_{1,1}, \ldots, \mathcal{C}_{n,k_n})$$

is the block matrix

$$[\tau]_{\mathcal{C}} = \begin{bmatrix} \mathcal{J}(\lambda_1, e_{1,1}) & & & & & \\ & \ddots & & & & \\ & & \mathcal{J}(\lambda_1, e_{1,k_1}) & & & \\ & & & \ddots & & \\ & & & & \ddots & \\ & & & & & \mathcal{J}(\lambda_n, e_{n,k_n}) \\ & & & & & & \ddots \\ & & & & & & & \mathcal{J}(\lambda_n, e_{n,k_n}) \end{bmatrix}_{block}$$

The matrix on the right is called the **Jordan canonical form** of τ. ∎

We leave it to the reader to show that, for an algebraically closed field F, the set of matrices that are in Jordan canonical form is indeed a set of canonical forms for similarity (at least up to the order of the Jordan blocks). In other words, every matrix over F is similar to exactly one matrix that is in Jordan canonical form (again up to order of the Jordan blocks).

Note that if τ has Jordan canonical form $\mathcal{J}$, then the diagonal elements of $\mathcal{J}$ are precisely the roots of the characteristic polynomial

$C_\tau(x)$, including multiplicities. In other words, the number of times each diagonal element appears in $\mathcal{J}$ is the multiplicity of that element as a root of the characteristic polynomial.

Geometric and Algebraic Multiplicities

If $\lambda \in F$ is an eigenvalue of a linear operator τ, then the multiplicity of λ, as a root of the characteristic polynomial $C_\tau(x)$, is called the **algebraic multiplicity** of λ. On the other hand, the dimension of the eigenspace $\mathcal{E}_\lambda$ is called the **geometric multiplicity** of λ.

Theorem 8.9 The geometric multiplicity of an eigenvalue λ of $\tau \in \mathcal{L}(V)$ is less than or equal to its algebraic multiplicity.

Proof. Suppose that λ is an eigenvalue of τ. Thus,

$$m_\tau(x) = (x-\lambda)^e p(x)$$

where $x-\lambda$ does not divide $p(x)$. Consider the rational canonical form for τ. In the primary decomposition of V, we have

$$V = V_{p_1} \oplus \cdots \oplus V_{p_n}$$

where we may assume that

$$V_{p_1} = \{\mathbf{v} \in V \mid (x-\lambda)^e \mathbf{v} = \mathbf{0}\}$$

The cyclic decomposition of this primary submodule is

$$V_{p_1} = \langle \mathbf{v}_1 \rangle \oplus \cdots \oplus \langle \mathbf{v}_k \rangle$$

with elementary divisors

$$min(\tau_j) = (x-\lambda)^{e_j}$$

where $\tau_j = \tau|_{\langle \mathbf{v}_j \rangle}$ and

$$e = e_1 \geq e_2 \geq \cdots \geq e_k$$

According to Theorem 8.4, the algebraic multiplicity of λ is

$$alg.\ mult. = \sum_{j=1}^{k} e_j = \sum_{j=1}^{k} dim(\langle \mathbf{v}_j \rangle) = dim(V_{p_1})$$

Now,

$$\begin{aligned} \mathbf{v} \in \mathcal{E}_\lambda &\Rightarrow \tau(\mathbf{v}) = \lambda \mathbf{v} \Rightarrow (\tau - \lambda)(\mathbf{v}) = \mathbf{0} \\ &\Rightarrow (\tau-\lambda)^e(\mathbf{v}) = \mathbf{0} \Rightarrow \mathbf{v} \in V_{p_1} \end{aligned}$$

and so $\mathcal{E}_\lambda \subset V_{p_1}$, that is, all eigenvectors associated to λ lie in V_{p_1}. In fact, we can say more. Recall that the set

$$\mathcal{C}_j = \left(\mathbf{v}_j, (\tau_j - \lambda)(\mathbf{v}_j), \ldots, (\tau_j - \lambda)^{e_j - 1}(\mathbf{v}_j)\right)$$

is a basis for $\langle \mathbf{v}_j \rangle$, for all $j = 1, \ldots, k$, and that

$$(\tau_j - \lambda)^{e_j}(\mathbf{v}_j) = \mathbf{0}$$

If

$$\mathbf{u} = \sum_{i=0}^{e_j - 1} r_i (\tau - \lambda)^i (\mathbf{v}_j)$$

is an eigenvector in V_{p_1}, then $(\tau - \lambda)(\mathbf{u}) = \mathbf{0}$, that is,

$$\mathbf{0} = \sum_{i=0}^{e_j - 1} r_i (\tau - \lambda)^{i+1} (\mathbf{v}_j) = \sum_{i=0}^{e_j - 2} r_i (\tau - \lambda)^{i+1} (\mathbf{v}_j)$$

and so $r_i = 0$ for $i = 0, \ldots, e_j - 2$, which implies that

$$\mathbf{u} = r_{e_j - 1} (\tau - \lambda)^{e_j - 1} (\mathbf{v}_j)$$

Hence, the only eigenvectors in $\langle \mathbf{v}_j \rangle$ are the scalar multiples of the vector

$$(\tau - \lambda)^{e_j - 1} (\mathbf{v}_j)$$

This shows that the geometric multiplicity of λ is the *number* k of τ-cyclic subspaces $\langle \mathbf{v}_j \rangle$ that form V_{p_1}. Since the algebraic multiplicity is the sum of the *dimensions* of these τ-cyclic subspaces, the theorem follows. ∎

Diagonalizable Operators

We are now in a position to give several different characterizations of diagonalizable linear operators, that is, operators that can be represented by a diagonal matrix. The first characterization amounts to little more than the definitions of the concepts involved.

Theorem 8.10 An operator $\tau \in \mathcal{L}(V)$ is diagonalizable if and only if there is a basis for V that consists entirely of eigenvectors of τ, that is, if and only if

$$\mathit{dim}(\mathcal{E}_{\lambda_1} \oplus \cdots \oplus \mathcal{E}_{\lambda_k}) = \mathit{dim}(V)$$

where $\lambda_1, \ldots, \lambda_k$ are the distinct eigenvalues of τ. ∎

The Jordan canonical form gives us another characterization of diagonalizable operators. For, suppose that τ is diagonalizable, and that

$$[\tau]_{\mathfrak{B}} = \begin{bmatrix} \lambda_1 & 0 & \cdots & 0 \\ 0 & \lambda_2 & \ddots & \vdots \\ \vdots & \ddots & \ddots & 0 \\ 0 & \cdots & 0 & \lambda_k \end{bmatrix}$$

where the diagonal elements are not necessarily distinct. Suppose further that

$$\lambda_{i_1}, \ldots, \lambda_{i_s}$$

are the *distinct* diagonal elements. Then the minimal polynomial of τ, which is the same as the minimal polynomial of $[\tau]_{\mathfrak{B}}$, is

$$m_\tau(x) = \prod_j (x - \lambda_{i_j})$$

Thus, $m_\tau(x)$ is the product of distinct linear factors.

Conversely, suppose that $\tau \in \mathcal{L}(V)$ has the property that $m_\tau(x)$ is the product of distinct linear factors, say

$$m_\tau(x) = (x - \lambda_1)\cdots(x - \lambda_s)$$

where the λ_i's are distinct. Then τ has a Jordan canonical form. Moreover, referring to Theorem 8.8, all of the elementary divisors have the form $x - \lambda_i$, and so

$$min(\tau_{i,j}) = x - \lambda_i$$

In other words, all vectors in $\langle \mathbf{v}_{i,j} \rangle$ are eigenvectors, and so the basis for V, constructed in Theorem 8.8, consists only of eigenvectors for V. Hence, τ is diagonalizable. We have established the following result.

Theorem 8.11 A linear operator $\tau \in \mathcal{L}(V)$ on a finite dimensional vector space is diagonalizable if and only if its minimal polynomial is the product of *distinct* linear factors. ∎

Projections

In order to obtain another characterization of diagonalizable operators, we turn to a discussion of a special type of operator.

Definition Let $V = S \oplus S^c$. The map $\rho: V \to V$ defined by

$$\rho(\mathbf{s} + \mathbf{s}^c) = \mathbf{s}$$

where $\mathbf{s} \in S$ and $\mathbf{s}^c \in S^c$, is a linear operator on V, called **projection on S along S^c**. □

Figure 8.1 illustrates the concept of a projection.

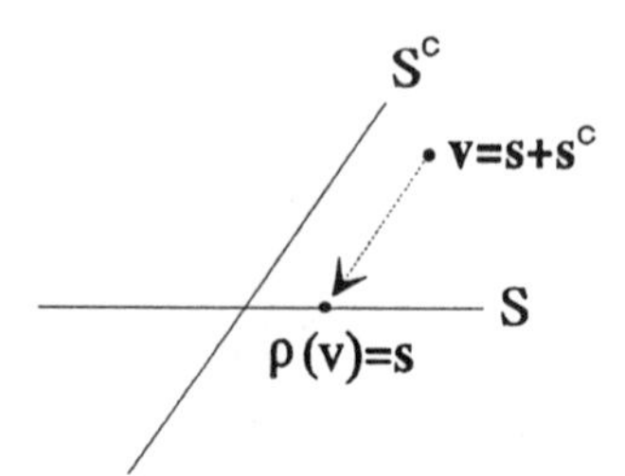

Figure 8.1

The following theorem describes projection operators.

Theorem 8.12
1) Let ρ be projection on S along S^c. Then

$$im(\rho) = S, \quad ker(\rho) = S^c$$

$$V = im(\rho) \oplus ker(\rho)$$

$$\mathbf{v} \in im(\rho) \Leftrightarrow \rho(\mathbf{v}) = \mathbf{v}$$

Note that the last condition says that a vector is in the image of ρ if and only if it is *fixed* by ρ.
2) Conversely, if $\sigma \in \mathcal{L}(V)$ has the property that

$$V = im(\sigma) \oplus ker(\sigma) \quad \text{and} \quad \sigma|_{im(\sigma)} = id$$

then σ is projection on $im(\sigma)$ along $ker(\sigma)$. ∎

Projection operators (or projections, for short) play a major role in the spectral theory of linear operators, which we will discuss in Chapter 10. Now we turn to some of the basic properties of these operators.

Theorem 8.13 A linear operator $\rho \in \mathcal{L}(V)$ is a projection if and only if it is **idempotent**, that is, if and only if $\rho^2 = \rho$.

Proof. To see that the projection operator ρ on S along S^c is idempotent, observe that, for any $\mathbf{s} \in S$ and $\mathbf{s}^c \in S^c$,

$$\rho^2(\mathbf{s} + \mathbf{s}^c) = \rho(\rho(\mathbf{s} + \mathbf{s}^c)) = \rho(\mathbf{s}) = \mathbf{s} = \rho(\mathbf{s} + \mathbf{s}^c)$$

and so $\rho^2 = \rho$. Conversely, if ρ is idempotent, let

$$S = \{\mathbf{v} \in V \mid \rho(\mathbf{v}) = \mathbf{v}\}$$

be the set of all vectors that are fixed by ρ. Then $S \subset im(\rho)$. Also, if $\mathbf{v} \in im(\rho)$, then $\mathbf{v} = \rho(\mathbf{w})$ for some $\mathbf{w} \in V$, and so

$$\rho(\mathbf{v}) = \rho^2(\mathbf{w}) = \rho(\mathbf{w}) = \mathbf{v}$$

Hence, $im(\rho) \subset S$, and so $S = im(\rho)$. In other words,

$$\rho \mid_{im(\rho)} = id$$

Now, if $\mathbf{v} \in im(\rho) \cap ker(\rho)$, we have

$$\rho(\mathbf{v}) = \mathbf{v} \quad \text{and} \quad \rho(\mathbf{v}) = \mathbf{0}$$

and so $\mathbf{v} = \mathbf{0}$. Hence, $im(\rho) \cap ker(\rho) = \{\mathbf{0}\}$.

Finally, observe that for $\mathbf{v} \in V$,

$$\mathbf{v} = \rho(\mathbf{v}) + (\mathbf{v} - \rho(\mathbf{v})) \in im(\rho) + ker(\rho)$$

and so

$$V = im(\rho) \oplus ker(\rho)$$

An appeal to Theorem 8.12 completes the proof. ∎

The Algebra of Projections

If ρ is a projection, then so is $\iota - \rho$, where ι is the identity operator on V, for we have

$$(\iota - \rho)^2 = \iota^2 - \iota\rho - \rho\iota + \rho^2 = \iota - \rho$$

It is not hard to see that $ker(\iota - \rho) = im(\rho)$ and $im(\iota - \rho) = ker(\rho)$. Hence, if ρ is projection on S along S^c, then $\iota - \rho$ is projection on S^c along S.

Definition Two projections $\rho, \sigma \in \mathcal{L}(V)$ are **orthogonal**, written $\rho \perp \sigma$, if $\rho\sigma = \sigma\rho = 0$. □

Note that $\rho \perp \sigma$ if and only if

$$im(\rho) \subset ker(\sigma) \quad \text{and} \quad im(\sigma) \subset ker(\rho)$$

The following example shows that it is not enough to have $\rho\sigma = 0$ in the definition of orthogonality, since it is possible for $\rho\sigma = 0$ and yet $\sigma\rho$ may not even be a projection.

Example 8.1 Let $V = F^2$, and let

$$D = \{(x,x) \mid x \in F\},\ X = \{(x,0) \mid x \in F\},\ \ Y = \{(0,y) \mid y \in F\}$$

Thus, D is the diagonal, X is the x-axis and Y is the y-axis in F^2. (The reader may wish to draw pictures in $\mathbb{R}^2$.) Using the notation $\rho_{A,B}$ for the projection on A along B, we have

$$\rho_{D,X}\rho_{D,Y} = \rho_{D,Y} \neq \rho_{D,X} = \rho_{D,Y}\rho_{D,X}$$

From this we deduce that if ρ and σ are projections, it may happen that both products $\rho\sigma$ and $\sigma\rho$ are projections, but that they are not equal.

We leave it to the reader to show that $\rho_{Y,X}\rho_{X,D} = 0$ (which is a projection), but that $\rho_{X,D}\rho_{Y,X}$ is not a projection. Thus, it may also happen that $\rho\sigma$ is a projection (even the zero projection) but that $\sigma\rho$ is not a projection. □

If ρ and σ are projections, it does not necessarily follow that $\rho+\sigma$, $\rho-\sigma$ or $\rho\sigma$ is a projection. The sum $\rho+\sigma$ is a projection if and only if

$$(\rho+\sigma)^2 = \rho+\sigma$$

or

$$\rho\sigma+\sigma\rho = 0 \tag{8.7}$$

Multiplying this on the left by ρ and on the right by ρ, gives

$$\rho\sigma+\rho\sigma\rho = 0$$

and

$$\rho\sigma\rho+\sigma\rho = 0$$

Hence,

$$\rho\sigma = \sigma\rho$$

which, together with (8.7), gives $2\rho\sigma = 0$. Therefore, if $\text{char}(F) \neq 2$ (so that $2 \neq 0$) then $\rho\sigma = 0$ and $\sigma\rho = 0$. This shows that if $\rho+\sigma$ is a projection, then $\rho \perp \sigma$. Conversely, if $\rho\sigma = \sigma\rho = 0$, then $(\rho+\sigma)^2 = \rho+\sigma$, and so $\rho+\sigma$ is a projection.

Now suppose that $\rho\sigma = \sigma\rho = 0$, and so $\rho+\sigma$ is a projection. To determine the kernel of $\rho+\sigma$, note that

$$(\rho+\sigma)(\mathbf{v}) = \mathbf{0} \Rightarrow \rho(\mathbf{v})+\sigma(\mathbf{v}) = \mathbf{0} \Rightarrow \rho^2(\mathbf{v})+\rho\sigma(\mathbf{v}) = \mathbf{0} \Rightarrow \rho(\mathbf{v}) = \mathbf{0}$$

and so $ker(\rho+\sigma) \subset ker(\rho)$. In a similar way, $ker(\rho+\sigma) \subset ker(\sigma)$, and so

$$ker(\rho+\sigma) \subset ker(\rho) \cap ker(\sigma)$$

But the reverse inclusion is obvious, and so

$$ker(\rho+\sigma) = ker(\rho) \cap ker(\sigma)$$

As to the image of $\rho+\sigma$, we know that $im(\rho+\sigma)$ is the set of vectors in V that are fixed by the projection $\rho+\sigma$. Hence,

$$\mathbf{v} \in im(\rho+\sigma) \Rightarrow \mathbf{v} = (\rho+\sigma)(\mathbf{v}) = \rho(\mathbf{v})+\sigma(\mathbf{v}) \in im(\rho)+im(\sigma)$$

and so $im(\rho+\sigma) \subset im(\rho)+im(\sigma)$. But notice that

$$\mathbf{x} \in im(\rho) \cap im(\sigma) \Rightarrow \rho(\mathbf{x}) = \mathbf{x} = \sigma(\mathbf{x}) \Rightarrow \mathbf{x} = \rho(\mathbf{x}) = \rho^2(\mathbf{x}) = \rho\sigma(\mathbf{x}) = \mathbf{0}$$

which implies that $im(\rho) \cap im(\sigma) = \{\mathbf{0}\}$. Therefore,

$$im(\rho+\sigma) \subset im(\rho) \oplus im(\sigma)$$

To establish the reverse inequality, observe that if $\mathbf{v} \in im(\rho) \oplus im(\sigma)$, then $\mathbf{v} = \mathbf{r} + \mathbf{s}$, for $\mathbf{r} \in im(\rho)$ and $\mathbf{s} \in im(\sigma)$, and so

$$(\rho+\sigma)(\mathbf{v}) = (\rho+\sigma)(\mathbf{r}) + (\rho+\sigma)(\mathbf{s}) = \mathbf{r} + \mathbf{s} = \mathbf{v}$$

This implies that $\mathbf{v} \in im(\rho+\sigma)$, and so

$$im(\rho+\sigma) = im(\rho) \oplus im(\sigma)$$

Let us summarize.

Theorem 8.14 Let $\rho,\sigma \in \mathcal{L}(V)$ be projections, where V is a vector space over a field of characteristic $\neq 2$. Then $\rho+\sigma$ is a projection if and only if $\rho \perp \sigma$, in which case $\rho+\sigma$ is projection on $im(\rho) \oplus im(\sigma)$ along $ker(\rho) \cap ker(\sigma)$. ∎

Let us next consider the difference $\rho - \sigma$. We know that $\rho - \sigma$ is a projection if and only if $\theta = \iota - (\rho - \sigma) = (\iota - \rho) + \sigma$ is a projection. Hence, we may apply the previous theorem to this case as well, to deduce that $\rho - \sigma$ is a projection if and only if

$$(\iota-\rho)\sigma = \sigma(\iota-\rho) = 0$$

or, equivalently,

$$\rho\sigma = \sigma\rho = \sigma$$

Moreover, in this case, $\rho - \sigma = \iota - \theta$ is projection on $ker(\theta)$ along $im(\theta)$. Again using Theorem 8.14, since $\theta = (\iota-\rho)+\sigma$, we have

$$im(\theta) = im(\iota-\rho) \oplus im(\sigma) = ker(\rho) \oplus im(\sigma)$$

and

$$ker(\theta) = ker(\iota-\rho) \cap ker(\sigma) = im(\rho) \cap ker(\sigma)$$

We have proved the following.

Theorem 8.15 Let $\rho,\sigma \in \mathcal{L}(V)$ be projections, where V is a vector space over a field of characteristic $\neq 2$. Then $\rho - \sigma$ is a projection if and only if

$$\rho\sigma = \sigma\rho = \sigma$$

in which case $\rho - \sigma$ is projection on $\mathrm{im}(\rho) \cap ker(\sigma)$ along $ker(\rho) \oplus im(\sigma)$. ∎

Finally, let us consider the product $\rho\sigma$ of two projections.

Theorem 8.16 Let $\rho,\sigma \in \mathcal{L}(V)$ be projections. If $\rho\sigma = \sigma\rho$ then $\rho\sigma$ is a projection. In this case, $\rho\sigma$ is projection on $im(\rho) \cap im(\sigma)$ along

$ker(\rho)+ker(\sigma)$.

Proof. If $\rho\sigma=\sigma\rho$, then

$$(\rho\sigma)^2=\rho\sigma\rho\sigma=\rho^2\sigma^2=\rho\sigma$$

and so $\rho\sigma$ is a projection. To find the image of $\rho\sigma$, we observe that if $\mathbf{v}=\rho\sigma(\mathbf{v})$, then

$$\rho(\mathbf{v})=\rho(\rho\sigma(\mathbf{v}))=\rho\sigma(\mathbf{v})=\mathbf{v}$$

and so $\mathbf{v}\in im(\rho)$. Similarly, $\mathbf{v}\in im(\sigma)$ and so

$$im(\rho\sigma)\subset im(\rho)\cap im(\sigma)$$

The reverse inclusion is clear, and so

$$im(\rho\sigma)=im(\rho)\cap im(\sigma)$$

Next, we observe that if $\mathbf{v}\in ker(\rho\sigma)$, then $\rho\sigma(\mathbf{v})=\mathbf{0}$ and so $\sigma(\mathbf{v})\in ker(\rho)$. Hence,

$$\mathbf{v}=\sigma(\mathbf{v})+(\iota-\sigma(\mathbf{v}))\in ker(\rho)+ker(\sigma)$$

Moreover, if $\mathbf{v}=\mathbf{r}+\mathbf{s}\in ker(\rho)+ker(\sigma)$, then

$$\rho\sigma(\mathbf{v})=\rho\sigma(\mathbf{r}+\mathbf{s})=\sigma\rho(\mathbf{r})+\rho\sigma(\mathbf{s})=\mathbf{0}+\mathbf{0}=\mathbf{0}$$

and so $\mathbf{v}\in ker(\rho\sigma)$. Thus,

$$ker(\rho\sigma)=ker(\rho)+ker(\sigma) \tag{8.8}$$ ∎

We should remark that if $\rho=\sigma$, then $ker(\rho)=ker(\sigma)$, and so the sum in (8.8) need not be direct.

Resolutions of the Identity

If ρ is a projection, then

$$\rho\perp(\iota-\rho) \quad\text{and}\quad \rho+(\iota-\rho)=\iota$$

Let us generalize this to more than two projections.

Definition If $\rho_1,\ldots,\rho_k$ are projections for which
1) $\rho_i\perp\rho_j$ for $i\neq j$
2) $\rho_1+\cdots+\rho_k=\iota$

then we refer to the sum in (2) as a **resolution of the identity.** □

The next theorem displays a correspondence between direct sum decompositions of V and resolutions of the identity.

Theorem 8.17

1) If $\rho_1 + \cdots + \rho_k = \iota$ is a resolution of the identity, then

$$V = im(\rho_1) \oplus \cdots \oplus im(\rho_k)$$

2) Conversely, if $V = S_1 \oplus \cdots \oplus S_k$, and ρ_i is projection on S_i along $S_1 \oplus \cdots \oplus \widehat{S_i} \oplus \cdots \oplus S_k$, where the hat ^ means that the corresponding term is missing from the direct sum. Then

$$\rho_1 + \cdots + \rho_k = \iota$$

is a resolution of the identity.

Proof. To prove (1) suppose that $\rho_1 + \cdots + \rho_k = \iota$ is a resolution of the identity. Then for any $\mathbf{v} \in V$, we have

$$\mathbf{v} = \iota\mathbf{v} = \rho_1(\mathbf{v}) + \cdots + \rho_k(\mathbf{v})$$

which shows that $V = im(\rho_1) + \cdots + im(\rho_k)$. Now, since the projections ρ_i are orthogonal, for each i, we have $im(\rho_j) \subset ker(\rho_i)$ for all $j \neq i$. Hence,

$$\sum_{j \neq i} im(\rho_j) \subset ker(\rho_i)$$

which shows that

$$im(\rho_i) \cap \sum_{j \neq i} im(\rho_j) = \{\mathbf{0}\}$$

and so

$$V = im(\rho_1) \oplus \cdots \oplus im(\rho_k)$$

To prove part (2), observe that for $i \neq j$,

$$im(\rho_i) = S_i \subset ker(\rho_j)$$

and so $\rho_i \perp \rho_j$. Also,

$$\iota\mathbf{v} = \mathbf{s}_1 + \cdots + \mathbf{s}_k = \rho_1(\mathbf{v}) + \cdots + \rho_k(\mathbf{v}) = \sum_u \rho_u(\mathbf{v})$$

and so $\iota = \rho_1 + \cdots + \rho_k$ is a resolution of the identity. ∎

Projections and Diagonalizability

Now let us consider a linear combination

$$\tau = \lambda_1\rho_1 + \cdots + \lambda_k\rho_k$$

where $\rho_1 + \cdots + \rho_k = \iota$ is a resolution of the identity. Since any vector $\mathbf{v} \in V$ has the form

$$\mathbf{v} = \mathbf{s}_1 + \cdots + \mathbf{s}_k$$

where $\mathbf{s}_i \in im(\rho_i)$, we have

$$\tau(\mathbf{v}) = \sum_i \lambda_i \rho_i(\mathbf{v}) = \sum_i \lambda_i \mathbf{s}_i$$

Thus, the action of τ has the particularly simple form

$$\mathbf{v} = \mathbf{s}_1 + \cdots + \mathbf{s}_k \Rightarrow \tau(\mathbf{v}) = \lambda_1 \mathbf{s}_1 + \cdots + \lambda_k \mathbf{s}_k \tag{8.9}$$

Moreover, we have the following.

Theorem 8.18 A linear operator $\tau \in \mathcal{L}(V)$ is diagonalizable if and only if it has the form

$$\tau = \lambda_1 \rho_1 + \cdots + \lambda_k \rho_k \tag{8.10}$$

for some resolution of the identity $\rho_1 + \cdots + \rho_k = \iota$. Moreover, if τ has the form (8.10), where the λ_i's are distinct and the ρ_i's are nonzero, then the λ_i's are the eigenvalues of τ, and $im(\rho_i)$ is the eigenspace of τ associated with λ_i.

Proof. Suppose that τ is diagonalizable, and that $\lambda_1, \ldots, \lambda_k$ are the distinct eigenvalues of τ. Then we have

$$V = \mathcal{E}_{\lambda_1} \oplus \cdots \oplus \mathcal{E}_{\lambda_k}$$

According to Theorem 8.17, the projections ρ_i on $\mathcal{E}_{\lambda_i}$ along

$$\mathcal{E}_{\lambda_1} \oplus \cdots \oplus \widehat{\mathcal{E}}_{\lambda_i} \oplus \cdots \oplus \mathcal{E}_{\lambda_k}$$

form a resolution of the identity. Moreover, if $\mathbf{v} \in \mathcal{E}_{\lambda_i}$, then

$$\tau(\mathbf{v}) = \lambda_i \mathbf{v} = (\lambda_1 \rho_1 + \cdots + \lambda_k \rho_k)(\mathbf{v})$$

from which we deduce that $\tau = \lambda_1 \rho_1 + \cdots + \lambda_k \rho_k$.

Conversely, suppose that τ has the form (8.10). Then, using Theorem 8.14, we may assume that the λ_i's are distinct. Also,

$$V = im(\rho_1) \oplus \cdots \oplus im(\rho_k)$$

In view of (8.9), we have, for any $\mathbf{s}_i \in im(\rho_i)$

$$\tau(\mathbf{s}_i) = \lambda_i \mathbf{s}_i$$

which shows that λ_i is an eigenvalue of τ, and that $im(\rho_i) \subset \mathcal{E}_{\lambda_i}$.

To see the reverse inclusion, let $\mathbf{v} = \mathbf{s}_1 + \cdots + \mathbf{s}_k \in \mathcal{E}_{\lambda_i}$, where $\mathbf{s}_i \in im(\rho_i)$. Then

$$\lambda_i(\mathbf{s}_1 + \cdots + \mathbf{s}_k) = \lambda_i \mathbf{v} = \tau(\mathbf{v}) = \lambda_1 \mathbf{s}_1 + \cdots + \lambda_k \mathbf{s}_k$$

which implies that $(\lambda_i - \lambda_j)\mathbf{s}_j = \mathbf{0}$ for all j, and since the λ_j's are distinct, we have $\mathbf{s}_j = \mathbf{0}$ for $j \neq i$. Thus, $\mathbf{v} = \mathbf{s}_i \in im(\rho_i)$, and so $im(\rho_i) = \mathcal{E}_{\lambda_i}$. This shows that

$$V = \mathcal{E}_{\lambda_1} \oplus \cdots \oplus \mathcal{E}_{\lambda_k}$$

and so τ is diagonalizable.

Finally, we observe that if λ is an eigenvalue of τ, with eigenvector $\mathbf{v} = \mathbf{s}_1 + \cdots + \mathbf{s}_k$. Then

$$\lambda(\mathbf{s}_1 + \cdots + \mathbf{s}_k) = \lambda\mathbf{v} = \tau(\mathbf{v}) = \lambda_1\mathbf{s}_1 + \cdots + \lambda_k\mathbf{s}_k$$

and so $(\lambda - \lambda_i)\mathbf{s}_i = \mathbf{0}$ for all i. Therefore, if $\lambda \neq \lambda_i$ for all i, we have $\mathbf{s}_i = 0$ for all i, that is, $\mathbf{v} = \mathbf{0}$, which is not possible. Hence, all eigenvalues are among the coefficients λ_i. ∎

Definition The set of eigenvalues of a linear operator on a finite dimensional vector space V is called the **spectrum** of τ. If τ is diagonalizable, and

$$\tau = \lambda_1\rho_1 + \cdots + \lambda_k\rho_k \tag{8.11}$$

where $\rho_1 + \cdots + \rho_k = \iota$ is a resolution of the identity, the λ_i's are distinct, and the ρ_i's are nonzero, then (8.11) is called the **spectral resolution** of the operator τ. □

Projections and Invariance

There is a connection between projections and the notions of invariant subspace and reducing pair for a linear operator τ.

Theorem 8.19 Let $\tau \in \mathcal{L}(V)$.

1) If S is an invariant subspace under τ, then $\rho\tau\rho = \tau\rho$ for all projections ρ on S.
2) If S is a subspace of V, and if $\rho\tau\rho = \tau\rho$ for any projection ρ on S, then S is invariant under τ.

Proof. To prove part (1), let S be invariant under τ, and let ρ be projection on S along T, whence $V = S \oplus T$. Now, let $\mathbf{v} = \mathbf{s} + \mathbf{t} \in V$, where $\mathbf{s} \in S$ and $\mathbf{t} \in T$. Then, since ρ fixes S,

$$\rho\tau\rho(\mathbf{v}) = \rho\tau(\mathbf{s}) = \tau(\mathbf{s}) = \tau\rho(\mathbf{v})$$

and so $\rho\tau\rho = \tau\rho$. As to part (2), suppose that $\rho\tau\rho = \tau\rho$, where ρ is projection on S along T. Let $\mathbf{s} \in S$. Then, since ρ fixes S,

$$\rho\tau(\mathbf{s}) = \rho\tau\rho(\mathbf{s}) = \tau\rho(\mathbf{s}) = \tau(\mathbf{s})$$

and so ρ fixes $\tau(\mathbf{s})$, which implies that $\tau(\mathbf{s}) \in S$. Thus, S is invariant under τ. ∎

Theorem 8.20 Let $V = S \oplus T$. Then a linear operator $\tau \in \mathcal{L}(V)$ is reduced by the pair (S,T) if and only if $\tau\rho = \rho\tau$, where ρ is projection on S along T.

Proof. Suppose first that $\tau\rho = \rho\tau$, where ρ is projection on S along T. Since $\mathbf{v} \in S$ if and only if $\rho(\mathbf{v}) = \mathbf{v}$, we have for $\mathbf{s} \in S$,

$$\rho\tau(\mathbf{s}) = \tau\rho(\mathbf{s}) = \tau(\mathbf{s})$$

and so ρ fixes $\tau(\mathbf{s})$, which implies that $\tau(\mathbf{s}) \in S$. Hence, S is invariant under τ. Also, $\mathbf{v} \in T$ if and only if $\rho(\mathbf{v}) = \mathbf{0}$, and so, for $\mathbf{t} \in T$,

$$\rho\tau(\mathbf{t}) = \rho\tau(\mathbf{t}) = 0$$

implies that $\tau(\mathbf{t}) \in T$. Hence, T is invariant under τ.

Conversely, suppose that (S,T) reduces τ. Then the projection operator ρ fixes vectors in S, and sends vectors in T to $\mathbf{0}$. Hence, for $\mathbf{s} \in S$ and $\mathbf{t} \in T$, since $\tau(\mathbf{s}) \in S$, we have

$$\rho\tau(\mathbf{s}) = \tau(\mathbf{s}) = \tau\rho(\mathbf{s})$$

and

$$\rho\tau(\mathbf{t}) = 0 = \tau\rho(\mathbf{t})$$

which imply that $\rho\tau = \tau\rho$. ∎

EXERCISES

1. A linear operator $\tau \in \mathcal{L}(V)$ is said to be **nonderogatory** if its minimal polynomial is equal to its characteristic polynomial. Prove that τ is nonderogatory if and only if V is a cyclic module.
2. Show that the eigenspace of an eigenvalue λ is a subspace of V.
3. Prove *directly* that the eigenvalues of a matrix are invariants under similarity.
4. Prove that the eigenvalues of a matrix do not form a complete set of invariants under similarity.
5. Show that not all matrices (hence linear operators) have eigenvalues in the base field.
6. Show that the set $\mathfrak{C}_{i,j}$ in (8.6) is an ordered basis for $\langle \mathbf{v}_{i,j} \rangle$.
7. Show that $\tau \in \mathcal{L}(V)$ is invertible if and only if 0 is *not* an eigenvalue of τ.
8. Let A be an $n \times n$ matrix over an *algebraically closed* field F, such as the complex field $\mathbb{C}$. Thus, all of the roots of the characteristic polynomial lie in F. Prove that $\det(A)$ is the product of the eigenvalues of A. Formulate a statement in this regard about linear operators.
9. Show that is λ is an eigenvalue of τ, then $p(\lambda)$ is an eigenvalue of $p(\tau)$, for any polynomial $p(x)$. Also, if $\lambda \neq 0$, then λ^{-1} is an eigenvalue for τ^{-1}.
10. An operator $\tau \in \mathcal{L}(V)$ is **nilpotent** if $\tau^n = 0$ for some $n \in \mathbb{N}$.

a) Show that if τ is nilpotent, then 0 is an eigenvalue of τ, and it is the *only* eigenvalue of τ.
b) Find an operator τ that has 0, and only 0 as an eigenvalue, but is not nilpotent.

11. Show that if $\tau,\sigma \in \mathcal{L}(V)$, then $\tau\sigma$ and $\sigma\tau$ have the same eigenvalues.
12. Suppose that $\tau, \sigma \in \mathcal{L}(V)$. Show that if $\tau\sigma = \sigma\tau$, then τ and σ have a common eigenvector.
13. Let ρ be a projection. Show that $ker(\iota - \rho) = im(\rho)$ and $im(\iota - \rho) = ker(\rho)$.
14. Complete the details of Example 8.1.
15. Find projections ρ and σ for which $\rho\sigma$ is a *nonzero* projection, but $\sigma\rho$ is not a projection.
16. (Halmos)
    a) Find a linear operator τ that is *not* idempotent, but for which $\tau^2(\iota - \tau) = 0$.
    b) Find a linear operator τ that is *not* idempotent, but for which $\tau(\iota - \tau)^2 = 0$.
    c) Prove that if $\tau^2(\iota - \tau) = \tau(\iota - \tau)^2 = 0$, then τ is idempotent.
17. An **involution** is a linear operator θ for which $\theta^2 = \iota$. If τ is an idempotent, what can you say about $2\tau - \iota$? Construct a one-to-one correspondence between the set of idempotents on V and the set of involutions.

The Trace of a Matrix

18. Let A be an $n \times n$ matrix over an *algebraically closed* field F, such as the complex field $\mathbb{C}$. Thus, all of the roots of the characteristic polynomial lie in F. The **trace** of A, denoted $tr(A)$, is the sum of the elements on the main diagonal of A. Verify the following statements.
    a) $tr(rA) = r\, tr(A)$, for $r \in F$
    b) $tr(A + B) = tr(A) + tr(B)$
    c) $tr(AB) = tr(BA)$
    d) the trace is an invariant under similarity
    e) the trace of A is the sum of the eigenvalues of A.

    Formulate a definition of the trace of a linear operator, show that it is well-defined, and relate this concept to the eigenvalues of the operator.
19. Use the concept of the trace of a matrix, as defined in the previous exercise, to prove that there are no matrices $A, B \in \mathcal{M}_n(\mathbb{C})$ for which $AB - BA = I$.
20. Let $T:\mathcal{M}_n(F) \to F$ be a function with the following properties. For all matrices $A, B \in \mathcal{M}_n(F)$, and $r \in F$,
    1) $T(rA) = rT(A)$

2) $T(A+B) = T(A) + T(B)$
3) $T(AB) = T(BA)$

Show that there exists an $s \in F$ for which $T(A) = s\, tr(A)$, for all $A \in \mathcal{M}_n(F)$.

Simultaneous Diagonalizability

A pair of linear operators $\sigma, \tau \in \mathcal{L}(V)$ are **simultaneously diagonalizable** if there is an ordered basis $\mathcal{B}$ for V for which $[\tau]_{\mathcal{B}}$ and $[\sigma]_{\mathcal{B}}$ are both diagonal.

21. The purpose of this exercise is to prove that two *diagonalizable* operators σ and τ are simultaneously diagonalizable if and only if they commute, that is, $\sigma\tau = \tau\sigma$.
    a) To prove necessity, suppose that $\mathcal{B}$ is a basis of eigenvectors for both σ and τ. Show that $\tau\sigma$ and $\sigma\tau$ agree on the vectors in $\mathcal{B}$.
    b) To prove sufficiency, suppose that $\sigma\tau = \tau\sigma$. Show that the eigenspaces of τ are invariant under σ.
    c) If σ_i is the restriction of σ to the ith eigenspace of τ, show that σ_i is diagonalizable. *Hint*: consider the minimal polynomials of σ and σ_i.
    d) Use the results of part (b) and (c) to complete the proof.

CHAPTER 9

Real and Complex Inner Product Spaces

Contents: *Introduction. Norm and Distance. Isometries. Orthogonality. Orthogonal and Orthonormal Sets. The Projection Theorem. The Gram-Schmidt Orthogonalization Process. The Riesz Representation Theorem. Exercises.*

Introduction

We now turn to a discussion of real or complex vector spaces that have an additional function defined on them, called an *inner product*, as described in the upcoming definition. *Thus, in this chapter,* F *will denote either the real or complex field.* If r is a complex number, then $\bar{r}$ denotes the complex conjugate of r.

Definition Let V be a vector space over F, where $F = \mathbb{R}$ or $F = \mathbb{C}$. An **inner product** on V is a function $\langle , \rangle : V \times V \to F$ with the following properties.

1) **(Positive definiteness)** For all $\mathbf{v} \in V$,

$$\langle \mathbf{v},\mathbf{v} \rangle \geq 0 \text{ and } \langle \mathbf{v},\mathbf{v} \rangle = 0 \text{ if and only if } \mathbf{v} = \mathbf{0}$$

2) For $F = \mathbb{C}$: **(Conjugate symmetry**, or **Hermitian symmetry)**

$$\langle \mathbf{u},\mathbf{v} \rangle = \overline{\langle \mathbf{v},\mathbf{u} \rangle}$$

For $F = \mathbb{R}$: **(Symmetry)**

$$\langle \mathbf{u},\mathbf{v} \rangle = \langle \mathbf{v},\mathbf{u} \rangle$$

3) **(Linearity in the first coordinate)** For all $\mathbf{u},\mathbf{v} \in V$ and $r,s \in F$

$$\langle r\mathbf{u} + s\mathbf{v},\mathbf{w} \rangle = r\langle \mathbf{u},\mathbf{w} \rangle + s\langle \mathbf{v},\mathbf{w} \rangle$$

A real (or complex) vector space V, together with an inner product defined on V, is called a real (or complex) **inner product space.** ▯

We will study bilinear forms ("inner products") on vector spaces over fields other than $\mathbb{R}$ or $\mathbb{C}$ in Chapter 13. Note that property (1) implies that the quantity $\langle \mathbf{v},\mathbf{v}\rangle$ is always *real*, even if V is a complex vector space.

Combining properties (2) and (3), we get, in the complex case

$$\langle \mathbf{w},r\mathbf{u}+s\mathbf{v}\rangle = \overline{\langle r\mathbf{u}+s\mathbf{v},\mathbf{w}\rangle} = \bar{r}\overline{\langle \mathbf{u},\mathbf{w}\rangle} + \bar{s}\overline{\langle \mathbf{v},\mathbf{w}\rangle} = \bar{r}\langle \mathbf{w},\mathbf{u}\rangle + \bar{s}\langle \mathbf{w},\mathbf{v}\rangle$$

This is referred to as **conjugate linearity**. Thus, a complex inner product is *linear* in its first coordinate and *conjugate linear* in its second coordinate. This is often described by saying that the inner product is **sesquilinear.** (*Sesqui* means *one and a half times.*)

In the real case $(F=\mathbb{R})$, the inner product is linear in both coordinates – a property referred to as **bilinearity.**

Example 9.1

1) The vector space $\mathbb{R}^n$ is an inner product space under the **standard inner product**, or **dot product**, defined by

$$\langle (r_1,\ldots,r_n),(s_1,\ldots,s_n)\rangle = r_1s_1+\cdots+r_ns_n$$

The inner product space $\mathbb{R}^n$ is often called n-dimensional **Euclidean space.**

2) The vector space $\mathbb{C}^n$ is an inner product space under the **standard inner product**, defined by

$$\langle (r_1,\ldots,r_n),(s_1,\ldots,s_n)\rangle = r_1\bar{s}_1+\cdots+r_n\bar{s}_n$$

This inner product space is often called n-dimensional **unitary space.**

3) The vector space $V(n,2)$ of all binary n-tuples is an inner product space, under the inner product

$$\langle (r_1,\ldots,r_n),(s_1,\ldots,s_n)\rangle = (r_1s_1+\cdots+r_ns_n) \bmod 2$$

4) The vector space $C[a,b]$ of all continuous complex-valued functions on the closed interval $[a,b]$ is an inner product space under the inner product

$$\langle f,g\rangle = \int_a^b f(x)\overline{g(x)}\,dx$$ ▯

Example 9.2 One of the most important inner product spaces is the vector space ℓ^2 of all complex sequences (s_n) with the property that $\sum |s_n|^2 < \infty$, under the inner product

$$\langle(s_n),(t_n)\rangle = \sum_{n=0}^{\infty} s_n\overline{t_n}$$

Of course, for this inner product to make sense, the sum on the right must converge. To see this, observe that since (s_n), $(t_n) \in \ell^2$,

$$s = \sum_{n=0}^{\infty} |s_n|^2 < \infty \quad \text{and} \quad t = \sum_{n=0}^{\infty} |t_n|^2 < \infty$$

Now,

$$0 \le (|s_n| - |t_n|)^2 = |s_n|^2 - 2|s_n|\,|t_n| + |t_n|^2$$

and so

$$2|s_n t_n| \le |s_n|^2 + |t_n|^2$$

which gives

$$2\left|\sum_{n=0}^{\infty} s_n\overline{t_n}\right| \le 2\sum_{n=0}^{\infty} |s_n t_n| \le \sum_{n=0}^{\infty} |s_n|^2 + \sum_{n=0}^{\infty} |t_n|^2 = s + t < \infty$$

We leave it to the reader to verify that ℓ^2 is an inner product space. □

The following simple result is quite useful and easy to prove.

Lemma 9.1 If V is an inner product space, and $\langle \mathbf{u},\mathbf{x}\rangle = \langle \mathbf{v},\mathbf{x}\rangle$ for all $\mathbf{x} \in V$, then $\mathbf{u} = \mathbf{v}$. ∎

Note that a vector subspace S of an inner product space V is also an inner product space under the restriction of the inner product of V to S.

Norm and Distance

If V is an inner product space, then we can define the **norm**, or **length**, of each $\mathbf{v} \in V$ by

$$\|\mathbf{v}\| = \sqrt{\langle \mathbf{v},\mathbf{v}\rangle} \tag{9.1}$$

A vector $\mathbf{v} \in V$ is a **unit vector** if $\|\mathbf{v}\| = 1$.

Here are the basic properties of the norm.

Theorem 9.2

1) $\|\mathbf{v}\| \ge 0$ and $\|\mathbf{v}\| = 0$ if and only if $\mathbf{v} = \mathbf{0}$.
2) $\|r\mathbf{v}\| = |r|\,\|\mathbf{v}\|$, for all $r \in F$, $\mathbf{v} \in V$
3) **(The Cauchy-Schwarz inequality)** For all $\mathbf{u},\mathbf{v} \in V$,

$$|\langle \mathbf{u},\mathbf{v}\rangle| \le \|\mathbf{u}\|\,\|\mathbf{v}\|$$

with equality if and only if $\mathbf{u} = r\mathbf{v}$ for some $r \in F$.

4) **(The triangle inequality)** For all $\mathbf{u},\mathbf{v} \in V$,

$$\|\mathbf{u}+\mathbf{v}\| \leq \|\mathbf{u}\| + \|\mathbf{v}\|$$

with equality if and only if $\mathbf{u} = r\mathbf{v}$ for some $r \in F$.

5) For all $\mathbf{u},\mathbf{v},\mathbf{x} \in V$,

$$\|\mathbf{u}-\mathbf{v}\| \leq \|\mathbf{u}-\mathbf{x}\| + \|\mathbf{x}-\mathbf{v}\|$$

6) For all $\mathbf{u},\mathbf{v} \in V$,

$$\Big| \|\mathbf{u}\| - \|\mathbf{v}\| \Big| \leq \|\mathbf{u}-\mathbf{v}\|$$

7) **(The parallelogram law)** For all $\mathbf{u},\mathbf{v} \in V$,

$$\|\mathbf{u}+\mathbf{v}\|^2 + \|\mathbf{u}-\mathbf{v}\|^2 = 2\|\mathbf{u}\|^2 + 2\|\mathbf{v}\|^2$$

Proof. We prove only (3) and (4). To prove (3), we proceed as follows. If either $\mathbf{u}$ or $\mathbf{v}$ is zero, the result follows. Assume that $\mathbf{u},\mathbf{v} \neq \mathbf{0}$. Then, for any *real* number $r \in \mathbb{R}$,

$$\begin{aligned}
0 &\leq \|\mathbf{u}+r\mathbf{v}\|^2 \\
&= \langle \mathbf{u}+r\mathbf{v},\mathbf{u}+r\mathbf{v}\rangle \\
&= \langle \mathbf{u},\mathbf{u}\rangle + r\langle \mathbf{u},\mathbf{v}\rangle + r\langle \mathbf{v},\mathbf{u}\rangle + r^2\langle \mathbf{v},\mathbf{v}\rangle \\
&= \langle \mathbf{u},\mathbf{u}\rangle + r\overline{\langle \mathbf{v},\mathbf{u}\rangle} + r\langle \mathbf{v},\mathbf{u}\rangle + r^2\langle \mathbf{v},\mathbf{v}\rangle \\
&\leq \langle \mathbf{u},\mathbf{u}\rangle + 2r\,|\langle \mathbf{v},\mathbf{u}\rangle| + r^2\langle \mathbf{v},\mathbf{v}\rangle = f(r)
\end{aligned}$$

This implies that the quadratic polynomial $f(r)$ must have nonpositive discriminant, that is,

$$4\,|\langle \mathbf{v},\mathbf{u}\rangle|^2 - 4\langle \mathbf{v},\mathbf{v}\rangle\langle \mathbf{u},\mathbf{u}\rangle \leq 0$$

from which the Cauchy-Schwarz inequality follows. Furthermore, if equality holds, then there exists an $r \in F$ for which $f(r) = 0$, that is, $0 = \|\mathbf{u}+r\mathbf{v}\|^2$, and so $\mathbf{u}+r\mathbf{v} = \mathbf{0}$, which implies that $\mathbf{u}$ is a scalar multiple of $\mathbf{v}$. (If $\mathbf{u}$ is a scalar multiple of $\mathbf{v}$, then equality is easily seen to hold.)

To prove the triangle inequality, we use the Cauchy-Schwarz inequality to get

$$\begin{aligned}
\|\mathbf{u}+\mathbf{v}\|^2 &= \langle \mathbf{u}+\mathbf{v},\mathbf{u}+\mathbf{v}\rangle \\
&= \langle \mathbf{u},\mathbf{u}\rangle + \langle \mathbf{u},\mathbf{v}\rangle + \langle \mathbf{v},\mathbf{u}\rangle + \langle \mathbf{v},\mathbf{v}\rangle \\
&\leq \|\mathbf{u}\|^2 + 2|\langle \mathbf{u},\mathbf{v}\rangle| + \|\mathbf{v}\|^2 \\
&\leq \|\mathbf{u}\|^2 + 2\|\mathbf{u}\|\,\|\mathbf{v}\| + \|\mathbf{v}\|^2 \\
&= (\|\mathbf{u}\| + \|\mathbf{v}\|)^2
\end{aligned}$$

from which the triangle inequality follows. The proof of the statement concerning equality is left to the reader. ∎

Any vector space V, together with a function $\| \ \|: V \to \mathbb{R}$ that satisfies properties (1), (2) and (4) of Theorem 9.2, is called a **normed linear space.** (And the function $\| \ \|$ is called a **norm.**) Thus, any inner product space is a normed linear space, under the norm given by (9.1).

It is interesting to observe that the inner product on V can be recovered from the norm.

Theorem 9.3

1) If V is a real inner product space, then

$$\langle \mathbf{u},\mathbf{v}\rangle = \tfrac{1}{4}(\|\mathbf{u}+\mathbf{v}\|^2 - \|\mathbf{u}-\mathbf{v}\|^2)$$

2) If V is a complex inner product space, then

$$\langle \mathbf{u},\mathbf{v}\rangle = \tfrac{1}{4}(\|\mathbf{u}+\mathbf{v}\|^2 - \|\mathbf{u}-\mathbf{v}\|^2) + \tfrac{1}{4}i(\|\mathbf{u}+i\mathbf{v}\|^2 - \|\mathbf{u}-i\mathbf{v}\|^2)$$

The formulas in Theorem 9.3 are known as the **polarization identities.** The norm can be used to define the distance between any two vectors in an inner product space.

Definition Let V be an inner product space. We define the **distance** $d(\mathbf{u},\mathbf{v})$ between any two vectors $\mathbf{u}$ and $\mathbf{v}$ in V by

$$d(\mathbf{u},\mathbf{v}) = \|\mathbf{u}-\mathbf{v}\| \tag{9.2}$$ □

Here are the basic properties of distance.

Theorem 9.4

1) $d(\mathbf{u},\mathbf{v}) \geq 0$ and $d(\mathbf{u},\mathbf{v}) = 0$ if and only if $\mathbf{u} = \mathbf{v}$
2) (**Symmetry**) $d(\mathbf{u},\mathbf{v}) = d(\mathbf{v},\mathbf{u})$
3) (**Triangle inequality**)

$$d(\mathbf{u},\mathbf{v}) \leq d(\mathbf{u},\mathbf{w}) + d(\mathbf{w},\mathbf{v})$$ □

Any nonempty set V, together with a function $d: V \times V \to \mathbb{R}$ that satisfies the properties of Theorem 9.4, is called a **metric space,** and the function d is called a **metric** on V. Thus, any inner product space is a metric space under the metric (9.2).

Before continuing, we should make a few remarks about our goals in this and the next chapter. The presence of an inner product, and hence a metric, raises a host of *topological issues,* related to the notion of convergence. We say that a sequence $(\mathbf{v}_n)$ of vectors in an inner product space **converges** to $\mathbf{v} \in V$ if

$$\lim_{n\to\infty} d(\mathbf{v}_n,\mathbf{v}) = 0$$

or, equivalently, if

$$\lim_{n\to\infty} \| \mathbf{v}_n - \mathbf{v} \| = 0$$

Some of the more important concepts related to convergence are closedness and closures, completeness, and the continuity of linear operators and linear functionals.

In the *finite dimensional* case, the situation is very straightforward – all subspaces are closed, all inner product spaces are complete, and all linear operators and functionals are continuous. However, in the infinite dimensional case, things are not as simple.

Our goals in this chapter and the next are to describe some of the basic properties of inner product spaces – both finite and infinite dimensional, and then to discuss certain special types of operators (normal, unitary and self-adjoint), in the *finite dimensional* case only. To achieve the latter goal as rapidly as possible, we will postpone a discussion of topological properties until Chapter 13. This means that we must describe *some* results for the finite dimensional case only in this chapter, deferring the infinite dimensional case to Chapter 13.

Isometries

An isomorphism of vector spaces preserves the vector space operations. The corresponding concept for inner product spaces is the following.

Definition Let V and W be inner product spaces, and let $\tau \in \mathcal{L}(V,W)$.

1) τ is an **isometry** if it preserves the inner product, that is, if

$$\langle \tau(\mathbf{u}),\tau(\mathbf{v})\rangle = \langle \mathbf{u},\mathbf{v}\rangle$$

for all $\mathbf{u},\mathbf{v} \in V$.

2) A bijective isometry is called an **isometric isomorphism**. When $\tau:V\to W$ is an isomorphism, we say that V and W are **isometrically isomorphic**. □

It is not hard to show that an isometry is injective, and so it is an isometric isomorphism provided it is also surjective. Moreover, if $dim(V) = dim(W) < \infty$, injectivity implies surjectivity, and so the concepts of isometry and isometric isomorphism are equivalent. On the other hand, the following example shows that this is not the case for infinite dimensional inner product spaces.

Example 9.3 Let $\tau:\ell^2 \to \ell^2$ be defined by

$$\tau(x_1,x_2,x_3\ldots) = (0,x_1,x_2,\ldots)$$

(This is the *right shift* operator.) Then τ is an isometry, but it is clearly not surjective. □

Theorem 9.5 A linear transformation $\tau \in \mathcal{L}(V,W)$ is an isometry if and only if it preserves the norm, that is, if and only if

$$\| \tau(\mathbf{v}) \| = \| \mathbf{v} \|$$

for all $\mathbf{v} \in V$.

Proof. Clearly, an isometry preserves the norm. The converse follows from Theorem 9.3. In the real case, if τ preserves the norm, then

$$\begin{aligned}\langle \tau(\mathbf{u}),\tau(\mathbf{v})\rangle &= \tfrac{1}{4}(\| \tau(\mathbf{u}) + \tau(\mathbf{v}) \|^2 - \| \tau(\mathbf{u}) - \tau(\mathbf{v}) \|^2) \\ &= \tfrac{1}{4}(\| \tau(\mathbf{u}+\mathbf{v}) \|^2 - \| \tau(\mathbf{u}-\mathbf{v}) \|^2) \\ &= \tfrac{1}{4}(\| \mathbf{u}+\mathbf{v} \|^2 - \| \mathbf{u}-\mathbf{v} \|^2) \\ &= \langle \mathbf{v},\mathbf{w}\rangle\end{aligned}$$

and so τ is an isometry. The complex case is similar. ∎

The next result points out one of the main differences between real and complex inner product spaces.

Theorem 9.6 Let V be an inner product space, and let $\tau \in \mathcal{L}(V)$.

1) If $\langle \tau(\mathbf{v}),\mathbf{w}\rangle = 0$ for all $\mathbf{v}, \mathbf{w} \in V$, then $\tau = 0$.
2) If V is a *complex* inner product space, and $\langle \tau(\mathbf{v}),\mathbf{v}\rangle = 0$ for all $\mathbf{v} \in V$, then $\tau = 0$.
3) Part (2) does *not* hold in general for real inner product spaces.

Proof. Part (1) follows directly from Lemma 9.1. As for part (2), let $\mathbf{v} = r\mathbf{x} + \mathbf{y}$, for $\mathbf{x},\mathbf{y} \in V$ and $r \in F$. Then

$$\begin{aligned}0 &= \langle \tau(r\mathbf{x}+\mathbf{y}),r\mathbf{x}+\mathbf{y}\rangle \\ &= |r|^2\langle \tau(\mathbf{x}),\mathbf{x}\rangle + \langle \tau(\mathbf{y}),\mathbf{y}\rangle + r\langle \tau(\mathbf{x}),\mathbf{y}\rangle + \bar{r}\langle \tau(\mathbf{y}),\mathbf{x}\rangle \\ &= r\langle \tau(\mathbf{x}),\mathbf{y}\rangle + \bar{r}\langle \tau(\mathbf{y}),\mathbf{x}\rangle\end{aligned}$$

Setting $r = 1$ gives

$$\langle \tau(\mathbf{x}),\mathbf{y}\rangle + \langle \tau(\mathbf{y}),\mathbf{x}\rangle = 0$$

and setting $r = i$ gives

$$\langle \tau(\mathbf{x}),\mathbf{y}\rangle - \langle \tau(\mathbf{y}),\mathbf{x}\rangle = 0$$

These two equations imply that $\langle \tau(\mathbf{x}),\mathbf{y}\rangle = 0$ for all $\mathbf{x},\mathbf{y} \in V$, and so

$\tau = 0$ by part (1). As for part (3), consider the real inner product space $\mathbb{R}^2$, and let $\tau \in \mathcal{L}(V)$ be defined by $\tau(\mathbf{e}_1) = \mathbf{e}_2$ and $\tau(\mathbf{e}_2) = -\mathbf{e}_1$. Thus, τ is rotation by 90°, and $\langle \tau(\mathbf{v}), \mathbf{v} \rangle = 0$ for all $\mathbf{v}$, but $\tau \neq 0$. ∎

Orthogonality

The presence of an inner product allows us to define the concept of orthogonality, or perpendicularity.

Definition Let V be an inner product space.

1) Two vectors $\mathbf{u}, \mathbf{v} \in V$ are said to be **orthogonal** if $\langle \mathbf{u}, \mathbf{v} \rangle = 0$. In this case, we write $\mathbf{u} \perp \mathbf{v}$.
2) If S and T are subsets of V, and $\mathbf{s} \perp \mathbf{t}$ for all $\mathbf{s} \in S$ and $\mathbf{t} \in T$, we say that S is **orthogonal** to T, and write $S \perp T$.
3) The **orthogonal complement** of a *subset* $S \subset V$ is the set

$$S^\perp = \{\mathbf{v} \in V \mid \mathbf{v} \perp S\}$$ □

The following result is easily proved.

Theorem 9.7 Let V be an inner product space.

1) For any subset $S \subset V$, the orthogonal complement $S^\perp$ of S is a *subspace* of V.
2) For any subspace S of V, $S \cap S^\perp = \{\mathbf{0}\}$. ∎

Orthogonal and Orthonormal Sets

Definition A nonempty collection $\mathcal{O} = \{\mathbf{u}_i \mid i \in K\}$ of vectors in an inner product space is said to be an **orthogonal set** if $\mathbf{u}_i \perp \mathbf{u}_j$ for all $i \neq j \in K$. If, in addition, each vector $\mathbf{u}_i$ is a unit vector, the set $\mathcal{O}$ is an **orthonormal set**. Thus, a set is orthonormal if

$$\langle \mathbf{u}_i, \mathbf{u}_j \rangle = \delta_{i,j}$$

for all $i,j \in K$, where $\delta_{i,j}$ is the Kronecker delta function. □

Note that given any nonzero vector $\mathbf{v} \in V$, we may obtain a unit vector $\mathbf{u}$ by simply multiplying $\mathbf{v}$ by the reciprocal of the norm of $\mathbf{v}$

$$\mathbf{u} = \frac{1}{\|\mathbf{v}\|}\mathbf{v}$$

Thus, it is a simple matter to construct an orthonormal set from an orthogonal set of nonzero vectors.

Theorem 9.8 Any orthogonal set of nonzero vectors in V is linearly independent.

Proof. Let $\mathcal{O} = \{u_i \mid i \in K\}$ be an orthogonal set of nonzero vectors, and suppose that

$$r_1u_1 + \cdots + r_nu_n = 0$$

Then, for any $k = 1,\ldots,n$,

$$0 = \langle r_1u_1 + \cdots + r_nu_n, u_k\rangle = r_k\langle u_k, u_k\rangle$$

and so $r_k = 0$, for all k. Hence, $\mathcal{O}$ is linearly independent. ∎

Definition A maximal orthonormal set in an inner product space V is called a **Hilbert basis** for V. □

Zorn's lemma can be used to show that any nontrivial inner product space has a Hilbert basis. We leave the details to the reader.

Extreme care must be taken here not to confuse the concepts of a basis for a vector space and a Hilbert basis for an inner product space. To avoid confusion, a vector space basis, that is, a maximal *linearly independent* set of vectors, is referred to as a **Hamel basis.** The following example shows that, in general, the two concepts of basis are not the same.

Example 9.4 Let $V = \ell^2$, and let M be the set of all vectors of the form

$$e_i = (0,\ldots,0,1,0\ldots)$$

where e_i has a 1 in the ith coordinate, and 0s elsewhere. Clearly, M is an orthonormal set. Moreover, it is maximal. For if $x = (x_n) \in \ell^2$ has the property that $x \perp M$, then

$$x_i = \langle x, e_i\rangle = 0$$

for all i, and so $x = 0$. Hence, no nonzero vector $x \notin M$ is orthogonal to M. This shows that M is a Hilbert basis for the *inner product space* ℓ^2.

On the other hand, the vector space span of M is the subspace S of all sequences in ℓ^2 that have *finite support*, that is, have only a finite number of nonzero terms, and since $span(M) = S \neq \ell^2$, we see that M is *not* a Hamel basis for the *vector space* ℓ^2. □

We will show in Chapter 13 that all Hilbert bases for an inner product space have the same cardinality, and so we can define the **Hilbert dimension** of an inner product space to be that cardinality. Once again, to avoid confusion, the cardinality of any Hamel basis for

V is referred to as the **Hamel dimension** of V. The Hamel dimension is, in general, *not* the same as the Hilbert dimension. However, as we will now show, they are equal when the Hamel dimension is finite.

Definition Let V be an inner product space with *finite* Hamel dimension. A Hamel basis for V that is also an orthogonal set is called an **orthogonal Hamel basis** for V, and a Hamel basis for V that is also an orthonormal set is called an **orthonormal Hamel basis** for V. (These concepts are defined only for finite dimensional vector spaces.) $\square$

Theorem 9.9 Let V be an inner product space with finite Hamel dimension. Then any Hilbert basis is a Hamel basis. Hence, the Hilbert dimension of V is the same as the Hamel dimension.

Proof. Let V have Hamel dimension n. Since orthonormal sets of vectors in V are linearly independent, their size cannot exceed n. In particular, a maximal orthonormal set has size at most n.

If $\mathcal{O} = \{\mathbf{u}_1, \ldots, \mathbf{u}_k\}$ is a maximal orthogonal set with $k < n$, then there exists a vector $\mathbf{v} \in V$ for which $\mathcal{O} \cup \{\mathbf{v}\}$ is linearly independent. If

$$\mathbf{w} = \mathbf{v} + r_1\mathbf{u}_1 + \cdots + r_k\mathbf{u}_k$$

then $\langle \mathbf{w}, \mathbf{u}_i \rangle = 0$ if and only if

$$0 = \langle \mathbf{w}, \mathbf{u}_i \rangle = \langle \mathbf{v}, \mathbf{u}_i \rangle + r_i \langle \mathbf{u}_i, \mathbf{u}_i \rangle$$

or, equivalently

$$r_i = \frac{\langle \mathbf{v}, \mathbf{u}_i \rangle}{\langle \mathbf{u}_i, \mathbf{u}_i \rangle}$$

Thus, by defining r_i in this way, we obtain a vector $\mathbf{w}$ that is orthogonal to all vectors in $\mathcal{O}$. Hence, $\mathcal{O} \cup \{\mathbf{w}/\|\mathbf{w}\|\}$ is an orthonormal set that properly contains the maximal orthonormal set $\mathcal{O}$. This contradiction implies that $k = n$, and so $\mathcal{O}$ is a Hamel basis. ▮

It is also true that if an inner product space has finite Hilbert dimension, then this is also equal to its Hamel dimension. (This will follow from our upcoming discussion of Gram-Schmidt orthogonalization.) Therefore, the term *finite dimensional* can be applied unambiguously to an inner product space. We will use the term *orthonormal basis* to refer to an orthonormal *Hamel* basis.

Orthonormal bases have a great advantage over arbitrary bases. To see this, suppose that $\mathcal{B} = \{\mathbf{v}_1, \ldots, \mathbf{v}_n\}$ is a basis for V. Then each $\mathbf{v} \in V$ has the form

$$\mathbf{v} = r_1\mathbf{v}_1 + \cdots + r_n\mathbf{v}_n$$

In general, however, determining the coordinates r_i requires solving a system of linear equations of size $n \times n$.

On the other hand, suppose that $\mathcal{O} = \{u_1, \ldots, u_n\}$ is an orthonormal basis for V. As before, any vector $v \in V$ has the form

$$v = r_1 u_1 + \cdots + r_n u_n$$

But now we have

$$\langle v, u_i \rangle = \langle r_1 u_1 + \cdots + r_n u_n, u_i \rangle = r_i \langle u_i, u_i \rangle = r_i$$

Thus, in the case of an orthonormal basis, we have a very simple procedure for finding the coordinates of any vector $v \in V$.

Theorem 9.10 Let $\mathcal{O} = \{u_1, \ldots, u_n\}$ be an orthonormal basis for V.

1) For any $v \in V$,

$$v = \langle v, u_1 \rangle u_1 + \cdots + \langle v, u_n \rangle u_n$$

The coordinates $\langle v, u_i \rangle$ are called the **Fourier coefficients** of v with respect to $\mathcal{O}$, and the expression for v on the right is called the **Fourier expansion** of v with respect to $\mathcal{O}$.

2) (**Bessel's identity**) For any $v \in V$,

$$\|v\|^2 = |\langle v, u_1 \rangle|^2 + \cdots + |\langle v, u_n \rangle|^2$$

3) (**Parseval's identity**) For $v, w \in V$,

$$\langle v, w \rangle = \langle v, u_1 \rangle \overline{\langle w, u_1 \rangle} + \cdots + \langle v, u_n \rangle \overline{\langle w, u_n \rangle}$$ ∎

Theorem 9.10 shows clearly that orthonormal bases are a pleasure to work with. The following result is included primarily to establish an analogy with the infinite dimensional case, which we will discuss in Chapter 13.

Theorem 9.11 Let V be a finite dimensional inner product space. Let $\mathcal{O} = \{u_1, \ldots, u_k\}$ be an orthonormal set of vectors in V. For any $v \in V$, the vector

$$\hat{v} = \langle v, u_1 \rangle u_1 + \cdots + \langle v, u_k \rangle u_k$$

is called the **Fourier expansion** of v with respect to $\mathcal{O}$.

1) (**Bessel's inequality**) For all $v \in V$,

$$\|\hat{v}\| \le \|v\|$$

2) The set $\mathcal{O}$ is an (orthonormal) Hamel basis for V if and only if $\hat{v} = v$, for all $v \in V$.

3) The set $\mathcal{O}$ is an (orthonormal) Hamel basis for V if and only if Bessel's identity holds for all $v \in V$, that is, if and only if

$$\|\hat{v}\| = \|v\|$$

for all $\mathbf{v} \in V$.

4) The set $\mathcal{O}$ is an (orthonormal) Hamel basis for V if and only if Parseval's identity holds for all $\mathbf{v},\mathbf{w} \in V$, that is, if and only if

$$\langle \mathbf{v},\mathbf{w}\rangle = \langle \mathbf{v},\mathbf{u}_1\rangle\overline{\langle \mathbf{w},\mathbf{u}_1\rangle} + \cdots + \langle \mathbf{v},\mathbf{u}_k\rangle\overline{\langle \mathbf{w},\mathbf{u}_k\rangle}$$

for all $\mathbf{v},\mathbf{w} \in V$. ∎

The Projection Theorem

We have seen that if S is a subspace of an inner product space V, then $S \cap S^{\perp} = \{\mathbf{0}\}$. This raises the question of whether or not the orthogonal complement of a subspace S is a (vector space) complement of S, that is, whether or not $V = S \oplus S^{\perp}$.

If S is a *finite dimensional* subspace of V, the answer is yes, but for infinite dimensional subspaces, S must have the topological property of being complete. Hence, in accordance with our goals in this chapter, we will postpone a discussion of the general case to Chapter 13, contenting ourselves here with an example to show that, in general, $V \neq S \oplus S^{\perp}$.

Example 9.5 As in Example 9.4, let $V = \ell^2$, and let S be the subspace spanned by the vectors

$$\mathbf{e}_i = (0,\ldots,0,1,0\ldots)$$

where $\mathbf{e}_i$ has a 1 in the ith coordinate, and 0s elsewhere. Thus, S is the subspace of all sequences in ℓ^2 that have *finite support*, that is, have only a finite number of nonzero terms.

Now, if $\mathbf{x} = (x_n) \in S^{\perp}$, then $x_i = \langle \mathbf{x},\mathbf{e}_i\rangle = 0$ for all i, and so $\mathbf{x} = \mathbf{0}$. Therefore, $S^{\perp} = \{\mathbf{0}\}$, and

$$S \oplus S^{\perp} = S \neq \ell^2$$

□

As the next theorem shows, in the *finite dimensional case*, orthogonal complements are also vector space complements. This theorem is often called the *projection theorem*, for reasons that will become apparent when we discuss projection operators. (We will discuss the projection theorem in the infinite dimensional case in Chapter 13.)

Theorem 9.12 **(The projection theorem)** Let S be a *finite dimensional* subspace of an inner product space. Then $V = S \oplus S^{\perp}$. That is, for any $\mathbf{v} \in V$, there are unique vectors $\mathbf{s} \in S$ and $\mathbf{s}^{\perp} \in S^{\perp}$ for which

$$\mathbf{v} = \mathbf{s} + \mathbf{s}^{\perp}$$

Proof. Let $\mathcal{O} = \{\mathbf{u}_1,\ldots,\mathbf{u}_k\}$ be an orthonormal basis for S. For each $\mathbf{v} \in V$, consider the Fourier expansion

$$\hat{\mathbf{v}} = \langle \mathbf{v},\mathbf{u}_1\rangle \mathbf{u}_1 + \cdots + \langle \mathbf{v},\mathbf{u}_k\rangle \mathbf{u}_k$$

with respect to $\mathcal{O}$. We may write

$$\mathbf{v} = \hat{\mathbf{v}} + (\mathbf{v} - \hat{\mathbf{v}})$$

where $\hat{\mathbf{v}} \in S$. Moreover, $\mathbf{v} - \hat{\mathbf{v}} \in S^{\perp}$, since

$$\langle \mathbf{v} - \hat{\mathbf{v}},\mathbf{u}_i\rangle = \langle \mathbf{v},\mathbf{u}_i\rangle - \langle \hat{\mathbf{v}},\mathbf{u}_i\rangle = 0$$

Hence $V = S + S^{\perp}$. We have already observed that $S \cap S^{\perp} = \{\mathbf{0}\}$, and so $V = S \oplus S^{\perp}$. ∎

According to the proof of the projection theorem, the component of $\mathbf{v}$ that lies in S is just the Fourier expansion of $\mathbf{v}$ with respect to any orthonormal basis $\mathcal{O}$ for S. This is pictured in Figure 9.1.

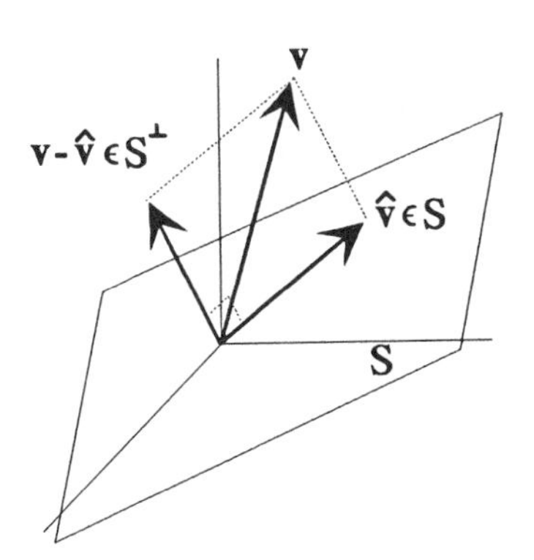

Figure 9.1

Definition Let V be an inner product space, and let $S_1,\ldots,S_n$ be subspaces of V. If

1) $V = S_1 \oplus \cdots \oplus S_n$
2) $S_i \perp S_j$ for $i \neq j$

then we say that V is the **orthogonal direct sum** of $S_1,\ldots,S_n$, and write $S = S_1 \odot \cdots \odot S_n$. □

Theorem 9.12 states that $V = S \odot S^{\perp}$, for any *finite dimensional* subspace S of V. The following simple result is very useful.

Theorem 9.13 Let V be an inner product space. The following are equivalent.

1) $V = S \odot T$

2) $V = S \oplus T$ and $T = S^{\perp}$
3) $V = S \oplus T$ and $T \subset S^{\perp}$

Proof. Suppose (1) holds. Then $V = S \oplus T$ and $S \perp T$, which implies that $T \subset S^{\perp}$. But if $\mathbf{w} \in S^{\perp}$, then $\mathbf{w} = \mathbf{s} + \mathbf{t}$ for $\mathbf{s} \in S$, $\mathbf{t} \in T$, and so

$$0 = \langle \mathbf{s}, \mathbf{w} \rangle = \langle \mathbf{s}, \mathbf{s} \rangle + \langle \mathbf{s}, \mathbf{t} \rangle = \langle \mathbf{s}, \mathbf{s} \rangle$$

showing that $\mathbf{s} = \mathbf{0}$, which implies that $\mathbf{w} \in T$. Thus, $S^{\perp} \subset T$ and so $S^{\perp} = T$. Hence, (2) holds. Of course, (2) implies (3). Finally, if (3) holds, then $T \subset S^{\perp}$, which implies that $S \perp T$, and so (1) holds. ∎

Theorem 9.14 Let V be an inner product space.
1) If $dim(V) < \infty$ and S is a subspace of V, then

$$dim(S^{\perp}) = dim(V) - dim(S)$$

2) If S is a finite dimensional subspace of V, then $S^{\perp\perp} = S$.
3) If S is a *subset* of V and $dim(span(S)) < \infty$, then $S^{\perp\perp} = span(S)$.

Proof. Since $V = S \oplus S^{\perp}$, we have $dim(V) = dim(S) + dim(S^{\perp})$, which proves part (1). As for part (2), it is clear that $S \subset S^{\perp\perp}$. On the other hand, if $\mathbf{v} \in S^{\perp\perp}$, then by the projection theorem

$$\mathbf{v} = \mathbf{s} + \mathbf{s}'$$

where $\mathbf{s} \in S$ and $\mathbf{s}' \in S^{\perp}$. But $\mathbf{v} \in S^{\perp\perp}$ implies that $0 = \langle \mathbf{v}, \mathbf{s}' \rangle = \langle \mathbf{s}', \mathbf{s}' \rangle$, and so $\mathbf{s}' = \mathbf{0}$, showing that $\mathbf{v} \in S$. Therefore, $S^{\perp\perp} \subset S$, and $S^{\perp\perp} = S$. We leave the proof of part (3) as an exercise. ∎

The Gram-Schmidt Orthogonalization Process

Given a linearly independent sequence $\mathcal{B} = (\mathbf{v}_1, \mathbf{v}_2, \ldots)$ in an inner product space V, we can easily construct an orthogonal sequence $\mathcal{O} = (\mathbf{u}_1, \mathbf{u}_2, \ldots)$ in V, with the property that

$$span\{\mathbf{u}_1, \ldots, \mathbf{u}_k\} = span\{\mathbf{v}_1, \ldots, \mathbf{v}_k\}$$

for all k. The following construction is known as the **Gram-Schmidt orthogonalization process.**

The first step is to let $\mathbf{u}_1 = \mathbf{v}_1$. Next, we search for a vector $\mathbf{u}_2$ of the form $\mathbf{u}_2 = \mathbf{v}_2 + r_1\mathbf{u}_1$ for which $\langle \mathbf{u}_2, \mathbf{u}_1 \rangle = 0$, that is, for which

$$0 = \langle \mathbf{u}_2, \mathbf{u}_1 \rangle = \langle \mathbf{v}_2 + r_1\mathbf{u}_1, \mathbf{u}_1 \rangle = \langle \mathbf{v}_2, \mathbf{u}_1 \rangle + r_1\langle \mathbf{u}_1, \mathbf{u}_1 \rangle$$

or, equivalently,

$$r_1 = -\frac{\langle \mathbf{v}_2, \mathbf{u}_1 \rangle}{\langle \mathbf{u}_1, \mathbf{u}_1 \rangle}$$

Hence, defining r_1 by this formula, we see that the set $\{u_1,u_2\}$ is orthogonal, and that $span\{u_1,u_2\} = span\{v_1,v_2\}$.

More generally, suppose that $\{u_1,\ldots,u_{k-1}\}$ is orthogonal, and that $span\{u_1,\ldots,u_{k-1}\} = span\{v_1,\ldots,v_{k-1}\}$. We want a vector u_k of the form

$$u_k = v_k + r_1u_1 + \cdots + r_{k-1}u_{k-1}$$

for which $\langle u_k,u_i\rangle = 0$ for all $i = 1,\ldots,k-1$, that is,

$$0 = \langle u_k,u_i\rangle = \langle v_k + r_1u_1 + \cdots + r_{k-1}u_{k-1},u_i\rangle = \langle v_k,u_i\rangle + r_i\langle u_i,u_i\rangle$$

or, equivalently,

$$r_i = -\frac{\langle v_k,u_i\rangle}{\langle u_i,u_i\rangle}$$

for all $i = 1,\ldots,j$. Defining the r_i's by this formula gives us the desired vector u_k. Let us summarize.

Theorem 9.15 (The Gram-Schmidt orthogonalization process) Suppose that $\mathcal{B} = (v_1,v_2,\ldots)$ is a sequence of linearly independent vectors in an inner product space V. If we define

$$u_k = v_k - \sum_{i=1}^{k-1} \frac{\langle v_k,u_i\rangle}{\langle u_i,u_i\rangle} u_i$$

then the sequence $\mathcal{O} = (u_1,u_2,\ldots)$ is an orthogonal sequence of linearly independent vectors, with the property that

$$span\{u_1,\ldots,u_k\} = span\{v_1,\ldots,v_k\}$$

for all $k = 1,2,\ldots$. ∎

Example 9.6 Consider the inner product space $F[x]$ of all polynomials over F, with inner product defined by

$$\langle p(x),q(x)\rangle = \int_{-1}^{1} p(x)q(x)dx$$

Applying the Gram-Schmidt process to the sequence $\mathcal{B} = (1,x,x^2,x^3,\ldots)$ gives

$$u_1(x) = 1$$

$$u_2(x) = x - \frac{\int_{-1}^{1} x\,dx}{\int_{-1}^{1} dx}\cdot 1 = x$$

$$u_3(x) = x^2 - \frac{\int_{-1}^{1} x^2\,dx}{\int_{-1}^{1} dx}\cdot 1 - \frac{\int_{-1}^{1} x^3\,dx}{\int_{-1}^{1} x\,dx}\cdot x = x^2 - \frac{1}{3}$$

$$u_4(x) = x^3 - \frac{\int_{-1}^{1} x^3\,dx}{\int_{-1}^{1} dx}\cdot 1 - \frac{\int_{-1}^{1} x^4\,dx}{\int_{-1}^{1} x\,dx}\cdot x - \frac{\int_{-1}^{1} x^3(x^2-\frac{1}{3})\,dx}{\int_{-1}^{1} (x^2-\frac{1}{3})^2 dx}\cdot (x^2 - \tfrac{1}{3})$$

$$= x^3 - \tfrac{3}{5}x$$

and so on. The polynomials in this sequence are (at least up to multiplicative constants) the **Legendre polynomials.** □

The Riesz Representation Theorem

If $\mathbf{x}$ is a vector in an inner product space V, then the function $\phi_{\mathbf{x}}:V\to F$ defined by

$$\phi_{\mathbf{x}}(\mathbf{v}) = \langle \mathbf{v},\mathbf{x}\rangle$$

is easily seen to be a linear functional on V. The following theorem shows that *all* linear functionals on a *finite dimensional* inner product space V have this form. (We will see in Chapter 13 that, in the infinite dimensional case, all *continuous* linear functionals on V have this form.)

Theorem 9.16 (The Riesz representation theorem) Let V be a *finite dimensional* inner product space, and let $f \in V^*$ be a linear functional on V. Then there exists a unique vector $\mathbf{x} \in V$ for which

$$f(\mathbf{v}) = \langle \mathbf{v},\mathbf{x}\rangle \tag{9.3}$$

for all $\mathbf{v} \in V$.

Proof. If f is the zero functional, we may take $\mathbf{x} = \mathbf{0}$, so let us assume that $f \neq 0$. By way of motivation, observe that if $\mathbf{x}$ has the desired property, then $\langle \mathbf{v},\mathbf{x}\rangle = 0$ for all $\mathbf{v} \in \mathit{ker}(f)$. Hence, we should look for an $\mathbf{x}$ in $\mathit{ker}(f)^{\perp}$.

Note that, if $\mathit{dim}(V) = n$, then $\mathit{dim}(\mathit{ker}(f)) = n - 1$. Hence, we can choose a unit vector $\mathbf{u} \in \mathit{ker}(f)^{\perp}$, and write

$$V = \langle \mathbf{u}\rangle \oplus \mathit{ker}(f)$$

Our goal is to find an $r \in F$ for which

$$f(\mathbf{v}) = \langle \mathbf{v}, r\mathbf{u}\rangle$$

for all $\mathbf{v} \in V$. In particular, for $\mathbf{v} = \mathbf{u}$, we want

$$f(\mathbf{u}) = \langle \mathbf{u}, r\mathbf{u}\rangle = \bar{r}\langle \mathbf{u},\mathbf{u}\rangle = \bar{r}$$

Therefore, let us take $r = \overline{f(\mathbf{u})}$, and so

$$\mathbf{x} = \overline{f(\mathbf{u})}\mathbf{u}$$

Any vector $\mathbf{v} \in V$ has the form $\mathbf{v} = a\mathbf{u} + b\mathbf{w}$, with $\mathbf{w} \in \mathit{ker}(f)$, and so

$$\langle \mathbf{v},\mathbf{x}\rangle = \langle \mathbf{v},\overline{f(\mathbf{u})}\mathbf{u}\rangle = f(\mathbf{u})\langle \mathbf{v},\mathbf{u}\rangle = f(\mathbf{u})a = f(a\mathbf{u}) = f(a\mathbf{u}+b\mathbf{w}) = f(\mathbf{v})$$

Proof of uniqueness is left as an exercise. ∎

Using the Riesz representation theorem, we can define a map $\phi:V^*\to V$ by $\phi(f) = \mathbf{x}$, where $\mathbf{x}$ is the unique vector in V for which (9.3) holds, that is, $\phi(f)$ is defined by

$$f(\mathbf{v}) = \langle \mathbf{v},\phi(f)\rangle$$

for all $\mathbf{v} \in V$. Since

$$\begin{aligned}
\langle \mathbf{v},\phi(rf+sg)\rangle &= (rf+sg)(\mathbf{v}) \\
&= rf(\mathbf{v}) + sg(\mathbf{v}) \\
&= \langle \mathbf{v},\bar{r}\phi(f)\rangle + \langle \mathbf{v},\bar{s}\phi(g)\rangle \\
&= \langle \mathbf{v},\bar{r}\phi(f) + \bar{s}\phi(g)\rangle
\end{aligned}$$

we have

$$\phi(rf+sg) = \bar{r}\phi(f) + \bar{s}\phi(g)$$

and so ϕ is **conjugate linear.** In addition, ϕ is clearly surjective, and it is injective, since $\phi(f) = \mathbf{0}$ implies that $f = 0$. Thus, the map $\phi:V^*\to V$ is a "conjugate isomorphism."

EXERCISES

1. Verify the statement concerning equality in the triangle inequality.
2. Prove the parallelogram law.
3. Prove **Appolonius' identity**

$$\|\mathbf{w}-\mathbf{u}\|^2 + \|\mathbf{w}-\mathbf{v}\|^2 = \tfrac{1}{2}\|\mathbf{u}-\mathbf{v}\|^2 + 2\|\mathbf{w}-\tfrac{1}{2}(\mathbf{u}+\mathbf{v})\|^2$$

4. Let V be an inner product space with basis $\mathfrak{B}$. Show that the inner product is uniquely defined by the values $\langle \mathbf{u},\mathbf{v}\rangle$, for all $\mathbf{u},\mathbf{v} \in \mathfrak{B}$.
5. Prove that two vectors $\mathbf{u}$ and $\mathbf{v}$ in a real inner product space V are orthogonal if and only if

$$\|\mathbf{u}+\mathbf{v}\|^2 = \|\mathbf{u}\|^2 + \|\mathbf{v}\|^2.$$

6. Show that an isometry is injective.
7. Use Zorn's lemma to show that any nontrivial inner product space has a Hilbert basis.
8. Prove Bessel's inequality.
9. Prove that an orthonormal set $\mathcal{O}$ is a basis for V if and only if $\hat{\mathbf{v}} = \mathbf{v}$, for all $\mathbf{v} \in V$.
10. Prove that an orthonormal set $\mathcal{O}$ is a basis for V if and only if Bessel's identity holds for all $\mathbf{v} \in V$, that is, if and only if

$$\|\hat{\mathbf{v}}\| = \|\mathbf{v}\|$$

for all $\mathbf{v} \in V$.

11. Prove that an orthonormal set $\mathcal{O}$ is a basis for V if and only if Parseval's identity holds for all $\mathbf{v},\mathbf{w} \in V$, that is, if and only if

$$\langle \mathbf{v},\mathbf{w}\rangle = \langle \mathbf{v},\mathbf{u}_1\rangle\langle \mathbf{w},\mathbf{u}_1\rangle + \cdots + \langle \mathbf{v},\mathbf{u}_k\rangle\langle \mathbf{w},\mathbf{u}_k\rangle$$

for all $\mathbf{v},\mathbf{w} \in V$.

12. Let V be an inner product space. Prove that $S \subset S^{\perp\perp}$ for any subspace $S \subset V$.
13. Let V be a finite dimensional inner product space. Prove that for any subset S of V, we have $S^{\perp\perp} = span(S)$.
14. Let $\mathcal{P}_3$ be the inner product of all polynomials of degree at most 3, under the inner product

$$\langle p(x),q(x)\rangle = \int_{-\infty}^{\infty} p(x)q(x)e^{-x^2}dx$$

Apply the Gram-Schmidt process to the basis $\{1,x,x^2,x^3\}$, thereby computing the first four **Hermite polynomials** (at least up to multiplicative constant).

15. Verify uniqueness in the Riesz representation theorem.

CHAPTER 10

The Spectral Theorem for Normal Operators

Contents: *The Adjoint of a Linear Operator. Orthogonal Diagonalizability. Motivation. Self-Adjoint Operators. Unitary Operators. Normal Operators. Orthogonal Diagonalization. Orthogonal Projections. Orthogonal Resolutions of the Identity. The Spectral Theorem. Functional Calculus. Positive Operators. The Polar Decomposition of an Operator. Exercises.*

The Adjoint of a Linear Operator

The purpose of this chapter is to study the structure of certain special types of linear operators on an inner product space. In order to define these operators, we introduce another type of adjoint (different from the operator adjoint of Chapter 3). We will define this adjoint in the finite dimensional case only, deferring the infinite dimensional case to Chapter 13.

Theorem 10.1 Let V and W be finite dimensional inner product spaces over F, and let $\tau \in \mathcal{L}(V,W)$. Then there is a unique function $\tau^*:W\to V$, defined by the condition

$$\langle \tau(\mathbf{v}),\mathbf{w}\rangle = \langle \mathbf{v},\tau^*(\mathbf{w})\rangle$$

for all $\mathbf{v}\in V$ and $\mathbf{w}\in W$. This function is in $\mathcal{L}(W,V)$, and is called the **adjoint** of τ.

Proof. For a fixed $\mathbf{w}\in W$, consider the function $\theta_{\mathbf{w}}:V\to F$ defined by

$$\theta_{\mathbf{w}}(\mathbf{v}) = \langle \tau(\mathbf{v}),\mathbf{w}\rangle$$

It is easy to verify that $\theta_{\mathbf{w}}$ is a linear functional on V, and so, by the Riesz representation theorem, there exists a *unique* vector $\mathbf{x} \in \mathrm{V}$ for which

$$\theta_{\mathbf{w}}(\mathbf{v}) = \langle \tau(\mathbf{v}),\mathbf{w} \rangle = \langle \mathbf{v},\mathbf{x} \rangle$$

for all $\mathbf{v} \in \mathrm{V}$. Hence, if we set $\tau^*(\mathbf{w}) = \mathbf{x}$, then

$$\langle \tau(\mathbf{v}),\mathbf{w} \rangle = \langle \mathbf{v},\tau^*(\mathbf{w}) \rangle$$

for all $\mathbf{v} \in \mathrm{V}$. This establishes the existence and uniqueness of τ^*. To show that τ^* is linear, observe that

$$\begin{aligned}
\langle \mathbf{v},\tau^*(\mathrm{r}\mathbf{w}+\mathrm{s}\mathbf{w}') \rangle &= \langle \tau(\mathbf{v}),\mathrm{r}\mathbf{w}+\mathrm{s}\mathbf{w}' \rangle \\
&= \bar{\mathrm{r}}\langle \tau(\mathbf{v}),\mathbf{w} \rangle + \bar{\mathrm{s}}\langle \tau(\mathbf{v}),\mathbf{w}' \rangle \\
&= \bar{\mathrm{r}}\langle \mathbf{v},\tau^*(\mathbf{w}) \rangle + \bar{\mathrm{s}}\langle \mathbf{v},\tau^*(\mathbf{w}') \rangle \\
&= \langle \mathbf{v},\mathrm{r}\tau^*(\mathbf{v}) \rangle + \langle \mathbf{v},\mathrm{s}\tau^*(\mathbf{w}') \rangle \\
&= \langle \mathbf{v},\mathrm{r}\tau^*(\mathbf{w}) + \mathrm{s}\tau^*(\mathbf{w}') \rangle
\end{aligned}$$

for all $\mathbf{v} \in \mathrm{V}$, and so

$$\tau^*(\mathrm{r}\mathbf{w}+\mathrm{s}\mathbf{w}') = \mathrm{r}\tau^*(\mathbf{w}) + \mathrm{s}\tau^*(\mathbf{w}')$$

Hence $\tau^* \in \mathcal{L}(\mathrm{V},\mathrm{W})$. ∎

We should make some remarks about the differences between the operator adjoint $\tau^\times$ of τ, as defined in Chapter 3, and the adjoint τ^* that we have just defined, which is sometimes called the **Hilbert space adjoint.** In the first place, if $\tau:\mathrm{V}\to\mathrm{W}$, then

$$\tau^\times:\mathrm{W}^*\to\mathrm{V}^*$$

but

$$\tau^*:\mathrm{W}\to\mathrm{V}$$

These maps are shown in Figure 10.1, where ϕ_1 and ϕ_2 are the conjugate linear maps that we discussed in Chapter 9, following our discussion of the Riesz representation theorem.

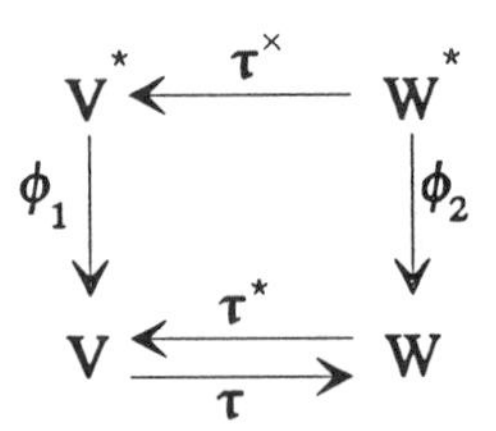

Figure 10.1

We can define a function $\sigma:W^* \to V^*$ by

$$\sigma = (\phi_1)^{-1}\tau^*\phi_2 \tag{10.1}$$

Because σ involves *two* conjugate linear maps (and one linear map), it is linear. Moreover, for all $f \in W^*$ and $\mathbf{v} \in V$

$$\begin{aligned}[\sigma(f)](\mathbf{v}) &= [(\phi_1)^{-1}\tau^*\phi_2(f)](\mathbf{v}) = (\phi_1)^{-1}[\tau^*\phi_2(f)](\mathbf{v}) \\ &= \langle \mathbf{v},\tau^*\phi_2(f)\rangle = \langle \tau(\mathbf{v}),\phi_2(f)\rangle = f(\tau(\mathbf{v})) = \tau^\times(f)(\mathbf{v})\end{aligned}$$

and so $\sigma = \tau^\times$. Hence, the relationship between $\tau^\times$ and τ^* is given by

$$\tau^\times = (\phi_1)^{-1}\tau^*\phi_2$$

In Chapter 3, we showed that the matrix of the operator adjoint $\tau^\times$ is the transpose of the matrix of the map τ. For Hilbert space adjoints, the situation is slightly different. Suppose that $\mathfrak{B} = (\mathbf{b}_1,\ldots,\mathbf{b}_n)$ is an ordered *orthonormal* basis for V, and $\mathfrak{C} = (\mathbf{c}_1,\ldots,\mathbf{c}_m)$ is an ordered *orthonormal* basis for W. If we let

$$[\tau]_{\mathfrak{B},\mathfrak{C}} = (a_{i,j})$$

then $a_{i,j}$ is the coordinate of $\mathbf{c}_i$ in $\tau(\mathbf{b}_j)$, that is

$$a_{i,j} = \langle \tau(\mathbf{b}_j),\mathbf{c}_i\rangle$$

On the other hand, if

$$[\tau^*]_{\mathfrak{C},\mathfrak{B}} = (\alpha_{i,j})$$

then $\alpha_{i,j}$ is the coordinate of $\mathbf{b}_i$ in $\tau^*(\mathbf{c}_j)$, that is

$$\alpha_{i,j} = \langle \tau^*(\mathbf{c}_j),\mathbf{b}_i\rangle = \overline{\langle \mathbf{b}_i,\tau^*(\mathbf{c}_j)\rangle} = \overline{\langle \tau(\mathbf{b}_i),\mathbf{c}_j\rangle} = \bar{a}_{j,i}$$

If $A = (a_{i,j})$ is a matrix over F, then the **conjugate transpose** of A is the matrix

$$A^* = (\bar{a}_{i,j})^T$$

With this terminology, we have proved the following.

Theorem 10.2 Let $\tau \in \mathcal{L}(V,W)$, where V and W are finite dimensional inner product spaces. Let $\mathfrak{B}$ and $\mathfrak{C}$ be ordered orthonormal bases for V and W, respectively. Then

$$[\tau^*]_{\mathfrak{C},\mathfrak{B}} = ([\tau]_{\mathfrak{B},\mathfrak{C}})^*$$

In words, the matrix of the adjoint τ^* is the *conjugate transpose* of the matrix of τ. ▮

Here are some of the basic properties of the adjoint.

Theorem 10.3 Let $\sigma,\tau \in \mathcal{L}(V,W)$, where V and W are finite dimensional.

1) $\langle \tau^*(\mathbf{v}),\mathbf{u}\rangle = \langle \mathbf{v},\tau(\mathbf{u})\rangle$
2) $(\sigma+\tau)^* = \sigma^* + \tau^*$
3) $(r\tau)^* = \bar{r}\tau^*$
4) $\tau^{**} = \tau^*$
5) $(\sigma\tau)^* = \tau^*\sigma^*$, if $V = W$
6) If τ is invertible, then $(\tau^{-1})^* = (\tau^*)^{-1}$

Proof. We prove (3) only:

$$\langle \mathbf{v},(r\tau)^*\mathbf{u}\rangle = \langle r\tau(\mathbf{v}),\mathbf{u}\rangle = r\langle \tau(\mathbf{v}),\mathbf{u}\rangle = r\langle \mathbf{v},\tau^*(\mathbf{u})\rangle = \langle \mathbf{v},\bar{r}\tau^*(\mathbf{u})\rangle$$

and so $(r\tau)^* = \bar{r}\tau^*$. ∎

Orthogonal Diagonalizability

Recall that a linear operator $\tau \in \mathcal{L}(V)$ on a finite dimensional vector space V is diagonalizable if and only if V has a basis consisting entirely of eigenvectors of τ, or equivalently, if and only if τ has a spectral resolution

$$\tau = \lambda_1\rho_1 + \cdots + \lambda_k\rho_k$$

where $\rho_1 + \cdots + \rho_k = \iota$ is a resolution of the identity, the λ_i's are the distinct eigenvalues of τ, and ρ_i is projection onto the eigenspace $\mathcal{E}_{\lambda_i}$. Since in this case

$$V = \mathcal{E}_{\lambda_1} \oplus \cdots \oplus \mathcal{E}_{\lambda_k}$$

the action of τ can be described in the simple form

$$\mathbf{v} = \mathbf{v}_1 + \cdots + \mathbf{v}_k \;\Rightarrow\; \tau(\mathbf{v}) = \lambda_1\mathbf{v}_1 + \cdots + \lambda_k\mathbf{v}_k$$

where $\mathbf{v}_i \in \mathcal{E}_{\lambda_i}$ for all i.

While this description of τ is simple, it does require finding the components of $\mathbf{v}$ that belong to each eigenspace $\mathcal{E}_{\lambda_i}$ which, in general, requires solving a system of equations.

However, suppose that V is a finite dimensional inner product space, and that $\mathcal{O}$ is an ordered *orthonormal* basis consisting entirely of eigenvalues of τ. If

$$\mathcal{O}_i = (\mathbf{u}_{i,1},\ldots,\mathbf{u}_{i,k_i})$$

is the subset of $\mathcal{O}$ consisting of the eigenvalues associated to λ_i, then it is not hard to see that $\mathcal{O}_i$ is an ordered orthonormal basis for $\mathcal{E}_{\lambda_i}$, and the component $\mathbf{v}_i$ of $\mathbf{v}$ in $\mathcal{E}_{\lambda_i}$ is the easily computed Fourier expansion

$$\hat{\mathbf{v}}_i = \langle \mathbf{v},\mathbf{u}_{i,1}\rangle\mathbf{u}_{i,1} + \cdots + \langle \mathbf{v},\mathbf{u}_{i,k_i}\rangle\mathbf{u}_{i,k_i}$$

of $\mathbf{v}$ with respect to $\mathcal{O}_i$. Hence, the action of τ has the *truly* simple form

$$\mathbf{v} = \hat{\mathbf{v}}_1 + \cdots + \hat{\mathbf{v}}_k \Rightarrow \tau(\mathbf{v}) = \lambda_1 \hat{\mathbf{v}}_1 + \cdots + \lambda_k \hat{\mathbf{v}}_k$$

Definition Let V be a finite dimensional inner product space, and let $\tau \in \mathcal{L}(V)$. If there is an *orthonormal* basis $\mathcal{O}$ for V for which $[\tau]_{\mathcal{O}}$ is a diagonal matrix, we say that τ is **orthogonally diagonalizable.** □

It is clear from this discussion that τ is orthogonally diagonalizable if and only if there is an orthonormal basis for V consisting entirely of eigenvectors of τ, that is, if and only if

$$V = \mathcal{E}_{\lambda_1} \odot \cdots \odot \mathcal{E}_{\lambda_k}$$

Thus, orthogonally diagonalizable operators are very well behaved indeed, and this leads us to seek a simple criterion for determining whether or not a given operator is orthogonally diagonalizable. Remarkably, there is a simple criterion.

Motivation

By way of motivation, suppose that V is a finite dimensional inner product space over F, and that all of the roots of the characteristic polynomial of $\tau \in \mathcal{L}(V)$ lie in F, that is, that the minimal polynomial of τ splits into a product of linear factors over F, say,

$$m_\tau(x) = (x - \lambda_1)^{e_1} \cdots (x - \lambda_k)^{e_k}$$

where the λ_i's are the *distinct* eigenvalues of τ. (This happens for all operators on a *complex* inner product space, for instance.) Then, according to the primary decomposition theorem, we may write V as the direct sum

$$V = V_1 \oplus \cdots \oplus V_k$$

where

$$V_i = \{\mathbf{v} \in V \mid (\tau - \lambda_i)^{e_i}(\mathbf{v}) = \mathbf{0}\}$$

If $\mathbf{v}$ is an eigenvector of τ associated with λ_i, then $(\tau - \lambda_i)(\mathbf{v}) = \mathbf{0}$, and so $\mathbf{v} \in V_i$. In other words, $\mathcal{E}_{\lambda_i} \subset V_i$. Thus, τ will be orthogonally diagonalizable if and only if

1) $\mathcal{E}_{\lambda_i} = V_i$, for all i, and
2) $\mathcal{E}_{\lambda_i} \perp \mathcal{E}_{\lambda_j}$ for $i \neq j$.

Let us consider property (2) first. This property is equivalent to

$$\tau(\mathbf{v}) = \lambda_i\mathbf{v} \text{ and } \tau(\mathbf{w}) = \lambda_j\mathbf{w} \Rightarrow \langle\mathbf{v},\mathbf{w}\rangle = 0$$

for $i \neq j$. Now, let us observe that

$$\lambda_i\langle\mathbf{v},\mathbf{w}\rangle = \langle\tau(\mathbf{v}),\mathbf{w}\rangle = \langle\mathbf{v},\tau^*(\mathbf{w})\rangle$$

and further, if it were true that $\tau^*(\mathbf{w}) = \overline{\lambda}_j\mathbf{w}$, then we could continue

$$= \langle\mathbf{v},\overline{\lambda}_j\mathbf{w}\rangle = \lambda_j\langle\mathbf{v},\mathbf{w}\rangle$$

which implies that $\langle\mathbf{v},\mathbf{w}\rangle = 0$ (since $\lambda_i \neq \lambda_j$). Thus, if τ has the property that

$$\tau(\mathbf{w}) = \lambda_j\mathbf{w} \Rightarrow \tau^*(\mathbf{w}) = \overline{\lambda}_j\mathbf{w}$$

for all j, then property (2) will hold. This is equivalent to

$$(\tau - \lambda_j)(\mathbf{w}) = \mathbf{0} \Rightarrow (\tau^* - \overline{\lambda}_j)(\mathbf{w}) = \mathbf{0}$$

or, since $\lambda_j^* = \overline{\lambda}_j$ (that is $(\lambda_j\iota)^* = \overline{\lambda}_j\iota$, where ι is the identity operator),

$$(\tau - \lambda_j)(\mathbf{w}) = \mathbf{0} \Rightarrow (\tau - \lambda_j)^*(\mathbf{w}) = \mathbf{0}$$

If we set $\sigma = \tau - \lambda_j$, then this is equivalent to

$$\sigma(\mathbf{w}) = \mathbf{0} \Rightarrow \sigma^*(\mathbf{w}) = \mathbf{0}$$

which in turn is equivalent to

$$\langle\sigma(\mathbf{w}),\sigma(\mathbf{w})\rangle = 0 \Rightarrow \langle\sigma^*(\mathbf{w}),\sigma^*(\mathbf{w})\rangle = 0$$

This will hold if $\sigma^*\sigma = \sigma\sigma^*$, for in this case,

$$\langle\sigma(\mathbf{w}),\sigma(\mathbf{w})\rangle = \langle\sigma^*\sigma(\mathbf{w}),\mathbf{w}\rangle = \langle\sigma\sigma^*(\mathbf{w}),\mathbf{w}\rangle = \langle\sigma^*(\mathbf{w}),\sigma^*(\mathbf{w})\rangle$$

But $\sigma^*\sigma = \sigma\sigma^*$ if and only if $\tau^*\tau = \tau\tau^*$, and so we conclude that

$$\tau^*\tau = \tau\tau^* \Rightarrow \text{property (2) holds}$$

As we will see, if $\tau^*\tau = \tau\tau^*$, then property (1) holds as well! In any case, we have motivated one of the following definitions.

Definition Let V be an inner product space, and let $\tau \in \mathcal{L}(V)$. Then
1) τ is **self-adjoint**, or **Hermitian**, if $\tau^* = \tau$.
2) τ is **unitary** if it is bijective and $\tau^* = \tau^{-1}$.
3) τ is **normal** if $\tau\tau^* = \tau^*\tau$. □

It is clear that self-adjoint and unitary operators are normal.

There are also matrix versions of these definitions, but the terminology differs for real and complex matrices. Recall that if $A = (a_{i,j})$ is a matrix over F, then $A^* = (\overline{a}_{i,j})^T$ is the conjugate transpose of A. (If $F = \mathbb{R}$, then $A^* = A^T$.)

Definition Let A be a *complex* matrix. Then
1) A is **Hermitian** if $A^* = A$.
2) A is **skew-Hermitian**, if $A^* = -A$.
3) A is **unitary** if it is invertible and $A^* = A^{-1}$.
4) A is **normal** if $AA^* = A^*A$.
Let A be a *real* matrix. Then $A^* = A^T$, and we say that
5) A is **symmetric** if $A^T = A$.
6) A is **skew-symmetric** if $A^T = -A$.
7) A is **orthogonal** if A is invertible and $A^T = A^{-1}$. □

In the finite dimensional case, we have seen that

$$[\tau^*]_{\mathcal{O}} = [\tau]_{\mathcal{O}}^*$$

for any ordered orthonormal basis $\mathcal{O}$ of V, and so if τ is normal, then

$$[\tau]_{\mathcal{O}}[\tau]_{\mathcal{O}}^* = [\tau]_{\mathcal{O}}[\tau^*]_{\mathcal{O}} = [\tau\tau^*]_{\mathcal{O}}$$

$$= [\tau^*\tau]_{\mathcal{O}} = [\tau^*]_{\mathcal{O}}[\tau]_{\mathcal{O}} = [\tau]_{\mathcal{O}}^*[\tau]_{\mathcal{O}}$$

which implies that the matrix $[\tau]_{\mathcal{O}}$ of τ is normal. The converse holds as well. In fact, we can say that τ is normal (*resp.* self-adjoint, unitary) if and only if any matrix that represents τ, with respect to an ordered *orthonormal* basis $\mathcal{O}$, is normal (*resp.* Hermitian, unitary).

Let us now turn to a discussion of the three types of operators that we have just defined.

Self-Adjoint Operators

By definition, an operator τ is self-adjoint if and only if

$$\langle \tau(\mathbf{v}),\mathbf{w}\rangle = \langle \mathbf{v},\tau(\mathbf{w})\rangle$$

for all $\mathbf{v},\mathbf{w} \in V$. Here are some of the basic properties of these extremely important operators.

Theorem 10.4 Let V be an inner product space, and let $\sigma,\tau \in \mathcal{L}(V)$.
1) If σ and τ are self-adjoint, so is $\sigma + \tau$.
2) If τ is self-adjoint and r is *real*, then $r\tau$ is self-adjoint.
3) If σ and τ are self-adjoint, then $\sigma\tau$ is self-adjoint if and only if $\sigma\tau = \tau\sigma$.
4) If τ is self-adjoint and invertible, then so is τ^{-1}.

Proof. We prove only (3). To this end, observe that $(\sigma\tau)^* = \tau^*\sigma^*$, and so

$$(\sigma\tau)^* = \sigma\tau \;\Leftrightarrow\; \tau^*\sigma^* = \sigma\tau \;\Leftrightarrow\; \tau\sigma = \sigma\tau$$ ∎

Theorem 10.5 Let V be an inner product space.

1) If τ is self-adjoint, then $\langle\tau(\mathbf{v}),\mathbf{v}\rangle$ is real, for all $\mathbf{v}\in V$.
2) If V is *complex* and $\langle\tau(\mathbf{v}),\mathbf{v}\rangle$ is real for all $\mathbf{v}\in V$, then τ is self-adjoint.
3) If τ is self-adjoint and $\langle\tau(\mathbf{v}),\mathbf{v}\rangle=0$ for all $\mathbf{v}\in V$, then $\tau=0$ (*cf.* Theorem 9.6).
4) If τ is self-adjoint then $\tau^k(\mathbf{v})=\mathbf{0}$ for any $k>0$ implies that $\tau(\mathbf{v})=\mathbf{0}$.
5) If τ is self-adjoint, then all complex roots of the characteristic polynomial (and hence minimal polynomial) of τ are real.
6) If λ, μ are *distinct* eigenvalues of a self-adjoint operator τ, then $\mathcal{E}_\lambda\perp\mathcal{E}_\mu$.

Proof.

1) For part (1), we have

$$\langle\tau(\mathbf{v}),\mathbf{v}\rangle=\langle\mathbf{v},\tau(\mathbf{v})\rangle=\overline{\langle\tau(\mathbf{v}),\mathbf{v}\rangle}$$

and so $\langle\tau(\mathbf{v}),\mathbf{v}\rangle$ must be real.

2) To prove part (2), we have

$$\begin{aligned}\langle(\tau-\tau^*)(\mathbf{v}),\mathbf{v}\rangle&=\langle\tau(\mathbf{v}),\mathbf{v}\rangle-\langle\tau^*(\mathbf{v}),\mathbf{v}\rangle\\&=\langle\tau(\mathbf{v}),\mathbf{v}\rangle-\langle\mathbf{v},\tau(\mathbf{v})\rangle\\&=\langle\tau(\mathbf{v}),\mathbf{v}\rangle-\overline{\langle\tau(\mathbf{v}),\mathbf{v}\rangle}\\&=0\quad(\text{since }\langle\tau(\mathbf{v}),\mathbf{v}\rangle\text{ is real})\end{aligned}$$

Hence, according to Theorem 9.6, $\tau-\tau^*=0$, which shows that τ is self-adjoint.

3) As for part (3), Theorem 9.6 implies that this is true for the complex case, so we need only consider the real case, for which we have

$$\begin{aligned}0&=\langle\tau(\mathbf{x}+\mathbf{y}),\mathbf{x}+\mathbf{y}\rangle\\&=\langle\tau(\mathbf{x}),\mathbf{x}\rangle+\langle\tau(\mathbf{y}),\mathbf{y}\rangle+\langle\tau(\mathbf{x}),\mathbf{y}\rangle+\langle\tau(\mathbf{y}),\mathbf{x}\rangle\\&=\langle\tau(\mathbf{x}),\mathbf{y}\rangle+\langle\tau(\mathbf{y}),\mathbf{x}\rangle\\&=\langle\tau(\mathbf{x}),\mathbf{y}\rangle+\langle\mathbf{x},\tau(\mathbf{y})\rangle\\&=\langle\tau(\mathbf{x}),\mathbf{y}\rangle+\langle\tau(\mathbf{x}),\mathbf{y}\rangle\\&=2\langle\tau(\mathbf{x}),\mathbf{y}\rangle\end{aligned}$$

and so $\tau=0$.

4) If $\tau^k(\mathbf{v})=\mathbf{0}$ for all $\mathbf{v}\in V$, then $\tau^{2^m}(\mathbf{v})=\mathbf{0}$ for some m. Thus,

$$0=\langle\tau^{2^m}(\mathbf{v}),\mathbf{v}\rangle=\langle\tau^{2^{m-1}}\tau^{2^{m-1}}(\mathbf{v}),\mathbf{v}\rangle=\langle\tau^{2^{m-1}}(\mathbf{v}),\tau^{2^{m-1}}(\mathbf{v})\rangle$$

and so $\tau^{2^{m-1}} = 0$. Repeating this process, we eventually get $\tau = 0$.

5) Suppose first that V is a complex vector space, and that λ is a root of $C_\tau(x)$. Then $\tau(\mathbf{v}) = \lambda\mathbf{v}$, for some $\mathbf{v} \neq \mathbf{0}$ and we have
$$\langle\tau(\mathbf{v}),\mathbf{v}\rangle = \langle\lambda\mathbf{v},\mathbf{v}\rangle = \lambda\langle\mathbf{v},\mathbf{v}\rangle$$
and
$$\langle\tau(\mathbf{v}),\mathbf{v}\rangle = \langle\mathbf{v},\tau(\mathbf{v})\rangle = \langle\mathbf{v},\lambda\mathbf{v}\rangle = \overline{\lambda}\langle\mathbf{v},\mathbf{v}\rangle$$
and so $\lambda = \overline{\lambda}$, which shows that λ is real.

If V is a real vector space, we must be careful, since if λ is a complex root of $C_\tau(x)$, it does not follow that $\tau(\mathbf{v}) = \lambda\mathbf{v}$ for some $\mathbf{0} \neq \mathbf{v} \in V$. However, we can proceed as follows. Let τ be represented by the matrix A, with respect to some ordered basis for V. Then $C_\tau(x) = C_A(x)$. Now, A is a real symmetric matrix, but can be thought of as a complex Hermitian matrix, that happens to have real entries. As such, it represents a self-adjoint linear operator on the *complex* space $\mathbb{C}^n$, and so, by what we have just shown, all (complex) roots of its characteristic polynomial are real. But the characteristic polynomial of A is the same, whether we think of A as a real or a complex matrix, and so the result follows.

6) Suppose that $\tau(\mathbf{v}) = \lambda\mathbf{v}$ and $\tau(\mathbf{w}) = \mu\mathbf{w}$, where $\mathbf{v},\mathbf{w} \neq \mathbf{0}$. Then
$$\lambda\langle\mathbf{v},\mathbf{w}\rangle = \langle\tau(\mathbf{v}),\mathbf{w}\rangle = \langle\mathbf{v},\tau(\mathbf{w})\rangle = \langle\mathbf{v},\mu\mathbf{w}\rangle = \mu\langle\mathbf{v},\mathbf{w}\rangle$$
and so $\lambda \neq \mu$ implies that $\langle\mathbf{v},\mathbf{w}\rangle = 0$. ∎

Of course, the fact that all complex eigenvalues of a self-adjoint operator are real implies that the minimal polynomial of τ factors into a product of *linear* factors.

Unitary Operators

We now turn to the basic properties of unitary operators. Note that τ is unitary if and only if
$$\langle\tau(\mathbf{v}),\mathbf{w}\rangle = \langle\mathbf{v},\tau^{-1}(\mathbf{w})\rangle$$
for all $\mathbf{v},\mathbf{w} \in V$.

Theorem 10.6 Let V be an inner product space, and let $\sigma,\tau \in \mathcal{L}(V)$.

1) If τ is unitary, so is τ^{-1}.
2) If σ,τ are unitary, so is $\sigma\tau$.
3) τ is unitary if and only it is a surjective isometry
4) If $dim(V) < \infty$, then τ is unitary if and only if τ takes an orthonormal basis to an orthonormal basis.

5) If τ is unitary, then the eigenvalues of τ have absolute value 1.

Proof. We leave the proofs of (1) and (2) to the reader.

3) For a bijective linear map τ, we have

$$\begin{aligned}
\tau \text{ is an isometry } &\Leftrightarrow \langle \tau(\mathbf{v}),\tau(\mathbf{w})\rangle = \langle \mathbf{v},\mathbf{w}\rangle \text{ for all } \mathbf{v},\mathbf{w} \in V \\
&\Leftrightarrow \langle \mathbf{v},\tau^*\tau(\mathbf{w})\rangle = \langle \mathbf{v},\mathbf{w}\rangle \text{ for all } \mathbf{v},\mathbf{w} \in V \\
&\Leftrightarrow \tau^*\tau(\mathbf{w}) = \mathbf{w} \text{ for all } \mathbf{w} \in V \\
&\Leftrightarrow \tau^*\tau = \iota \\
&\Leftrightarrow \tau^* = \tau^{-1} \\
&\Leftrightarrow \tau \text{ is unitary}
\end{aligned}$$

4) Suppose that τ is unitary, and that $\mathcal{O} = \{\mathbf{u}_1,\ldots,\mathbf{u}_n\}$ is an orthonormal basis for V. Then

$$\langle \tau(\mathbf{u}_i),\tau(\mathbf{u}_j)\rangle = \langle \mathbf{u}_i,\mathbf{u}_j\rangle = \delta_{i,j}$$

and so $\tau(\mathcal{O})$ is an orthonormal basis for V. Conversely, suppose that $\mathcal{O}$ and $\tau(\mathcal{O})$ are orthonormal bases for V. Then

$$\langle \tau(\mathbf{u}_i),\tau(\mathbf{u}_j)\rangle = \delta_{i,j} = \langle \mathbf{u}_i,\mathbf{u}_j\rangle$$

and so, if $\mathbf{v} = \Sigma r_i\mathbf{u}_i$ and $\mathbf{w} = \Sigma s_j\mathbf{u}_j$, we have

$$\begin{aligned}
\langle \tau(\mathbf{v}),\tau(\mathbf{w})\rangle &= \Big\langle \sum_i r_i\tau(\mathbf{u}_i), \sum_j s_j\tau(\mathbf{u}_j)\Big\rangle \\
&= \sum_{i,j} r_i\overline{s_j}\langle \tau(\mathbf{u}_i),\tau(\mathbf{u}_j)\rangle \\
&= \sum_{i,j} r_i\overline{s_j}\langle \mathbf{u}_i,\mathbf{u}_j\rangle \\
&= \Big\langle \sum_i r_i\mathbf{u}_i, \sum_j s_j\mathbf{u}_j\Big\rangle \\
&= \langle \mathbf{v},\mathbf{w}\rangle
\end{aligned}$$

and so τ is unitary.

5) If τ is unitary, and $\tau(\mathbf{v}) = \lambda\mathbf{v}$, then

$$\lambda\overline{\lambda}\langle \mathbf{v},\mathbf{v}\rangle = \langle \lambda\mathbf{v},\lambda\mathbf{v}\rangle = \langle \tau(\mathbf{v}),\tau(\mathbf{v})\rangle = \langle \mathbf{v},\mathbf{v}\rangle$$

and so $|\lambda|^2 = \lambda\overline{\lambda} = 1$, which implies that $|\lambda| = 1$. ∎

We also have the following theorem concerning unitary (and orthogonal) matrices.

Theorem 10.7 Let A be a matrix.

1) An $n \times n$ matrix A is unitary if and only if the columns of A form an orthonormal set in $\mathbb{C}^n$.
2) An $n \times n$ matrix A is unitary if and only if the rows of A form an orthonormal set in $\mathbb{C}^n$.
3) If A is unitary, then $|\det(A)| = 1$. In particular, if A is orthogonal, then $\det(A) = \pm 1$.

Proof. The matrix A is unitary if and only if $AA^* = I$, which is equivalent to saying that the rows of A are orthonormal. Similarly, A is unitary if and only if $A^*A = I$, which is equivalent to saying that the columns of A are orthonormal. As for part (3), we have

$$AA^* = I \Rightarrow \det(A)\det(A^*) = 1 \Rightarrow \det(A)\overline{\det(A)} = 1$$

from which the result follows. ∎

Normal Operators

Now let us discuss the properties of normal operators, including the key properties that we used to motivate the definition of normal operators.

Theorem 10.8 Let V be an inner product space, and let τ be a normal operator on V.

1) For any polynomial $p(x) \in F[x]$, the operator $p(\tau)$ is also normal.
2) $\tau(\mathbf{v}) = \mathbf{0} \Rightarrow \tau^*(\mathbf{v}) = \mathbf{0}$
3) $\tau^k(\mathbf{v}) = \mathbf{0}$ for any $k > 0 \Rightarrow \tau(\mathbf{v}) = \mathbf{0}$
4) For any $\lambda \in F$, $(\tau - \lambda)^k(\mathbf{v}) = \mathbf{0} \Rightarrow (\tau - \lambda)(\mathbf{v}) = \mathbf{0}$
5) If $\tau(\mathbf{v}) = \lambda\mathbf{v}$, then $\tau^*(\mathbf{v}) = \overline{\lambda}(\mathbf{v})$.
6) If λ, μ are *distinct* eigenvalues of τ, then $\mathcal{E}_\lambda \perp \mathcal{E}_\mu$.

Proof. We leave the proofs of parts (1) and (2) as exercises.

3) The operator $\sigma = \tau\tau^*$ is easily seen to be self adjoint, and since τ is normal, we have
$$\sigma^k(\mathbf{v}) = (\tau^*)^k(\tau)^k(\mathbf{v}) = \mathbf{0}$$
and so, according to Theorem 10.5, $\sigma(\mathbf{v}) = \mathbf{0}$, that is, $\tau\tau^*(\mathbf{v}) = \mathbf{0}$. But then
$$0 = \langle \tau\tau^*(\mathbf{v}), \mathbf{v} \rangle = \langle \tau(\mathbf{v}), \tau(\mathbf{v}) \rangle$$
and so $\tau(\mathbf{v}) = 0$.
4) Part (4) follows from parts (1) and (3).
5) Suppose that $\tau(\mathbf{v}) = \lambda\mathbf{v}$, where $\mathbf{v} \neq \mathbf{0}$. Then $(\tau - \lambda)(\mathbf{v}) = \mathbf{0}$. Hence, according to part (2), $(\tau - \lambda)^*(\mathbf{v}) = \mathbf{0}$. But $(\tau - \lambda)^* = \tau^* - \overline{\lambda}$, from which the result follows.

6) Suppose that $\tau(\mathbf{v}) = \lambda\mathbf{v}$ and $\tau(\mathbf{w}) = \mu\mathbf{w}$, where $\mathbf{v},\mathbf{w} \neq \mathbf{0}$. Then

$$\lambda\langle\mathbf{v},\mathbf{w}\rangle = \langle\tau(\mathbf{v}),\mathbf{w}\rangle = \langle\mathbf{v},\tau^*(\mathbf{w})\rangle = \langle\mathbf{v},\overline{\mu}\mathbf{w}\rangle = \mu\langle\mathbf{v},\mathbf{w}\rangle$$

and so $\lambda \neq \mu$ implies that $\langle\mathbf{v},\mathbf{w}\rangle = 0$. ∎

Orthogonal Diagonalization

We are now in a position to state one of the most beautiful theorems in linear algebra.

Theorem 10.9 Let V be a finite dimensional *complex* inner product space.

1) A linear operator τ on V is orthogonally diagonalizable if and only if it is normal.
2) Among all normal operators on V, we can characterize self-adjoint and unitary ones by their eigenvalues. To wit:
   a) A normal operator is self-adjoint if and only if all of its eigenvalues are real.
   b) A normal operator is unitary if and only if all of its eigenvalues have absolute value 1.

Proof. To prove part (1), let τ be a normal operator on a complex inner product space. If the prime factorization of the minimal polynomial of τ is

$$m_\tau(x) = (x - \lambda_1)^{e_1} \cdots (x - \lambda_k)^{e_k}$$

then the primary decomposition theorem gives

$$V = V_1 \oplus \cdots \oplus V_k$$

where, according to part (4) of Theorem 10.8,

$$V_i = \{\mathbf{v} \in V \mid (\tau - \lambda_i)^{e_i}(\mathbf{v}) = \mathbf{0}\} = \{\mathbf{v} \in V \mid (\tau - \lambda_i)(\mathbf{v}) = \mathbf{0}\} = \mathcal{E}_{\lambda_i}$$

Hence, the minimal polynomial of $\tau|_{V_i}$ is $x - \lambda_i$, and so $e_i = 1$ for all i. Thus

$$V = \mathcal{E}_{\lambda_1} \oplus \cdots \oplus \mathcal{E}_{\lambda_k}$$

Moreover, part (6) of Theorem 10.8 shows that V is the *orthogonal* direct sum

$$V = \mathcal{E}_{\lambda_1} \odot \cdots \odot \mathcal{E}_{\lambda_k}$$

Hence, we may construct an orthonormal basis of eigenvectors of τ by combining orthonormal bases for each eigenspace, and so τ is orthogonally diagonalizable.

For the converse, if τ is orthogonally diagonalizable, then there is an orthonormal basis $\mathcal{O} = \{\mathbf{u}_1,\ldots,\mathbf{u}_k\}$ for V consisting of eigenvectors of τ, say $\tau(\mathbf{u}_i) = \lambda_i\mathbf{u}_i$. Then

$$\langle \mathbf{u}_i, \tau^*(\mathbf{u}_j)\rangle = \langle \tau(\mathbf{u}_i), \mathbf{u}_j)\rangle = \lambda_i \langle \mathbf{u}_i, \mathbf{u}_j\rangle = \lambda_i \delta_{i,j} = \lambda_j \delta_{i,j} = \langle \mathbf{u}_i, \overline{\lambda}_j \mathbf{u}_j\rangle$$

and so $\tau^*(\mathbf{u}_j) = \overline{\lambda}_j \mathbf{u}_j$. Thus,

$$\tau\tau^*(\mathbf{u}_j) = \overline{\lambda}_j \tau(\mathbf{u}_j) = \overline{\lambda}_j \lambda_j \mathbf{u}_j = \lambda_j \overline{\lambda}_j \mathbf{u}_j = \lambda_j \tau^*(\mathbf{u}_j) = \tau^*\tau(\mathbf{u}_j)$$

and so τ is normal.

As for part (2a), we have already seen that a self-adjoint operator is normal, and has real eigenvalues. On the other hand, if τ is normal and has real eigenvalues, then for any eigenvector $\mathbf{u}_j$, associated to λ_j,

$$\tau^*(\mathbf{u}_j) = \overline{\lambda}_j \mathbf{u}_j = \lambda_j \mathbf{u}_j = \tau(\mathbf{u}_j)$$

and since there is a basis of eigenvectors, τ is self-adjoint. The proof of part (2b) is similar. ∎

Thus, on a finite dimensional *complex* inner product space, diagonal matrices form a set of canonical forms for the class of normal operators (at least up to order of the diagonal entries). For *real* inner product spaces, the situation is a bit different.

Theorem 10.10 A linear operator τ on a finite dimensional *real* inner product space is orthogonally diagonalizable if and only if it is self-adjoint.

Proof. Suppose that V is a real inner product space. If τ is self-adjoint, then according to part (5) of Theorem 10.5, the minimal polynomial of τ splits over $\mathbb{R}$. Moreover, parts (4) and (6) of Theorem 10.5 (with part (4) applied to the symmetric operator $\tau - \lambda$), show that V has an orthonormal basis of eigenvectors for τ. Hence, τ is orthogonally diagonalizable. (This is similar to the proof of Theorem 10.9.)

Here is a matrix proof of the converse. If τ is orthogonally diagonalizable, then there is an orthonormal basis $\mathcal{O}$ for V for which $[\tau]_{\mathcal{O}}$ is diagonal, and since $[\tau]_{\mathcal{O}}$ is real, it is symmetric. Hence,

$$[\tau^*]_{\mathcal{O}} = [\tau]_{\mathcal{O}}^* = [\tau]_{\mathcal{O}}^{\mathsf{T}} = [\tau]_{\mathcal{O}}$$

and so $\tau^* = \tau$. ∎

The matrix versions of Theorems 10.9 and 10.10 are as follows.

Theorem 10.11

1) Let A be a square complex matrix. Then there exists a unitary matrix U for which UAU^{-1} is diagonal if and only if A is normal.
2) Let A be a square real matrix. Then there exists an orthogonal matrix O for which OAO^{-1} is diagonal if and only if A is symmetric. ∎

We must work a little harder to find a canonical form for unitary operators over the real field. The problem is that the minimal polynomial $m_\tau(x)$ of a real unitary operator τ may not split over $\mathbb{R}$.

However, we can proceed as follows. If τ is real unitary, then $\sigma = \tau + \tau^* = \tau + \tau^{-1}$ is self-adjoint, and has a complete set of real eigenvalues, so we may decompose V as in the proof of Theorem 10.9,

$$V = \mathcal{E}_{\lambda_1} \oplus \cdots \oplus \mathcal{E}_{\lambda_k}$$

where

$$\mathcal{E}_{\lambda_i} = \{\mathbf{v} \in V \mid (\tau + \tau^{-1} - \lambda_i)(\mathbf{v}) = \mathbf{0}\}$$

or, multiplying by τ,

$$\mathcal{E}_{\lambda_i} = \{\mathbf{v} \in V \mid (\tau^2 - \lambda_i \tau + 1)(\mathbf{v}) = \mathbf{0}\}$$

If $\lambda_i = 2$, then since τ is normal, we have

$$\mathcal{E}_2 = \{\mathbf{v} \in V \mid (\tau - 1)^2(\mathbf{v}) = \mathbf{0}\} = \{\mathbf{v} \in V \mid (\tau - 1)(\mathbf{v}) = \mathbf{0}\}$$

and if $\lambda_i = -2$,

$$\mathcal{E}_{-2} = \{\mathbf{v} \in V \mid (\tau + 1)^2(\mathbf{v}) = \mathbf{0}\} = \{\mathbf{v} \in V \mid (\tau + 1)(\mathbf{v}) = \mathbf{0}\}$$

Thus, on the eigenspaces $\mathcal{E}_2$ and $\mathcal{E}_{-2}$ (if indeed they exist), the operator τ is just multiplication by 1 or -1, respectively.

We may decompose each $\mathcal{E}_{\lambda_i}$, for $\lambda_i \neq \pm 2$, as follows. Take $\mathbf{v} \in \mathcal{E}_{\lambda_i}$, and consider $span\{\mathbf{v}, \tau(\mathbf{v})\}$. This subspace of $\mathcal{E}_{\lambda_i}$ is invariant, since $\tau(\tau(\mathbf{v})) = \tau^2(\mathbf{v}) = \lambda_i \tau(\mathbf{v}) - \mathbf{v}$. Thus, we can write

$$\mathcal{E}_{\lambda_i} = span\{\mathbf{v}, \tau(\mathbf{v})\} \oplus span\{\mathbf{v}, \tau(\mathbf{v})\}^{\perp}$$

Continuing in this way, we can write each $\mathcal{E}_{\lambda_i}$ as the orthogonal direct sum of two-dimensional subspaces on which τ is real unitary. This gives

$$V = \mathcal{E}_2 \oplus \mathcal{E}_{-2} \oplus \mathcal{D}_1 \oplus \cdots \oplus \mathcal{D}_m$$

where $dim(\mathcal{D}_i) = 2$ and each summand is invariant under τ.

Hence, we need only determine the matrix of a real unitary operator τ on a two-dimensional space $\mathcal{D}$. The matrix of τ with respect to any orthonormal basis for $\mathcal{D}$ is orthogonal, and so if

$$[\tau] = \begin{bmatrix} a & b \\ c & d \end{bmatrix}$$

then it follows that

$$\begin{aligned} a^2 + b^2 &= 1 \\ c^2 + d^2 &= 1 \\ ac + bd &= 0 \end{aligned}$$

Moreover, since $\det([\tau])$ is the constant term of the minimal polynomial $m_\tau(x) = \tau^2 - \lambda_i\tau + 1 = 0$, we have $\det(\tau) = 1$, that is,

$$ad - bc = 1$$

Solving these equations gives $d = a$ and $c = -b$, and so

$$[\tau] = \begin{bmatrix} a & b \\ -b & a \end{bmatrix}$$

Since (a,b) is a unit vector in $\mathbb{R}^2$, we can write $(a,b) = (\cos\theta, \sin\theta)$, for some real θ, and so

$$[\tau] = \begin{bmatrix} \cos\theta & \sin\theta \\ -\sin\theta & \cos\theta \end{bmatrix}$$

Thus, we arrive at the following result.

Theorem 10.12 Let τ be a unitary operator on a finite dimensional *real* inner product space V. Then there is an orthonormal basis for V for which the matrix of τ has the block form

$$\begin{bmatrix} 1 & & & & & & & \\ & \ddots & & & & & & \\ & & 1 & & & & & \\ & & & -1 & & & & \\ & & & & \ddots & & & \\ & & & & & -1 & & \\ & & & & & & \begin{bmatrix} \cos\theta_1 & \sin\theta_1 \\ -\sin\theta_1 & \cos\theta_1 \end{bmatrix} & & \\ & & & & & & & \ddots & \\ & & & & & & & & \begin{bmatrix} \cos\theta_k & \sin\theta_k \\ -\sin\theta_k & \cos\theta_k \end{bmatrix} \end{bmatrix}_{block}$$

∎

Orthogonal Projections

We now wish to characterize orthogonal diagonalizability in terms of projection operators.

Definition Let $V = S \odot S^{\perp}$. The projection map $\rho{:}V \to S$ on S along $S^{\perp}$ is called **orthogonal projection** onto S. Put another way, a projection map ρ is an orthogonal projection if $V = im(\rho) \odot ker(\rho)$. □

Thus, orthogonal projection is just a special type of projection operator, where $ker(\rho) = im(\rho)^{\perp}$. Note that some care must be taken to avoid confusion between the term orthogonal projection and the concept of projections that are orthogonal to each other, that is, for which $\rho\sigma = \sigma\rho = 0$.

We saw in Chapter 8 that an operator ρ is a projection operator if and only if it is idempotent. Here is the analogous characterization of orthogonal projections.

Theorem 10.13 A linear operator $\rho \in \mathcal{L}(V)$ is an orthogonal projection if and only if it is idempotent and self-adjoint.

Proof. Suppose that ρ is idempotent and self-adjoint. Then ρ is projection on $im(\rho)$ along $ker(\rho)$, and $V = im(\rho) \oplus ker(\rho)$. Furthermore, if $\mathbf{x} \in ker(\rho)$, we have

$$\langle \rho(\mathbf{v}), \mathbf{x} \rangle = \langle \mathbf{v}, \rho(\mathbf{x}) \rangle = 0$$

and so $im(\rho) \perp ker(\rho)$. Hence, $V = im(\rho) \odot ker(\rho)$, which shows that ρ is orthogonal projection.

For the converse, suppose that ρ is orthogonal projection. Then ρ is idempotent, and we need only show that ρ is self-adjoint. Since ρ is orthogonal projection, we have $V = im(\rho) \odot ker(\rho)$. But if $\mathbf{v} \in im(\rho)$, then $\mathbf{v} = \rho(\mathbf{w})$ and so

$$\rho(\mathbf{v}) = \rho(\rho(\mathbf{w})) = \rho^2(\mathbf{w}) = \rho(\mathbf{w}) = \mathbf{v}$$

Hence all nonzero vectors in $im(\rho)$ are eigenvectors associated with the eigenvalue 1. Moreover, if $\mathbf{x} \in ker(\rho)$, then

$$\rho(\mathbf{x}) = \mathbf{0} = 0\mathbf{x}$$

and so all nonzero vectors in $ker(\rho)$ are eigenvectors associated with the eigenvalue 0. Therefore, we can find an orthonormal basis for V that consists entirely of eigenvectors for ρ, which means that ρ is normal. Finally, since the eigenvalues of ρ are real, ρ must be self-adjoint. ∎

Note that for an orthogonal projection ρ, we have

$$\langle \mathbf{v},\rho(\mathbf{v})\rangle = \langle \mathbf{v},\rho^2(\mathbf{v})\rangle = \langle \rho(\mathbf{v}),\rho(\mathbf{v})\rangle$$

The following theorem gives another characterization of orthogonal projections.

Theorem 10.14 A linear operator $\rho \in \mathcal{L}(V)$ is an orthogonal projection if and only if it is idempotent and $\| \rho(\mathbf{v}) \| \leq \| \mathbf{v} \|$ for all $\mathbf{v} \in V$.

Proof. We leave proof of the necessity as an exercise. Suppose that ρ is idempotent and that $\| \rho(\mathbf{v}) \| \leq \| \mathbf{v} \|$. We want to show that $V = ker(\rho) \oplus im(\rho)$, which can be done, according to Theorem 9.13, by showing that $im(\rho) \subset ker(\rho)^{\perp}$. Now, the key to this is the fact that $V = ker(\rho) \oplus ker(\rho)^{\perp}$, which holds for $dim(ker(\rho)) < \infty$ by the projection theorem. However, we will see in Chapter 13 that it is also true in general.

Proceeding under this assumption then, for any $\mathbf{w} \in im(\rho)$, we have $\mathbf{w} = \mathbf{x} + \mathbf{y}$, where $\mathbf{x} \in ker(\rho)$ and $\mathbf{y} \in ker(\rho)^{\perp}$, and since ρ is idempotent,

$$\mathbf{w} = \rho(\mathbf{w}) = \rho(\mathbf{x}) + \rho(\mathbf{y}) = \rho(\mathbf{y})$$

and so

$$\| \mathbf{x} \|^2 + \| \mathbf{y} \|^2 = \| \mathbf{w} \|^2 = \| \rho(\mathbf{y}) \|^2 \leq \| \mathbf{y} \|$$

which implies that $\| \mathbf{x} \| = 0$, and hence that $\mathbf{x} = \mathbf{0}$. Thus, $\mathbf{w} = \mathbf{y} \in ker(\rho)^{\perp}$, and so $im(\rho) \subset ker(\rho)^{\perp}$, as desired. ∎

The next three theorems gives some additional properties of orthogonal projections.

Theorem 10.15

1) If ρ and σ are both orthogonal projections, then $\rho\sigma = 0$ implies $\sigma\rho = 0$.
2) Two orthogonal projections ρ and σ are orthogonal to each other if and only if $im(\rho) \perp im(\sigma)$. ∎

Theorem 10.16 Let V be a vector space over a field of characteristic $\neq 2$.

1) Let ρ and σ both be orthogonal projections. Then $\rho + \sigma$ is an orthogonal projection if and only if $\rho \perp \sigma$, in which case $\rho + \sigma$ is projection on $im(\rho) \oplus im(\sigma)$ along $ker(\rho) \cap ker(\sigma)$.
2) Let $\rho_1, \ldots, \rho_k$ be orthogonal projections. Then $\rho = \rho_1 + \cdots + \rho_k$ is an orthogonal projection if and only if $\rho_i \perp \rho_j$ for all $i \neq j$.
3) Let ρ and σ both be orthogonal projections. Then $\rho - \sigma$ is an orthogonal projection if and only if

$$\rho\sigma = \sigma\rho = \sigma$$

in which case $\rho - \sigma$ is projection on $\mathrm{im}(\rho) \cap \mathit{ker}(\sigma)$ along $\mathit{ker}(\rho) \oplus \mathit{im}(\sigma)$.

4) Let ρ and σ both be orthogonal projections. If $\rho\sigma = \sigma\rho$ then $\rho\sigma$ is an orthogonal projection. In this case, $\rho\sigma$ is projection on $\mathit{im}(\rho) \cap \mathit{im}(\sigma)$ along $\mathit{ker}(\rho) \oplus \mathit{ker}(\sigma)$.

Proof. We prove only part (2). If the ρ_i's are orthogonal projections, and if $\rho_i \perp \rho_j$ for all $i \neq j$, then $\rho_i\rho_j = 0$ for all $i \neq j$, and so it is straightforward to check that $\rho^2 = \rho$ and that $\rho^* = \rho$. Hence, ρ is an orthogonal projection. Conversely, suppose that ρ is an orthogonal projection, and that $\mathbf{x} \in \mathit{im}(\rho_i)$. Then $\rho_i(\mathbf{x}) = \mathbf{x}$, and so

$$\|\mathbf{x}\|^2 \geq \|\rho(\mathbf{x})\|^2 = \langle \rho(\mathbf{x}),\rho(\mathbf{x})\rangle = \langle \rho(\mathbf{x}),\mathbf{x}\rangle$$

$$= \sum_j \langle \rho_j(\mathbf{x}),\mathbf{x}\rangle = \sum_j \|\rho_j(\mathbf{x})\|^2 \geq \|\rho_i(\mathbf{x})\|^2 = \|\mathbf{x}\|^2$$

which implies that $\rho_j(\mathbf{x}) = \mathbf{0}$ for $j \neq i$. In other words,

$$\mathit{im}(\rho_i) \subset \mathit{ker}(\rho_j) = \mathit{im}(\rho_j)^\perp$$

Therefore,

$$0 = \langle \rho_j(\mathbf{v}),\rho_i(\mathbf{w})\rangle = \langle \rho_i\rho_j(\mathbf{v}),\mathbf{w}\rangle$$

for all $\mathbf{v},\mathbf{w} \in V$, which shows that $\rho_i\rho_j = 0$, that is, $\rho_i \perp \rho_j$. ∎

Theorem 10.17 The following statements are equivalent for orthogonal projections ρ and σ.

1) $\langle (\rho - \sigma)(\mathbf{v}),\mathbf{v}\rangle \geq 0$ for all $\mathbf{v} \in V$
2) $\|\sigma(\mathbf{v})\| \leq \|\rho(\mathbf{v})\|$ for all $\mathbf{v} \in V$
3) $\mathit{im}(\sigma) \subset \mathit{im}(\rho)$
4) $\rho\sigma = \sigma$
5) $\sigma\rho = \sigma$

If any (and hence all) of these conditions obtain, we say that σ is **less than or equal to** ρ, and write $\sigma \leq \rho$.

Proof. Suppose that (1) holds. Then

$$0 \leq \langle (\rho - \sigma)(\mathbf{v}),\mathbf{v}\rangle = \langle \rho(\mathbf{v}),\mathbf{v}\rangle - \langle \sigma(\mathbf{v}),\mathbf{v}\rangle$$

$$= \langle \rho(\mathbf{v}),\rho(\mathbf{v})\rangle - \langle \sigma(\mathbf{v}),\sigma(\mathbf{v})\rangle = \|\rho(\mathbf{v})\|^2 - \|\sigma(\mathbf{v})\|^2$$

from which (2) follows. Next, suppose that (2) holds. Then for any $\mathbf{v} \in \mathit{im}(\sigma)$, we have $\mathbf{v} = \mathbf{x} + \mathbf{y}$, where $\mathbf{x} \in \mathit{im}(\rho) \perp \mathbf{y} \in \mathit{ker}(\rho)$. Then,

$$\|\mathbf{x}\|^2 + \|\mathbf{y}\|^2 = \|\mathbf{v}\|^2 = \|\sigma(\mathbf{v})\|^2 \leq \|\rho(\mathbf{v})\|^2 = \|\mathbf{x}\|^2$$

and so $\mathbf{y} = \mathbf{0}$, that is, $\mathbf{v} \in \mathit{im}(\rho)$. This proves (3). Now suppose that (3) holds. Then since $\sigma(\mathbf{v}) \in \mathit{im}(\sigma) \subset \mathit{im}(\rho)$ for any $\mathbf{v} \in V$, we have

$$\rho(\sigma(\mathbf{v})) = \sigma(\mathbf{v})$$

and so $\rho\sigma = \sigma$. Hence, (4) holds. If (4) holds, then $\sigma\rho = \sigma^*\rho^* = (\rho\sigma)^* = \sigma^* = \sigma$, and so (5) holds. Finally, suppose that (5) holds. Then so does (4), and so $\rho - \sigma$ is an orthogonal projection, from which it follows that

$$\langle(\rho-\sigma)(\mathbf{v}),\mathbf{v}\rangle = \langle(\rho-\sigma)(\mathbf{v}),(\rho-\sigma)(\mathbf{v})\rangle \geq 0$$

and so (1) holds. ∎

Orthogonal Resolutions of the Identity

Definition If $\rho_1,\ldots,\rho_k$ are *orthogonal* projections for which

$$\rho_1 + \cdots + \rho_k = \iota \tag{10.2}$$

is a resolution of the identity, then we refer to (10.2) as an **orthogonal resolution of the identity.** □

The following theorem displays a correspondence between orthogonal direct sum decompositions of V and orthogonal resolutions of the identity. It should be compared to Theorem 8.17.

Theorem 10.18

1) If $\rho_1 + \cdots + \rho_k = \iota$ is an *orthogonal* resolution of the identity, then
$$V = im(\rho_1) \odot \cdots \odot im(\rho_k)$$
2) Conversely, if $V = S_1 \odot \cdots \odot S_k$ and ρ_i is projection on S_i along $S_1 \odot \cdots \odot \widehat{S_i} \odot \cdots \odot S_k$, where the hat ˆ means that the corresponding term is missing from the direct sum. Then
$$\rho_1 + \cdots + \rho_k = \iota$$
is an orthogonal resolution of the identity.

Proof. To prove (1) suppose that $\rho_1 + \cdots + \rho_k = \iota$ is an orthogonal resolution of the identity. According to Theorem 8.17, we have

$$V = im(\rho_1) \oplus \cdots \oplus im(\rho_k)$$

However, since the ρ_i's are orthogonal, they are self-adjoint, and so for $i \neq j$,

$$\langle\rho_i(\mathbf{v}),\rho_j(\mathbf{w})\rangle = \langle\mathbf{v},\rho_i\rho_j(\mathbf{w})\rangle = 0$$

which shows that

$$V = im(\rho_1) \odot \cdots \odot im(\rho_k)$$

For the converse, we know from Theorem 8.17 that $\rho_1 + \cdots + \rho_k = \iota$ is a resolution of the identity, and we need only show that each ρ_i is an orthogonal projection. But this follows from the fact that

$$im(\rho_i) = S_i \perp (S_1 \oplus \cdots \oplus \widehat{S}_i \oplus \cdots \oplus S_k) = ker(\rho_i)$$ ∎

The Spectral Theorem

We can now characterize the orthogonally diagonalizable operators on a finite dimensional complex inner product space.

Theorem 10.19 (The spectral theorem for normal operators) Let $\tau \in \mathcal{L}(V)$, where V is a finite dimensional *complex* inner product space. The following statements are equivalent.

1) τ is orthogonally diagonalizable, that is,

$$V = \mathcal{E}_{\lambda_1} \oplus \cdots \oplus \mathcal{E}_{\lambda_k}$$

2) τ is normal, that is,

$$\tau\tau^* = \tau^*\tau$$

3) τ has the **orthogonal spectral resolution**

$$\tau = \lambda_1\rho_1 + \cdots + \lambda_k\rho_k \tag{10.3}$$

where $\lambda_i \in \mathbb{C}$ and where $\rho_1 + \cdots + \rho_k = \iota$ is an orthogonal resolution of the identity.

Moreover, if τ has the form (10.3), where the λ_i's are distinct and the ρ_i's are nonzero, then the λ_i's are the eigenvalues of τ and $im(\rho_i)$ is the eigenspace of τ associated with λ_i.

Proof. We have seen (Theorem 10.9) that (1) and (2) are equivalent. Suppose that τ is orthogonally diagonalizable. We know from Theorem 8.18 that (10.3) holds for some resolution of the identity, and we need only observe that since

$$V = \mathcal{E}_{\lambda_1} \oplus \cdots \oplus \mathcal{E}_{\lambda_k}$$

this is an orthogonal resolution. Hence, (3) holds.

Conversely, if (10.3) holds, we have

$$V = im(\rho_1) \oplus \cdots \oplus im(\rho_k)$$

But Theorem 8.18 implies that $im(\rho_i) = \mathcal{E}_{\lambda_i}$, and so τ is orthogonally diagonalizable. ∎

In the real case, we have the following.

Theorem 10.20 (The spectral theorem for self-adjoint operators) Let $\tau \in \mathcal{L}(V)$, where V is a finite dimensional *real* inner product space.

The following statements are equivalent.

1) τ is orthogonally diagonalizable, that is,
$$V = \mathcal{E}_{\lambda_1} \oplus \cdots \oplus \mathcal{E}_{\lambda_k}$$
2) τ is self-adjoint, that is,
$$\tau^* = \tau$$
3) τ has the **orthogonal spectral resolution**

(10.4)
$$\tau = \lambda_1 \rho_1 + \cdots + \lambda_k \rho_k$$
where $\lambda_i \in \mathbb{R}$ and $\rho_1 + \cdots + \rho_k = \iota$ is an orthogonal resolution of the identity.

Moreover, if τ has the form (10.4), where the λ_i's are distinct and the ρ_i's are nonzero, then the λ_i's are the eigenvalues of τ, and $im(\rho_i)$ is the eigenspace of τ associated with λ_i. ∎

Functional Calculus

Let us consider some applications of the spectral theorem. Recall that if V is a vector space over F, if $\tau \in \mathcal{L}(V)$, and if $p(x)$ is a polynomial over F, then the operator $p(\tau) \in \mathcal{L}(V)$ is well-defined. Now, suppose that V is a finite dimensional inner product space, and τ has spectral resolution $\tau = \lambda_1 \rho_1 + \cdots + \lambda_k \rho_k$. Then $\rho_i^m = \rho_i$ for $m \geq 1$, and $\rho_i \rho_j = 0$ for $i \neq j$. Thus,
$$\tau^n = (\lambda_1 \rho_1 + \cdots + \lambda_k \rho_k)^n = \lambda_1^n \rho_1 + \cdots + \lambda_k^n \rho_k$$
and, more generally, for any polynomial $p(x)$ over F,
$$p(\tau) = p(\lambda_1)\rho_1 + \cdots + p(\lambda_k)\rho_k$$

In fact, we can extend this further by *defining*, for any function $f:\{\lambda_1,\ldots,\lambda_k\} \to F$,
$$f(\tau) = f(\lambda_1)\rho_1 + \cdots + f(\lambda_k)\rho_k$$
Thus, we may define $\sqrt{\tau}$, τ^{-1}, e^τ, and so on. Notice, however, that since the spectral resolution of τ is a *finite* sum, we actually gain nothing (but convenience) by using functions other than polynomials. To see this, suppose that $f:\{\lambda_1,\ldots,\lambda_k\} \to F$ is any function, and let
$$f(\lambda_i) = \alpha_i$$
Then we can find a polynomial $p(x)$ for which $p(\lambda_i) = \alpha_i$ for $i = 1,\ldots,k$, and so
$$f(\tau) = f(\lambda_1)\rho_1 + \cdots + f(\lambda_k)\rho_k = p(\lambda_1)\rho_1 + \cdots + p(\lambda_k)\rho_k = p(\tau)$$
The study of the properties of functions of an operator τ is referred to as the *functional calculus* of τ.

According to the spectral theorem, if V is complex ($F = \mathbb{C}$) and τ is normal, then $f(\tau)$ is a normal operator whose eigenvalues are $f(\lambda_i)$. Similarly, if V is real ($F = \mathbb{R}$), and τ is self-adjoint, then $f(\tau)$ is self-adjoint, with eigenvalues $f(\lambda_i)$. Let us consider some special cases of this construction.

For each $j = 1,\ldots,k$, if $p_j(x)$ is a polynomial for which

$$p_j(\lambda_j) = 1, \quad p_j(\lambda_i) = 0 \quad \text{for } i \neq j$$

then

$$p_j(\tau) = \rho_j$$

and so we see that each projection ρ_j is a polynomial function of τ.

If τ is invertible, then $\lambda_i \neq 0$ for all i, and so we may let $f(x) = x^{-1}$, giving

$$\tau^{-1} = \lambda_1^{-1}\rho_1 + \cdots + \lambda_k^{-1}\rho_k$$

as can easily be verified by direct calculation.

If $f(\lambda_i) = \overline{\lambda}_i$ and if τ is normal, then each ρ_i is self-adjoint, and so

$$f(\tau) = \overline{\lambda}_1\rho_1 + \cdots + \overline{\lambda}_k\rho_k = \tau^*$$

The functional calculus can be applied to the study of the commutativity properties of operators. Here are two simple examples.

Theorem 10.21 Let τ have spectral resolution

$$\tau = \lambda_1\rho_1 + \cdots + \lambda_k\rho_k$$

Then an operator σ commutes with τ if and only if it commutes with each ρ_i.

Proof. If σ commutes with each ρ_i, then clearly σ commutes with τ. For the converse, we simply observe that ρ_i is a polynomial in τ, and since σ commutes with τ, it commutes with any polynomial in τ. ∎

Theorem 10.22 Let V be a finite dimensional complex inner product space, and let $\tau,\sigma \in \mathcal{L}(V)$ be normal operators. Then τ and σ commute if and only if they have the form $\tau = p(\theta)$, $\sigma = q(\theta)$, where p and q are polynomials, and $\theta = r(\tau,\sigma)$ is a polynomial in τ and σ.

Proof. If τ and σ are polynomials in θ, then they clearly commute. For the converse, suppose that $\tau\sigma = \sigma\tau$, and let

$$\tau = \lambda_1\rho_1 + \cdots + \lambda_k\rho_k$$

and

$$\sigma = \mu_1 v_1 + \cdots + \mu_m v_m$$

be the orthogonal spectral resolutions of τ and σ. Then according to Theorem 10.21, $\rho_i \nu_j = \nu_j \rho_i$. Now, let us choose any polynomial $r(x,y)$ with the property that $\alpha_{i,j} = r(\lambda_i, \mu_j)$ are distinct. Since each ρ_i and ν_j is self-adjoint, we may set $\theta = r(\tau,\sigma)$ and deduce that

$$\theta = r(\tau,\sigma) = \sum_{i,j} \alpha_{i,j} \rho_i \nu_j$$

We also choose $p(x)$ and $q(x)$ so that $p(\alpha_{i,j}) = \lambda_i$ for all j and $q(\alpha_{i,j}) = \mu_j$ for all i. Then

$$p(\theta) = \sum_{i,j} p(\alpha_{i,j}) \rho_i \nu_j = \sum_{i,j} \lambda_i \rho_i \nu_j = \Big(\sum_i \lambda_i \rho_i\Big)\Big(\sum_j \nu_j\Big) = \sum_i \lambda_i \rho_i = \tau$$

and similarly, $q(\theta) = \sigma$. ∎

Positive Operators

One of the most important cases of the functional calculus is when $f(x) = \sqrt{x}$. First, we need some definitions.

Definition A *self-adjoint* linear operator $\tau \in \mathcal{L}(V)$ is **nonnegative** if $\langle \tau(\mathbf{v}), \mathbf{v} \rangle \geq 0$ for all $\mathbf{v} \in V$ and **positive** if it is nonnegative and $\langle \tau(\mathbf{v}), \mathbf{v} \rangle > 0$ for $\mathbf{v} \neq \mathbf{0}$. □

Theorem 10.23 A self-adjoint operator τ on a finite dimensional inner product space is

1) nonnegative if and only if all of its eigenvalues are nonnegative
2) positive if and only if all of its eigenvalues are positive.

Proof. If $\langle \tau(\mathbf{v}), \mathbf{v} \rangle \geq 0$ and $\tau(\mathbf{v}) = \lambda \mathbf{v}$, then $0 \leq \langle \tau(\mathbf{v}), \mathbf{v} \rangle = \lambda \langle \mathbf{v}, \mathbf{v} \rangle$, and so $\lambda \geq 0$. Conversely, if all eigenvalues of τ are nonnegative, then we have

$$\tau = \lambda_1 \rho_1 + \cdots + \lambda_k \rho_k, \quad \lambda_i \geq 0$$

and since $\iota = \rho_1 + \cdots + \rho_k$,

$$\langle \tau(\mathbf{v}), \mathbf{v} \rangle = \sum_{i,j} \lambda_i \langle \rho_i(\mathbf{v}), \rho_j(\mathbf{v}) \rangle = \sum_i \lambda_i \| \rho_i(\mathbf{v}) \|^2 \geq 0$$

and so τ is nonnegative. Part (2) is proved similarly. ∎

If τ is a nonnegative operator, with spectral resolution

$$\tau = \lambda_1 \rho_1 + \cdots + \lambda_k \rho_k, \quad \lambda_i \geq 0$$

then we may take the **nonnegative square root** of τ,

$$\sqrt{\tau} = \sqrt{\lambda_1} \rho_1 + \cdots + \sqrt{\lambda_k} \rho_k$$

where $\sqrt{\lambda_i}$ is the nonnegative square root of λ_i.

It is clear that

$$(\sqrt{\tau})^2 = \tau$$

and it is not hard to see that $\sqrt{\tau}$ is the *only* nonnegative operator whose square is τ. In other words, every nonnegative operator has a *unique* nonnegative square root. Conversely, if τ has a nonnegative square root, that is, if $\tau = \sigma^2$, for some nonnegative operator σ, then τ is nonnegative. Hence, an operator τ is nonnegative if and only if it has a nonnegative square root.

Here is an application of square roots.

Theorem 10.24 If τ and σ are nonnegative operators, and $\tau\sigma = \sigma\tau$, then $\tau\sigma$ is nonnegative.

Proof. Since τ is a nonnegative operator, it has a nonnegative square root $\sqrt{\tau}$, which is a polynomial in τ, and similarly for σ. Therefore, since τ and σ commute, so do $\sqrt{\tau}$ and $\sqrt{\sigma}$. Hence,

$$(\sqrt{\tau}\sqrt{\sigma})^2 = (\sqrt{\tau})^2(\sqrt{\sigma})^2 = \tau\sigma$$

Since $\sqrt{\tau}$ and $\sqrt{\sigma}$ are self-adjoint and commute, their product is self-adjoint, and so $\tau\sigma$ is nonnegative. ∎

The Polar Decomposition of an Operator

It is well-known that any nonzero complex number z can be written in the *polar form* $z = re^{i\theta}$, where r is a positive number, and θ is real. We can do the same for any nonzero linear operator τ on a finite dimensional complex inner product space.

Theorem 10.25 Let τ be a nonzero linear operator on a finite dimensional complex inner product space V. Then there exists a unique positive operator ρ, and a unitary operator ν for which $\tau = \nu\rho$. Moreover, if τ is invertible, then ν is also unique.

Proof. Let us suppose for a moment that $\tau = \nu\rho$. Then $\tau^* = (\nu\rho)^* = \rho^*\nu^* = \rho\nu^{-1}$ and so

$$\tau^*\tau = \rho\nu^{-1}\nu\rho = \rho^2$$

Also, if $\mathbf{v} \in V$, then

$$\tau(\mathbf{v}) = \nu\rho(\mathbf{v})$$

These equations give us the clue as to how to define ρ and ν.

Let us *define* ρ to be the unique nonnegative square root of the nonnegative operator $\tau^*\tau$. Then

(10.5)
$$\| \rho(\mathbf{v}) \|^2 = \langle \rho(\mathbf{v}),\rho(\mathbf{v})\rangle = \langle \rho^2(\mathbf{v}),\mathbf{v}\rangle = \langle \tau^*\tau(\mathbf{v}),\mathbf{v}\rangle = \| \tau(\mathbf{v}) \|^2$$

Let us define ν on the image $im(\rho)$ by

(10.6)
$$\nu(\rho(\mathbf{v})) = \tau(\mathbf{v})$$

for all $\mathbf{v} \in V$. To see that this is well-defined, observe that (10.5) gives

$$\rho(\mathbf{v}) = \rho(\mathbf{w}) \Rightarrow \rho(\mathbf{v}-\mathbf{w}) = \mathbf{0} \Rightarrow \| \rho(\mathbf{v}-\mathbf{w}) \| = 0$$
$$\Rightarrow \| \tau(\mathbf{v}-\mathbf{w}) \| = 0 \Rightarrow \tau(\mathbf{v}) = \tau(\mathbf{w})$$

Moreover, ν is an isometry (on its domain $im(\rho)$), since (10.5) again gives

$$\| \nu(\rho(\mathbf{v})) \| = \| \tau(\mathbf{v}) \| = \| \rho(\mathbf{v}) \|$$

Since $\nu{:}im(\rho)\to im(\nu)$ is injective, we have

$$dim(im(\rho)) = dim(im(\nu))$$

and so

$$dim(im(\rho)^{\perp}) = dim(im(\nu)^{\perp})$$

which means that we may extend ν to a unitary map (perhaps in many ways) ν on V. Equation (10.6) then shows that $\tau = \nu\rho$.

As for the uniqueness, suppose that $\tau = \nu\rho = \nu'\rho'$. Then

$$\tau^*\tau = \rho^*\nu^*\nu\rho = \rho^2 \quad \text{and} \quad \tau^*\tau = \rho'^*\nu'^*\nu'\rho' = (\rho')^2$$

and so $\rho^2 = (\rho')^2$, and since ρ^2 has a unique nonnegative square root, we deduce that $\rho = \rho'$. Thus, ρ is unique. Finally, if τ is invertible, then (10.5) shows that ρ is also invertible. Hence, ρ is a bijection, and so (10.6) uniquely determines ν. ∎

Applying the previous theorem to the map τ^*, we get

$$\tau = (\tau^*)^* = (\nu\rho)^* = \rho\nu^{-1} = \rho\mu$$

We leave it as an exercise to show that any unitary operator μ has the form $\mu = e^{i\sigma}$, where σ is a self-adjoint operator. This gives the following corollary.

Corollary 10.26 **(Polar decomposition)** Let τ be a nonzero linear operator on a finite dimensional complex inner product space. Then there is a unique positive operator ρ and a self-adjoint operator σ for which τ has the **polar decomposition**

$$\tau = \rho e^{i\sigma}$$ ∎

Normal operators can be characterized using the polar decomposition.

Theorem 10.27 Let $\tau = \rho e^{i\sigma}$ be a polar decomposition of a nonzero linear operator τ. Then τ is normal if and only if $\rho\sigma = \sigma\rho$.

Proof. Since

$$\tau\tau^* = \rho e^{i\sigma} e^{-i\sigma} \rho = \rho^2$$

and

$$\tau^*\tau = e^{-i\sigma} \rho\rho e^{i\sigma} = e^{-i\sigma} \rho^2 e^{i\sigma}$$

we see that τ is normal if and only if

$$e^{-i\sigma} \rho^2 e^{i\sigma} = \rho^2$$

or equivalently,

$$\rho^2 e^{i\sigma} = e^{i\sigma} \rho^2 \tag{10.7}$$

Now, ρ is a function of ρ^2, and σ is a function of $e^{i\sigma}$, and so (10.7) holds if and only if $\rho\sigma = \sigma\rho$. ∎

EXERCISES

1. Prove that τ is self-adjoint (unitary) if and only if any matrix that represents τ, with respect to an ordered orthonormal basis $\mathcal{O}$, is Hermitian (unitary). (Substitute the correct terms when $F = \mathbb{R}$.)
2. Show that if τ is self-adjoint, then so is τ^n for any $n \in \mathbb{N}$.
3. Let $\tau \in \mathcal{L}(V)$, and let
$$\tau_1 = \frac{1}{2}(\tau + \tau^*) \quad \text{and} \quad \tau_2 = \frac{1}{2i}(\tau - \tau^*)$$
Show that τ_1 and τ_2 are self-adjoint, and that
$$\tau = \tau_1 + i\tau_2 \quad \text{and} \quad \tau^* = \tau_1 - i\tau_2$$
What can you say about the uniqueness of these representations of τ and τ^*?
4. Show that a nonzero self-adjoint operator cannot be nilpotent.
5. Prove that all of the roots of the characteristic polynomial of a skew-Hermitian matrix are pure imaginary.
6. Prove that if τ is unitary, then so is τ^{-1}.
7. Prove that if σ, τ are unitary, then so is $\sigma\tau$.
8. Prove that a normal operator is unitary if and only if all of its eigenvalues have absolute value 1.
9. Let τ be a unitary operator on a finite dimensional inner product space V. Show that if a subspace S of V is invariant under τ, then so is $S^\perp$.
10. Give an example of a normal operator that is neither self-adjoint nor unitary.
11. Prove that if $\|\tau(\mathbf{v})\| = \|\tau^*(\mathbf{v})\|$ for all $\mathbf{v} \in V$, where V is

complex, then τ is normal.

12. Show that if τ is a normal operator on a finite dimensional inner product space, then $\tau^* = p(\tau)$, for some polynomial $p(x) \in F[x]$.
13. Show that a linear operator τ on a finite dimensional inner product space V is normal if and only if whenever S is an invariant subspace under τ, so is $S^{\perp}$.
14. Let V be a finite dimensional inner product space, and let τ be a normal operator on V.
    a) Prove that if τ is idempotent, then it is also self-adjoint.
    b) Prove that if τ is nilpotent, then $\tau = 0$.
    c) Prove that if $\tau^2 = \tau^3$, then τ is idempotent.
15. Show that if τ is a normal operator on a finite dimensional complex inner product space, then the algebraic multiplicity is equal to the geometric multiplicity for all eigenvalues of τ.
16. Use the results of the previous exercise to show that if τ is normal, and if $\sigma\tau = \tau\sigma$, then $\sigma\tau^* = \tau^*\sigma$. In other words, τ^* commutes with all operators that commute with τ.
17. Recall that it is possible for two projections to have the property that $\sigma\rho$ is a projection, but $\rho\sigma$ is not. Show that this cannot happen if ρ and σ are both orthogonal projections.
18. Show that two orthogonal projections σ and ρ are orthogonal to each other if and only if $im(\sigma) \perp im(\rho)$.
19. Show that the spectral resolution of a normal operator is unique.
20. If ν is a unitary operator on a complex inner product space, show that there exists a self-adjoint operator σ for which $\nu = e^{i\sigma}$.
21. Show that, in the complex case, we need not specify that τ is self-adjoint in defining nonnegative operators.
22. Show that a nonnegative operator has a unique nonnegative square root.
23. Let α_i, β_i be complex numbers, for $i = 1, \ldots, k$. Construct a polynomial $p(x)$ for which $p(\alpha_i) = \beta_i$ for all i.
24. Prove that if τ has a square root, that is, if $\tau = \sigma^2$, for some nonnegative operator σ, then τ is nonnegative.
25. Prove that a self-adjoint operator on a finite dimensional inner product space is positive if and only if all of its eigenvalues are positive.
26. Prove that if $\sigma \leq \tau$ and if θ is a positive operator that commutes with both σ and τ, then $\sigma\theta \leq \tau\theta$.
27. Does every self-adjoint operator on a finite dimensional real inner product space have a square root?
28. Let τ be a liner operator on $\mathbb{C}^n$, and let $\lambda_1, \ldots, \lambda_n$ be the eigenvalues of τ, each one written a number of times equal to its algebraic multiplicity. Show that

$$\sum_i |\lambda_i|^2 \leq tr(\tau^*\tau)$$

where tr is the trace, defined in the exercises in Chapter 8. Show also that equality holds if and only if τ is normal.

Part 2

Topics

CHAPTER 11

Metric Vector Spaces

Contents: *Symmetric, Skew-symmetric and Alternate Forms. The Matrix of a Bilinear Form. Quadratic Forms. Linear Functionals. Orthogonality. Orthogonal Complements. Orthogonal Direct Sums. Quotient Spaces. Symplectic Geometry–Hyperbolic Planes. Orthogonal Geometry–Orthogonal Bases. The Structure of an Orthogonal Geometry. Isometries. Symmetries. Witt's Cancellation Theorem. Witt's Extension Theorem. Maximum Hyperbolic Subspaces. Exercises.*

Symmetric, Skew-Symmetric and Alternate Forms

In this chapter, we study vector spaces over arbitrary fields that have a *bilinear form* defined on them. As we will see, the study of such vector spaces has a very geometric flavor, and hence so does the terminology.

Unless otherwise mentioned, all vector spaces are assumed to be finite dimensional. The symbol F *denotes an arbitrary field, and* F_q *denotes a finite field of size* q.

Definition Let V be a vector space over F. A mapping $\langle,\rangle: V \times V \to F$ is called a **bilinear form** if it is a linear function of each coordinate, that is, if

$$\langle \alpha\mathbf{x} + \beta\mathbf{y}, \mathbf{z}\rangle = \alpha\langle \mathbf{x}, \mathbf{z}\rangle + \beta\langle \mathbf{y}, \mathbf{z}\rangle$$

and

$$\langle \mathbf{z}, \alpha\mathbf{x} + \beta\mathbf{y}\rangle = \alpha\langle \mathbf{z}, \mathbf{x}\rangle + \beta\langle \mathbf{z}, \mathbf{y}\rangle$$

A bilinear form is

1) **symmetric** if
$$\langle \mathbf{x},\mathbf{y}\rangle = \langle \mathbf{y},\mathbf{x}\rangle$$
for all $\mathbf{x}, \mathbf{y} \in V$.
2) **skew-symmetric** if
$$\langle \mathbf{x},\mathbf{y}\rangle = -\langle \mathbf{y},\mathbf{x}\rangle$$
for all $\mathbf{x},\mathbf{y} \in V$.
3) **alternate** if
$$\langle \mathbf{x},\mathbf{x}\rangle = 0$$
for all $\mathbf{x} \in V$. □

Definition A bilinear form that is either symmetric, skew-symmetric, or alternate is referred to as an **inner product**, and a pair $(V,\langle,\rangle)$, where V is a vector space and $\langle,\rangle$ is an inner product on V, is called a **metric vector space**. □

Notice that the real inner products discussed in Chapter 9 are inner products in the present sense and have the additional property of being positive definite. On the other hand, the complex inner products of Chapter 9, being sesquilinear, are *not* inner products in the present sense. Note also that metric vector spaces should not be confused with *metric spaces*, which we will study in the next chapter.

As is traditional, when the inner product is understood, we will use the phrase "let V be a metric vector space."

Definition Let V be a metric vector space over a field F. If $\langle,\rangle$ is symmetric, then V is called an **orthogonal geometry** over F, and if $\langle,\rangle$ is alternate, then V is called a **symplectic geometry** over F. □

Thus, a real inner product space is an orthogonal geometry, but a complex inner product space is *not* an orthogonal geometry.

As we will see, not all metric vector spaces behave as nicely as the real inner product spaces, and this necessitates the introduction of a new set of terminology to cover various types of behavior. Here is one example.

Definition A metric vector space is **nonsingular** (or **nondegenerate**) if
$$\langle \mathbf{x},\mathbf{v}\rangle = 0 \text{ for all } \mathbf{v} \in V \Rightarrow \mathbf{x} = \mathbf{0}$$
□

Example 11.1 **Minkowski space** M_4 is the four-dimensional nonsingular real orthogonal geometry $\mathbb{R}^4$ with inner product defined by
$$\langle \mathbf{e}_1,\mathbf{e}_1\rangle = \langle \mathbf{e}_2,\mathbf{e}_2\rangle = \langle \mathbf{e}_3,\mathbf{e}_3\rangle = 1$$

$$\langle \mathbf{e}_4,\mathbf{e}_4\rangle = -1, \text{ and } \langle \mathbf{e}_i,\mathbf{e}_j\rangle = 0 \text{ for } i \neq j$$

where $\mathbf{e}_1,\ldots,\mathbf{e}_4$ is the standard basis for $\mathbb{R}^4$. ∎

The concepts of being symmetric, skew-symmetric and alternate are not independent. However, their relationship depends on the characteristic of the base field F.

Theorem 11.1 Let V be a vector space over a field F.

1) If $\text{char}(F) = 2$, then a bilinear form on V is skew-symmetric if and only if it is symmetric. Furthermore, an alternate bilinear form is symmetric (and skew-symmetric).
2) If $\text{char}(F) \neq 2$, then a bilinear form on V is skew-symmetric if and only if it is alternate.

Proof. First, we observe that for any field, if $\langle,\rangle$ is alternate, then

$$0 = \langle \mathbf{x}+\mathbf{y},\mathbf{x}+\mathbf{y}\rangle = \langle \mathbf{x},\mathbf{x}\rangle + \langle \mathbf{x},\mathbf{y}\rangle + \langle \mathbf{y},\mathbf{x}\rangle + \langle \mathbf{y},\mathbf{y}\rangle = \langle \mathbf{x},\mathbf{y}\rangle + \langle \mathbf{y},\mathbf{x}\rangle$$

Thus,

$$\langle \mathbf{x},\mathbf{y}\rangle + \langle \mathbf{y},\mathbf{x}\rangle = 0$$

or

$$\langle \mathbf{x},\mathbf{y}\rangle = -\langle \mathbf{y},\mathbf{x}\rangle$$

which shows that $\langle,\rangle$ is skew-symmetric. Thus, alternate implies skew-symmetric.

Now, if $\text{char}(F) = 2$, then $-\alpha = \alpha$ for all $\alpha \in F$, and so the definitions of symmetric and skew-symmetric are equivalent. Suppose that $\text{char}(F) \neq 2$. Then if $\langle,\rangle$ is skew-symmetric, for any $\mathbf{x} \in V$, we have

$$\langle \mathbf{x},\mathbf{x}\rangle = -\langle \mathbf{x},\mathbf{x}\rangle$$

or

$$2\langle \mathbf{x},\mathbf{x}\rangle = 0$$

which implies that $\langle \mathbf{x},\mathbf{x}\rangle = 0$. Hence, $\langle,\rangle$ is alternate. ∎

Theorem 11.1 tells us that we do not need to consider skew-symmetric forms *per se*, since skew-symmetric is always equivalent to either symmetric or alternate.

Example 11.2 The standard inner product on $V(n,q)$, defined by

$$(x_1,\ldots,x_n)\cdot(y_1,\ldots,y_n) = x_1y_1 + \cdots + x_ny_n$$

is symmetric, but not alternate, since

$$(1,0,\ldots,0)\cdot(1,0,\ldots,0) = 1 \neq 0$$

□

The Matrix of a Bilinear Form

If $\mathcal{B} = (\mathbf{b}_1, \ldots, \mathbf{b}_n)$ is an ordered basis for a metric vector space V, then the form $\langle , \rangle$ is completely determined by the $n \times n$ matrix of values

$$M_{\mathcal{B}} = (a_{i,j}) = (\langle \mathbf{b}_i, \mathbf{b}_j \rangle)$$

which is referred to as the **matrix of the form** $\langle , \rangle$ with respect to the ordered basis $\mathcal{B}$.

Observe that if $\mathbf{x} = \Sigma x_i \mathbf{b}_i$ and $\mathbf{y} = \Sigma y_j \mathbf{b}_j$, then

$$\langle \mathbf{x}, \mathbf{y} \rangle = \sum_i \sum_j x_i y_j \langle \mathbf{b}_i, \mathbf{b}_j \rangle = \sum_i x_i \Big(\sum_j a_{i,j} y_j \Big) = [\mathbf{x}]_{\mathcal{B}}^{\mathsf{T}} \ M_{\mathcal{B}} [\mathbf{y}]_{\mathcal{B}}$$

where $[\mathbf{x}]_{\mathcal{B}}$ and $[\mathbf{y}]_{\mathcal{B}}$ are the coordinate matrices of $\mathbf{x}$ and $\mathbf{y}$, respectively.

Notice also that a form is symmetric if and only if the matrix $M_{\mathcal{B}} = (a_{i,j})$ of the form satisfies

$$a_{i,j} = a_{j,i}$$

for all $1 \le i,j \le n$, that is, if and only if $M_{\mathcal{B}}$ is a symmetric matrix. Similarly, a form is alternate if and only if the matrix $M_{\mathcal{B}} = (a_{i,j})$ of the form satisfies

$$a_{i,i} = 0, \qquad a_{i,j} = -a_{j,i} \ (i \neq j)$$

Such a matrix is referred to as **alternate.**

Now let us see how the matrix of a form behaves with respect to a change of basis. Let $\mathcal{C} = (\mathbf{c}_1, \ldots, \mathbf{c}_n)$ be an ordered basis for V. Recall from Chapter 2 that the change of basis matrix $M_{\mathcal{C},\mathcal{B}}$, whose ith column is $[\mathbf{c}_i]_{\mathcal{B}}$, satisfies

$$[\mathbf{v}]_{\mathcal{B}} = M_{\mathcal{C},\mathcal{B}} [\mathbf{v}]_{\mathcal{C}}$$

Hence,

$$\begin{aligned} \langle \mathbf{x}, \mathbf{y} \rangle &= [\mathbf{x}]_{\mathcal{B}}^{\mathsf{T}} \ M_{\mathcal{B}} [\mathbf{y}]_{\mathcal{B}} \\ &= ([\mathbf{x}]_{\mathcal{C}}^{\mathsf{T}} \ M_{\mathcal{C},\mathcal{B}}^{\mathsf{T}} \ ) M_{\mathcal{B}} (M_{\mathcal{C},\mathcal{B}} [\mathbf{y}]_{\mathcal{C}} \ ) \\ &= [\mathbf{x}]_{\mathcal{C}}^{\mathsf{T}} \ (M_{\mathcal{C},\mathcal{B}}^{\mathsf{T}} \ M_{\mathcal{B}} M_{\mathcal{C},\mathcal{B}}) [\mathbf{y}]_{\mathcal{C}} \end{aligned}$$

and so

$$M_{\mathcal{C}} = M_{\mathcal{C},\mathcal{B}}^{\mathsf{T}} \ M_{\mathcal{B}} M_{\mathcal{C},\mathcal{B}}$$

This prompts the following definition.

Definition Two matrices $A, B \in \mathcal{M}_n(F)$ are said to be **congruent** if there exists an invertible matrix P for which

$$A = PBP^{\mathsf{T}}$$

□

(Setting $Q = P^T$ gives $PBP^T = Q^TBQ$, and so it doesn't matter whether we use PBP^T or P^TBP in the preceding definition.) Let us summarize.

Theorem 11.2 If the matrix of a bilinear form on V with respect to an ordered basis $\mathfrak{B} = (\mathbf{b}_1,\ldots,\mathbf{b}_n)$ is

$$M_{\mathfrak{B}} = (\langle \mathbf{b}_i,\mathbf{b}_j\rangle)$$

then

$$\langle \mathbf{x},\mathbf{y}\rangle = [\mathbf{x}]_{\mathfrak{B}}^T\, M_{\mathfrak{B}}[\mathbf{y}]_{\mathfrak{B}}$$

Furthermore, if $\mathfrak{C} = (\mathbf{c}_1,\ldots,\mathbf{c}_n)$ is also an ordered basis for V, then we have

$$M_{\mathfrak{C}} = M_{\mathfrak{C},\mathfrak{B}}^T\, M_{\mathfrak{B}} M_{\mathfrak{C},\mathfrak{B}}$$

where $M_{\mathfrak{C},\mathfrak{B}}$ is the change of basis matrix from $\mathfrak{C}$ to $\mathfrak{B}$, whose ith column is $[\mathbf{c}_i]_{\mathfrak{B}}$. ∎

Thus, if two matrices represent the same bilinear form on V, they must be congruent. Conversely, congruent matrices represent the same bilinear form on V. For suppose that $B = M_{\mathfrak{B}}$ represents a bilinear form on V, with respect to the ordered basis $\mathfrak{B}$, and that

$$A = P^TBP$$

where P is nonsingular. We saw in Chapter 2 (see the discussion following Theorem 2.12) that there is an ordered basis $\mathfrak{C}$ for V with the property that

$$P = M_{\mathfrak{C},\mathfrak{B}}$$

and so

$$A = M_{\mathfrak{C},\mathfrak{B}}^T\, M_{\mathfrak{B}} M_{\mathfrak{C},\mathfrak{B}}$$

Thus, $A = M_{\mathfrak{C}}$ represents the same form with respect to $\mathfrak{C}$.

Theorem 11.3 Two matrices A and B represent the same bilinear forms on V if and only if they are congruent. ∎

In view of the fact that congruent matrices have the same rank, we may define the **rank** of a bilinear form to be the rank of any matrix that represents that form.

Note that a metric vector space V is nonsingular if and only if the matrix $M_{\mathfrak{B}}$ is nonsingular, for any ordered basis $\mathfrak{B}$.

If A and B are congruent matrices, then

$$\det(A) = \det(PAP^T) = \det(P)^2\det(B)$$

and so $\det(A)$ and $\det(B)$ differ by a square factor. The **discriminant** of a bilinear form is the set of all determinants of the matrices that represent the form under all choices of ordered bases. Thus, if $\det(A) = d$ for some matrix A representing the form, then the discriminant of the form is the set $\{r^2 d \mid 0 \neq r \in F\}$. While it is true that the discriminant often does not give us much information about the matrix of the form in question, we will see a case where the discriminant is a *complete* invariant for congruence of matrices.

Quadratic Forms

There is a close link between symmetric bilinear forms and another important type of function defined on a vector space.

Definition A **quadratic form** on a vector space V is a map $Q:V \to F$ with the following properties

1) $Q(r\mathbf{v}) = r^2 Q(\mathbf{v})$ for all $r \in F$, $\mathbf{v} \in V$
2) The map $\langle \mathbf{u},\mathbf{v} \rangle_Q = Q(\mathbf{u}+\mathbf{v}) - Q(\mathbf{u}) - Q(\mathbf{v})$ is a (symmetric) bilinear form. ▯

Every quadratic form Q defines a symmetric bilinear form, by (2). On the other hand, if $\mathrm{char}(F) \neq 2$, and if $\langle , \rangle$ is a symmetric bilinear form on V, then we can define a quadratic form Q by

$$Q(\mathbf{x}) = \tfrac{1}{2}\langle \mathbf{x},\mathbf{x} \rangle$$

We leave it to the reader to verify that this is indeed a quadratic form. Moreover, if Q is defined from a bilinear form in this way, then the bilinear form associated with Q is

$$\begin{aligned} \langle \mathbf{u},\mathbf{v} \rangle_Q &= Q(\mathbf{u}+\mathbf{v}) - Q(\mathbf{u}) - Q(\mathbf{v}) \\ &= \tfrac{1}{2}\langle \mathbf{u}+\mathbf{v},\mathbf{u}+\mathbf{v} \rangle - \tfrac{1}{2}\langle \mathbf{u},\mathbf{u} \rangle - \tfrac{1}{2}\langle \mathbf{v},\mathbf{v} \rangle \\ &= \tfrac{1}{2}\langle \mathbf{u},\mathbf{v} \rangle + \tfrac{1}{2}\langle \mathbf{v},\mathbf{u} \rangle = \langle \mathbf{u},\mathbf{v} \rangle \end{aligned}$$

which is the original bilinear form. In other words, the maps $\langle , \rangle \to Q$ and $Q \to \langle , \rangle_Q$ are inverses, and so there is a one-to-one correspondence between symmetric bilinear forms on V and quadratic forms on V.

Again assuming that $\mathrm{char}(F) \neq 2$, if $\mathcal{B} = (\mathbf{b}_1, \ldots, \mathbf{b}_n)$ is an ordered basis for an orthogonal geometry V, and if the matrix of the symmetric form on V is $M_{\mathcal{B}} = (a_{i,j})$, then for $\mathbf{x} = \Sigma x_i \mathbf{b}_i$,

$$Q(\mathbf{x}) = \tfrac{1}{2}\langle \mathbf{x},\mathbf{x} \rangle = \tfrac{1}{2}[\mathbf{x}]_{\mathcal{B}}^{\mathsf{T}}\, M_{\mathcal{B}}[\mathbf{x}]_{\mathcal{B}} = \sum_{i,j} \tfrac{1}{2}\, a_{i,j} x_i x_j$$

and so $Q(\mathbf{x}) = Q(x_1, \ldots, x_n)$ is a *homogeneous* polynomial of degree 2 in the coordinates x_i. (The term *form* means *homogeneous polynomial*— hence the term *quadratic form*.)

Linear Functionals

Let V be a metric vector space over F. The vector space of all linear functionals on V is known as the **algebraic dual space** of V and is denoted by V^*. Moreover, for finite dimensional vector spaces, we have $dim(V) = dim(V^*)$.

Now let $\mathbf{x} \in V$, and consider the map $\phi_{\mathbf{x}}:V \to F$ defined by

$$\phi_{\mathbf{x}}(\mathbf{v}) = \langle \mathbf{v},\mathbf{x} \rangle$$

which is easily seen to be a linear functional. Hence, we can define a function $\tau:V \to V^*$ by

$$\tau(\mathbf{x}) = \phi_{\mathbf{x}}$$

This function is easily seen to be linear, and its kernel is

$$\{\mathbf{x} \in V \mid \phi_{\mathbf{x}} = 0\} = \{\mathbf{x} \in V \mid \langle \mathbf{v},\mathbf{x} \rangle = 0 \text{ for all } \mathbf{v} \in V\}$$

Hence, if V is nonsingular, the kernel of τ is the zero subspace, and τ is injective. Moreover, since $dim(V) = \dim(V^*)$, we deduce that τ is surjective, and so it is an isomorphism from V onto V^*. This implies that *every* linear functional on V has the form $\phi_{\mathbf{x}}$, for some $\mathbf{x} \in V$. We have proved the Riesz representation theorem for nonsingular metric vector spaces.

Theorem 11.4 **(The Riesz representation theorem)** Let V be a *nonsingular* metric vector space, and let $f \in V^*$ be a linear functional on V. Then there exists a unique vector $\mathbf{x} \in V$ for which

$$f(\mathbf{v}) = \langle \mathbf{v},\mathbf{x} \rangle$$

for all $\mathbf{v} \in V$. ∎

Orthogonality

A vector $\mathbf{x}$ is **orthogonal** to a vector $\mathbf{y}$, written $\mathbf{x} \perp \mathbf{y}$, if $\langle \mathbf{x},\mathbf{y} \rangle = 0$. Any *nonzero* vector $\mathbf{x}$ that is orthogonal to itself is called a **null vector**, or an **isotropic vector.**

The following result explains why we restrict attention to symmetric or alternate forms (which includes skew-symmetric forms).

Theorem 11.5 Let $\langle , \rangle$ be a bilinear form on V. Then orthogonality is a symmetric relation, that is,

$$\mathbf{x} \perp \mathbf{y} \Leftrightarrow \mathbf{y} \perp \mathbf{x} \tag{11.1}$$

if and only if $\langle , \rangle$ is either symmetric or alternate. Thus, in these cases, we may use the phrase "$\mathbf{x}$ and $\mathbf{y}$ are orthogonal."

Proof. It is clear that (11.1) holds if $\langle , \rangle$ is symmetric. If $\langle , \rangle$ is alternate, then it is skew-symmetric, and so (11.1) also holds. For the converse, suppose that (11.1) holds. For $\mathbf{x}, \mathbf{y}$ and $\mathbf{z}$ in V, let

$$\mathbf{w} = \langle \mathbf{x},\mathbf{y}\rangle \mathbf{z} - \langle \mathbf{x},\mathbf{z}\rangle \mathbf{y}$$

Then $\mathbf{x} \perp \mathbf{w}$, and so by assumption, $\mathbf{w} \perp \mathbf{x}$. But this is equivalent to

$$\langle \mathbf{x},\mathbf{y}\rangle\langle \mathbf{z},\mathbf{x}\rangle - \langle \mathbf{x},\mathbf{z}\rangle\langle \mathbf{y},\mathbf{x}\rangle = 0 \tag{11.2}$$

Setting $\mathbf{y} = \mathbf{x}$ gives

$$\langle \mathbf{x},\mathbf{x}\rangle\Big(\langle \mathbf{z},\mathbf{x}\rangle - \langle \mathbf{x},\mathbf{z}\rangle\Big) = 0 \tag{11.3}$$

for all vectors $\mathbf{x}$ and $\mathbf{z}$ in V. Exchanging $\mathbf{x}$ and $\mathbf{z}$, and multiplying by -1, give

$$\langle \mathbf{z},\mathbf{z}\rangle\Big(\langle \mathbf{z},\mathbf{x}\rangle - \langle \mathbf{x},\mathbf{z}\rangle\Big) = 0$$

Thus, we deduce that, for any vectors $\mathbf{u}$ and $\mathbf{v}$ in V, if $\langle \mathbf{u},\mathbf{v}\rangle \neq \langle \mathbf{v},\mathbf{u}\rangle$, then $\mathbf{u}$ and $\mathbf{v}$ are null vectors. Equivalently, if $\mathbf{u}$ is nonnull, then $\langle \mathbf{u},\mathbf{v}\rangle = \langle \mathbf{v},\mathbf{u}\rangle$ for all $\mathbf{v} \in V$.

Now, suppose that $\langle , \rangle$ is not symmetric. Then there exists vectors $\mathbf{u}$ and $\mathbf{v}$ for which $\langle \mathbf{u},\mathbf{v}\rangle \neq \langle \mathbf{v},\mathbf{u}\rangle$. Hence $\langle \mathbf{u},\mathbf{u}\rangle = \langle \mathbf{v},\mathbf{v}\rangle = 0$. We wish to show that $\langle \mathbf{a},\mathbf{a}\rangle = 0$ for any $\mathbf{a} \in V$, which will show that $\langle , \rangle$ is alternate.

Since $\mathbf{u}$ is null,

$$\langle \mathbf{u}+\mathbf{a},\mathbf{u}+\mathbf{a}\rangle = \langle \mathbf{u},\mathbf{a}\rangle + \langle \mathbf{a},\mathbf{u}\rangle + \langle \mathbf{a},\mathbf{a}\rangle \tag{11.4}$$

Now, if $\mathbf{a}$ were nonnull, then $\langle \mathbf{a},\mathbf{x}\rangle = \langle \mathbf{x},\mathbf{a}\rangle$ for all $\mathbf{x} \in V$; in particular, $\langle \mathbf{a},\mathbf{u}\rangle = \langle \mathbf{u},\mathbf{a}\rangle$ and $\langle \mathbf{a},\mathbf{v}\rangle = \langle \mathbf{v},\mathbf{a}\rangle$. Furthermore, setting $\mathbf{y} = \mathbf{a}$, $\mathbf{x} = \mathbf{u}$, $\mathbf{z} = \mathbf{v}$ in (11.2) gives

$$\langle \mathbf{u},\mathbf{a}\rangle\langle \mathbf{v},\mathbf{u}\rangle - \langle \mathbf{u},\mathbf{v}\rangle\langle \mathbf{a},\mathbf{u}\rangle = 0$$

which is equivalent to

$$\langle \mathbf{u},\mathbf{a}\rangle\Big(\langle \mathbf{v},\mathbf{u}\rangle - \langle \mathbf{u},\mathbf{v}\rangle\Big) = 0$$

and since $\langle \mathbf{u},\mathbf{v}\rangle \neq \langle \mathbf{v},\mathbf{u}\rangle$, we must have

$$\langle \mathbf{a},\mathbf{u}\rangle = \langle \mathbf{u},\mathbf{a}\rangle = 0$$

Similarly,

$$\langle \mathbf{a},\mathbf{v}\rangle = \langle \mathbf{v},\mathbf{a}\rangle = 0$$

Hence, (11.4) becomes

$$\langle \mathbf{u}+\mathbf{a},\mathbf{u}+\mathbf{a}\rangle = \langle \mathbf{a},\mathbf{a}\rangle$$

But

$$\langle \mathbf{u}+\mathbf{a},\mathbf{v}\rangle = \langle \mathbf{u},\mathbf{v}\rangle \neq \langle \mathbf{v},\mathbf{u}\rangle = \langle \mathbf{v},\mathbf{u}+\mathbf{a}\rangle$$

and so $\mathbf{u}+\mathbf{a}$ is also null, showing that $\langle \mathbf{a},\mathbf{a}\rangle = 0$, which contradicts the assumption about $\mathbf{a}$. Hence, all $\mathbf{a} \in V$ are null, and $\langle , \rangle$ is alternate. ∎

Orthogonal Complements

If S is a subset of a metric vector space V, then S inherits the metric structure from V. With this structure, we refer to S as a **subspace** of V.

Definition Two subspaces S and T of a metric vector space V are **orthogonal**, denoted by $S \perp T$, if $\langle \mathbf{s},\mathbf{t}\rangle = 0$ for all $\mathbf{s} \in S$ and $\mathbf{t} \in T$. The **orthogonal complement** of S, denoted by $S^{\perp}$, is the set

$$S^{\perp} = \{\mathbf{v} \in V \mid \mathbf{v} \perp S\}$$ ▯

Definition If V is a metric vector space, then $V^{\perp}$ is called the **radical** of V, and denoted by $Rad(V)$. ▯

Thus, V is nonsingular if and only if $Rad(V) = \{\mathbf{0}\}$. Note that if S is a subspace of V, then the radical of S is $Rad(S) = S \cap S^{\perp}$.

It should be emphasized that the properties of orthogonality can be quite different for arbitrary base fields than for the familiar case of the real base field. For instance, in the case of real metric vector spaces, we have $S \cap S^{\perp} = \{\mathbf{0}\}$, whereas in the case of metric vector spaces over finite fields, for instance, we may even have $S = S^{\perp}$, as the next example shows.

Example 11.3 It is easy to see that the subspace

$$S = \{0000, 1100, 0011, 1111\}$$

of $V(4,2)$ has the property that $S = S^{\perp}$. Note that $V(4,2)$ is nonsingular, and yet the subspace S is quite singular. ▯

The previous example notwithstanding, we do have the following important result concerning dimensions.

Theorem 11.6 If S is a subspace of a nonsingular metric vector space V, then

$$dim(S) + dim(S^{\perp}) = dim(V)$$

Proof. For each $\mathbf{v} \in V$, let $\phi_{\mathbf{v}}$ be the linear functional in S^* defined by

$$\phi_{\mathbf{v}}: S \to F, \quad \phi_{\mathbf{v}}(\mathbf{u}) = \langle \mathbf{u},\mathbf{v}\rangle$$

We define a map $\tau: V \to S^*$ by

$$\tau(\mathbf{v}) = \phi_{\mathbf{v}}$$

This map is linear, and its kernel is

$$ker(\tau) = \{\mathbf{v} \in V \mid \phi_{\mathbf{v}} = 0\} = \{\mathbf{v} \in V \mid \langle \mathbf{u},\mathbf{v}\rangle = 0 \text{ for all } \mathbf{u} \in S\} = S^{\perp}$$

Moreover, by the Riesz representation theorem, the restriction $\tau|_S: S \to S^*$ is surjective, and so *a fortiori*, τ is surjective, that is,

$$im(\tau) = S$$

The theorem then follows from the fact that

$$dim(im(\tau)) + dim(ker(\tau)) = dim(V)$$ ∎

Theorem 11.7 If S is a subspace of a nonsingular metric vector space V, then
1) $S^{\perp\perp} = S$
2) $Rad(S) = S \cap S^{\perp} = Rad(S^{\perp})$ □

Let us summarize the terminology related to orthogonality in metric vector spaces. (Unfortunately, authors vary somewhat on their use of the term isotropic.)

Definition Let V be a metric vector space.
1) A nonzero $\mathbf{x} \in V$ is **null**, or **isotropic**, if $\langle \mathbf{x},\mathbf{x} \rangle = 0$.
2) The **radical** of V is $Rad(V) = V^{\perp}$.
3) V is **nonsingular**, or **nondegenerate**, if $V^{\perp} = \{\mathbf{0}\}$.
4) V is **null** if $\langle \mathbf{u},\mathbf{v} \rangle = 0$ for all $\mathbf{u},\mathbf{v} \in V$, that is, if $V^{\perp} = V$.
5) V is **isotropic** if V contains at least one isotropic vector.
6) V is **anisotropic** if V contains *no* isotropic vectors.
7) V is **totally isotropic** if all vectors in V are isotropic. □

Orthogonal Direct Sums

Definition Let V be a metric vector space. If S and T are subspaces of V with the property that $V = S \oplus T$ and $S \perp T$, then we say that V is the **orthogonal direct sum** of S and T and write $V = S \odot T$. □

In view of Example 11.3, it is reasonable to ask under what conditions on a subspace S is it true that $V = S \odot S^{\perp}$. The answer is given by the following theorem.

Theorem 11.8 Let S be a subspace of a nonsingular metric vector space V. The following statements are equivalent.
1) S is nonsingular
2) $S^{\perp}$ is nonsingular
3) $S \cap S^{\perp} = \{\mathbf{0}\}$
4) $V = S + S^{\perp}$
5) $V = S \odot S^{\perp}$

Proof. According to Theorem 11.7, statements 1, 2 and 3 are

equivalent. For any subspaces S and T of a vector space V,

$$dim(S+T) = dim(S) + dim(T) - dim(S \cap T)$$

and so

$$\begin{aligned} dim(S+S^{\perp}) &= dim(S) + dim(S^{\perp}) - dim(S \cap S^{\perp}) \\ &= dim(V) - \dim(S \cap S^{\perp}) \end{aligned}$$

which shows that 3 is equivalent to 4, and that 4 implies 5. Since 5 clearly implies 4, the proof is complete. ∎

Most of the important results that we have established so far require that the space be nonsingular. Fortunately, the following theorem says that we may restrict attention to such spaces, without loosing any important structure.

Theorem 11.9 Let V be a metric vector space. Then

$$V = Rad(V) \odot S$$

where $Rad(V)$ is null and S is nonsingular. ∎

Proof. Let S be a complement of $Rad(V)$, that is, $V = Rad(V) \oplus S$. Since all vectors are orthogonal to $Rad(V)$, we have $Rad(V) \perp S$, and so $V = Rad(V) \odot S$. Now, if $\mathbf{v} \in Rad(S)$, then $\mathbf{v} \perp S$, and so $\mathbf{v} \perp V$, which implies that $\mathbf{v} \in Rad(V) \cap S = \{\mathbf{0}\}$, that is, $\mathbf{v} = \mathbf{0}$. Hence, $Rad(S) = \{\mathbf{0}\}$, that is, S is nonsingular. ∎

Quotient Spaces

In general, if S is a subspace of V, the quotient space V/S does not inherit a metric structure from V. However, if $S = Rad(V) = V^{\perp}$, then $V/Rad(V)$ does inherit the metric structure of V as follows. Let

$$\langle \mathbf{u} + Rad(V), \mathbf{v} + Rad(V) \rangle = \langle \mathbf{u}, \mathbf{v} \rangle$$

To show that this inner product is well defined, we observe that if

$$\mathbf{u} + Rad(V) = \mathbf{u}' + Rad(V)$$

then $\mathbf{u} = \mathbf{u}' + \mathbf{r}$, where $\mathbf{r} \in Rad(V)$. Hence,

$$\begin{aligned} \langle \mathbf{u} + Rad(V), \mathbf{v} + Rad(V) \rangle &= \langle \mathbf{u}, \mathbf{v} \rangle \\ &= \langle \mathbf{u}' + \mathbf{r}, \mathbf{v} \rangle = \langle \mathbf{u}', \mathbf{v} \rangle = \langle \mathbf{u}' + Rad(V), \mathbf{v} + Rad(V) \rangle \end{aligned}$$

and similarly for the second component.

Symplectic Geometry – Hyperbolic Planes

Let us consider a nonsingular symplectic geometry V. Thus, by definition, every vector in V is null. Given $\mathbf{u} \in V$, there must exist a $\mathbf{v} \in V$ for which $\langle \mathbf{u},\mathbf{v} \rangle \neq 0$, since V is assumed nonsingular. Consider a two-dimensional subspace H with basis $\{\mathbf{u},\mathbf{v}\}$. Then

$$\langle \mathbf{u},\mathbf{u} \rangle = \langle \mathbf{v},\mathbf{v} \rangle = 0$$

and $\langle \mathbf{u},\mathbf{v} \rangle = a \neq 0$. Replacing $\mathbf{v}$ by $a^{-1}\mathbf{v}$, we can assume that

$$\langle \mathbf{u},\mathbf{v} \rangle = 1, \quad \langle \mathbf{v},\mathbf{u} \rangle = -1$$

The subspace H, thought of as a metric vector space, has matrix with respect to the basis $\{\mathbf{u},\mathbf{v}\}$

$$M_1 = \begin{bmatrix} 0 & 1 \\ -1 & 0 \end{bmatrix}$$

We pause for a definition.

Definition Let V be a metric vector space. If $\mathbf{u},\mathbf{v} \in V$ have the property that

$$\langle \mathbf{u},\mathbf{u} \rangle = \langle \mathbf{v},\mathbf{v} \rangle = 0, \quad \langle \mathbf{u},\mathbf{v} \rangle = 1$$

the ordered pair $(\mathbf{u},\mathbf{v})$ is called a **hyperbolic pair**, and the subspace $H = span\{\mathbf{u},\mathbf{v}\}$ is called a **hyperbolic plane.** Any space of the form $H_1 \oplus \cdots \oplus H_k$, where each H_i is a hyperbolic plane, is called a **hyperbolic space.** □

Note that in an orthogonal geometry, if $(\mathbf{u},\mathbf{v})$ is a hyperbolic pair, then $\langle \mathbf{v},\mathbf{u} \rangle = 1$, but in a symplectic space, $\langle \mathbf{v},\mathbf{u} \rangle = -1$.

Now let us return to the discussion at hand. Since H is nonsingular, we have $V = H \oplus H^{\perp}$, where $H^{\perp}$ is also nonsingular. Hence, we may repeat the preceding construction in $H^{\perp}$, to obtain an orthogonal decomposition of V of the form

$$V = H_1 \oplus H_2 \oplus \cdots \oplus H_k$$

where each H_i is a hyperbolic plane. This proves the following result.

Theorem 11.10 Any nonsingular symplectic geometry V is a hyperbolic space, that is,

$$V = H_1 \oplus H_2 \oplus \cdots \oplus H_k$$

where each H_i is a hyperbolic plane. Thus, there is a basis for V for which the matrix of the form is

$$M = \begin{bmatrix} 0 & 1 & & & & & \\ -1 & 0 & & & & & \\ & & 0 & 1 & & & \\ & & -1 & 0 & & & \\ & & & & \ddots & & \\ & & & & & 0 & 1 \\ & & & & & -1 & 0 \end{bmatrix}$$

In particular, the dimension of V is even. ▯

Corollary 11.11 Any symplectic geometry V has the form

$$V = H_1 \oplus H_2 \oplus \cdots \oplus H_k \oplus N$$

where each H_i is a hyperbolic space, and N is a null space. ▯

Orthogonal Geometry – Orthogonal Bases

The structure of orthogonal geometries is more closely tied to the characteristic of the base field than is the case for symplectic geometries.

Definition Let V be an orthogonal geometry. A basis $\mathcal{B} = \{u_1, \ldots, u_n\}$ for V is said to be **orthogonal** if $\langle u_i, u_j \rangle = 0$ for $i \neq j$. ▯

A basis $\mathcal{B}$ for V is orthogonal if and only if the matrix $M_{\mathcal{B}}$ of the form is diagonal. It happens that any orthogonal geometry has an orthogonal basis, provided that in case $\text{char}(F) = 2$, we exclude the case where V is both orthogonal and symplectic, since no nonnull symplectic geometry can have an orthogonal basis. (The matrix with respect to such a basis would have 0s off the diagonal, by orthogonality of the basis and 0s on the diagonal, by virtue of V being symplectic.)

Clearly, we may exclude from consideration the case where V is null, since in this case, all bases are orthogonal.

Let us consider first the case where V is nonsingular, orthogonal, and $\text{char}(F) \neq 2$. Let $u \in V$ have the property that $\langle u, u \rangle \neq 0$. Such a vector must exist, for if not, then V would be symplectic, and for $\text{char}(F) \neq 2$, there are no nonnull metric vector spaces that are both orthogonal and symplectic. Since the subspace $S = \mathit{span}\{u\}$ is nonsingular, we have

$$V = S \oplus S^{\perp}$$

where $S^{\perp}$ is nonsingular and orthogonal. Hence, we can repeat the argument on $S^{\perp}$, to get

$$V = S \oplus T \oplus T^{\perp}$$

where S and T are one-dimensional subspaces. Continuing in this way, we get

$$V = S_1 \oplus \cdots \oplus S_n$$

where S_i is spanned by a vector $\mathbf{u}_i$ for which $\langle \mathbf{u}_i, \mathbf{u}_i \rangle \neq 0$. Hence, the basis $\{\mathbf{u}_1, \ldots, \mathbf{u}_n\}$ is an orthogonal basis for V. Theorem 11.9 then implies that any orthogonal metric vector space (singular or nonsingular), with $\mathrm{char}(F) \neq 2$, has an orthogonal basis.

As to the case where V is nonsingular, orthogonal and $\mathrm{char}(F) = 2$, assuming that V is not symplectic implies that there is a nonnull vector $\mathbf{u}$ in V, and so we have

$$V = S \oplus S^{\perp}$$

just as before. Now, we know that $S^{\perp}$ is nonsingular and orthogonal. If it is not symplectic, then we may choose another nonnull vector and repeat the process. This will continue until we meet a nonsingular, orthogonal, symplectic subspace T of V, which is the orthogonal sum of hyperbolic planes, according to Theorem 11.10. Hence, we have

$$V = S_1 \oplus \cdots \oplus S_k \oplus H_1 \oplus \cdots \oplus H_m$$

Now, we leave it to the reader to show that a matrix of the form

$$M = \begin{bmatrix} a & 0 & 0 \\ 0 & 0 & 1 \\ 0 & 1 & 0 \end{bmatrix}$$

where $a \neq 0$, is congruent to a diagonal matrix. Hence, we can replace the basis vectors for S_k and H_1 by basis vectors that will replace $S_k \oplus H_1$ by $T_k \oplus T_{k+1} \oplus T_{k+2}$, where each summand has dimension 1. Continuing with this process, we eventually get V as an orthogonal sum of one-dimensional subspaces, and so V has an orthogonal basis. Another appeal to Theorem 11.9 handles the general (singular and nonsingular) case.

Let us summarize.

Theorem 11.12 Let V be an orthogonal geometry. Provided that V is not symplectic as well when $\mathrm{char}(F) = 2$, then V has an ordered orthogonal basis $\mathcal{B} = (\mathbf{u}_1, \ldots, \mathbf{u}_k, \mathbf{z}_1, \ldots, \mathbf{z}_m)$ for which $\langle \mathbf{u}_i, \mathbf{u}_i \rangle = a_i \neq 0$ and $\langle \mathbf{z}_i, \mathbf{z}_i \rangle = 0$. Hence, $M_{\mathcal{B}}$ has the diagonal form

$$M_{\mathcal{B}} = \begin{bmatrix} a_1 & & & & & \\ & \ddots & & & & \\ & & a_k & & & \\ & & & 0 & & \\ & & & & \ddots & \\ & & & & & 0 \end{bmatrix}$$

with k ones and m zeros on the diagonal. Furthermore, the number k is the rank of the bilinear form, and so k is uniquely determined by V. ∎

Corollary 11.13 Let M be a symmetric matrix. If $\text{char}(F) = 2$, we assume that M has a nonzero entry on the diagonal. Then M is congruent to a diagonal matrix. □

The Structure of an Orthogonal Geometry

According to Corollary 11.13, for $\text{char}(F) \neq 2$, any symmetric matrix over F is congruent to a diagonal matrix. However, since two distinct diagonal matrices can be congruent, we cannot say that the diagonal matrices form a set of canonical forms for congruence.

It should come as no surprise that the determination of a set of canonical forms for congruence depends on the properties of the base field. To see this more clearly, suppose that $\mathcal{B} = (b_1, \ldots, b_n)$ is an ordered orthogonal basis for V, and so the matrix of the form has the diagonal form

$$M_{\mathcal{B}} = \begin{bmatrix} a_1 & & & \\ & a_2 & & \\ & & \ddots & \\ & & & a_n \end{bmatrix}$$

If $r_1, \ldots, r_n$ are nonzero scalars, the set $\mathcal{C} = (r_1 b_1, \ldots, r_n b_n)$ is also an ordered orthogonal basis for V, and

$$\langle r_i b_i, r_j b_j \rangle = r_i r_j \langle b_i, b_j \rangle = r_i^2 \delta_{i,j}$$

Hence the matrix of the bilinear form with respect to $\mathcal{C}$ is

$$M_{\mathcal{C}} = \begin{bmatrix} r_1^2 a_1 & & & \\ & r_2^2 a_2 & & \\ & & \ddots & \\ & & & r_n^2 a_n \end{bmatrix} \tag{11.5}$$

Thus, by a simple change of basis, we can multiply any diagonal entry by a nonzero *square* in F.

Before considering some possibilities, we have the following definition.

Definition An orthogonal basis $\{u_1, \ldots, u_n\}$ for V is an **orthonormal basis** if $\langle u_i, u_i \rangle = 1$ for all i. □

Algebraically Closed Fields

A field F with the property that every polynomial $p(x) \in F[x]$ splits into linear factors over F is said to be **algebraically closed**. For example, the field of complex numbers is algebraically closed. However, the field of real numbers is not algebraically closed.

If F is algebraically closed, then the polynomial $x^2 - r = 0$ has a solution in F, that is, every element of F has a square root in F. Therefore, we may choose $r_i = \sqrt{a_i}$ in (11.5), which leads to the following result.

Theorem 11.14 Let V be an orthogonal geometry over an algebraically closed field F. Provided that V is not symplectic as well when $\mathrm{char}(F) = 2$, then V has an ordered orthogonal basis $\mathfrak{B} = (u_1, \ldots, u_k, z_1, \ldots, z_m)$ for which $\langle u_i, u_i \rangle = 1$ and $\langle z_i, z_i \rangle = 0$, and so $M_{\mathfrak{B}}$ has the diagonal form

$$M_{\mathfrak{B}} = Z_{k,m} = \begin{bmatrix} 1 & & & & & \\ & \ddots & & & & \\ & & 1 & & & \\ & & & 0 & & \\ & & & & \ddots & \\ & & & & & 0 \end{bmatrix}$$

with k ones and m zeros on the diagonal. Furthermore, the number k is the rank of the bilinear form, and so k is uniquely determined by V. In particular, if V is nonsingular as well, then V has an orthonormal basis. ∎

The matrix version of Theorem 11.14 follows.

Theorem 11.15 Let $\mathcal{S}$ be the set of all $n \times n$ symmetric matrices over an algebraically closed field F. In case $\mathrm{char}(F) = 2$, we restrict $\mathcal{S}$ to be the set of all symmetric matrices with at least one nonzero entry on the main diagonal.

1) Any matrix in $\mathcal{S}$ is congruent to a unique matrix of the form $Z_{k,m}$, for some $k = 0, \ldots, n$ and $m = n - k$.
2) The set of all matrices of the form $Z_{k,m}$ for $k + m = n$, is a set of canonical forms for congruence on $\mathcal{S}$.
3) The rank of a matrix is a complete invariant under congruence on $\mathcal{S}$. ∎

The Real Field $\mathbb{R}$

As we have remarked, the real field $\mathbb{R}$ is not algebraically closed. However, referring to (11.5), we can choose $r_i = \sqrt{|a_i|}$, so that all diagonal elements will be either 0, 1 or −1.

Theorem 11.16 (Sylvester's law of inertia) Any orthogonal geometry V over the real field $\mathbb{R}$, has an ordered orthogonal basis $\mathcal{B} = (\mathbf{u}_1,\ldots,\mathbf{u}_k,\mathbf{v}_1,\ldots,\mathbf{v}_m,\mathbf{z}_1,\ldots,\mathbf{z}_p)$ for which $\langle\mathbf{u}_i,\mathbf{u}_i\rangle = 1$, $\langle\mathbf{v}_i,\mathbf{v}_i\rangle = -1$ and $\langle\mathbf{z}_i,\mathbf{z}_i\rangle = 0$. Hence, the matrix $M_{\mathcal{B}}$ has the diagonal form

$$M_{\mathcal{B}} = Z_{k,m,p} = \begin{bmatrix} 1 & & & & & & & \\ & \ddots & & & & & & \\ & & 1 & & & & & \\ & & & -1 & & & & \\ & & & & \ddots & & & \\ & & & & & -1 & & \\ & & & & & & 0 & \\ & & & & & & & \ddots \\ & & & & & & & & 0 \end{bmatrix}$$

with k ones, m negative ones, and p zeros on the diagonal. Moreover, the numbers k, m and p are uniquely determined by V.

Proof. We need only prove the uniqueness statement. Let

$$\mathcal{P} = span\{\mathbf{u}_1,\ldots,\mathbf{u}_k\},\quad \mathcal{N} = span\{\mathbf{v}_1,\ldots,\mathbf{v}_m\},\quad \mathcal{Z} = span\{\mathbf{z}_1,\ldots,\mathbf{z}_p\}$$

Then if $\mathbf{v} = \Sigma r_i\mathbf{u}_i \in \mathcal{P}$, we have

$$\langle\mathbf{v},\mathbf{v}\rangle = \left\langle \sum r_i\mathbf{u}_i, \sum r_j\mathbf{u}_j\right\rangle = \sum_{i,j} r_ir_j\langle\mathbf{u}_i,\mathbf{u}_j\rangle = \sum_{i,j} r_ir_j\delta_{i,j} = \sum r_i^2 \geq 0$$

and so the bilinear form $\langle,\rangle$ is *positive definite* on $\mathcal{P}$. Similarly, the form is *negative definite* on $\mathcal{N}$, that is, $\langle\mathbf{v},\mathbf{v}\rangle \leq 0$ for all $\mathbf{v} \in \mathcal{N}$. Finally, the form is zero on $\mathcal{Z}$. Now suppose that $\mathcal{C}$ is an ordered basis of a similar type to $\mathcal{B}$, and

$$\overline{\mathcal{P}} = span\{\overline{\mathbf{u}}_1,\ldots,\overline{\mathbf{u}}_{\overline{k}}\},\quad \overline{\mathcal{N}} = span\{\overline{\mathbf{v}}_1,\ldots,\overline{\mathbf{v}}_{\overline{m}}\},\quad \overline{\mathcal{Z}} = span\{\overline{\mathbf{z}}_1,\ldots,\overline{\mathbf{z}}_{\overline{p}}\}$$

Then

$$\mathcal{P} \cap span\{\overline{\mathcal{N}},\overline{\mathcal{Z}}\} = \{\mathbf{0}\}$$

for if $\mathbf{v} \in \mathcal{P}$ then $\langle\mathbf{v},\mathbf{v}\rangle \geq 0$ and if $\mathbf{v} \in span\{\overline{\mathcal{N}},\overline{\mathcal{Z}}\}$, then $\langle\mathbf{v},\mathbf{v}\rangle \leq 0$, and so $\mathbf{v} \in \mathcal{P} \cap span\{\overline{\mathcal{N}},\overline{\mathcal{Z}}\}$ implies that $\langle\mathbf{v},\mathbf{v}\rangle = 0$, that is, $\mathbf{v} = \mathbf{0}$. Thus, if $dim(V) = n$, then

$$dim(\mathcal{P}) + dim(span\{\overline{\mathcal{N}},\overline{\mathcal{Z}}\}) \leq dim(V)$$

that is,

$$p + (n - \overline{p}) \leq n$$

Hence $p \leq \overline{p}$. By symmetry, $\overline{p} \leq p$ and so $p = \overline{p}$. In a similar way, we deduce that $n = \overline{n}$ and $z = \overline{z}$. ∎

Here is the matrix version of Theorem 11.16.

Theorem 11.17 Let $\mathcal{S}$ be the set of all $n \times n$ symmetric matrices over the real field $\mathbb{R}$.

1) Any matrix in $\mathcal{S}$ is congruent to a unique matrix of the form $Z_{k,m,p}$, for some k, m and $p = n - k - m$.
2) The set of all matrices of the form $Z_{k,m,p}$ for $k + m + p = n$ is a set of canonical forms for congruence on $\mathcal{S}$.
3) The pair (k,m), or equivalently the pair $(k + m,\ k - m)$, is a complete invariant under congruence on $\mathcal{S}$. The number $k + m$ is the rank of the form, and $k - m$ is called the **signature** of the form. ∎

Finite Fields

To deal with the case of finite fields, we need two preliminary results.

Theorem 11.18 Let F_q be a finite field with q elements.

1) If $char(F_q) = 2$, then every element of F_q is a square.
2) If $char(F) \neq 2$, then exactly half of the nonzero elements of F_q are squares. Moreover, if x is any nonsquare in F_q, then all nonsquares have the form r^2x, for some $r \in F$.

Proof. We first remark that, in any field F, the equation $x^2 = 1$ has two solutions $x = 1$ and -1, which are distinct if and only if $char(F) \neq 2$. Now, let $F = F_q$, and let F^* be the set of all nonzero elements in F. Consider the set

$$(F^*)^2 = \{a^2 \mid a \in F^*\}$$

of all nonzero squares in F. Observe that, for $a,b \in F^*$

$$a^2 = b^2 \Leftrightarrow (ab^{-1})^2 = 1 \Leftrightarrow ab^{-1} = \pm 1 \Leftrightarrow a = \pm b$$

Thus, if $char(F) = 2$,

$$a^2 = b^2 \Leftrightarrow a = b$$

and so $(F^*)^2 = F^*$, which proves part (1). On the other hand, if $char(F) \neq 2$, then the map $\{a,-a\} \to a^2$ is a one-to-one correspondence between the set of pairs of (distinct) elements of F and $(F^*)^2$, and so $|F^*| = 2|(F^*)^2|$. We leave proof of the last statement to the reader. ∎

Definition A bilinear form on V is **universal** if for any $0 \neq r \in F$, there exists a vector $\mathbf{v}$ for which $\langle \mathbf{v},\mathbf{v} \rangle = r$. □

Theorem 11.19 Let V be a nonsingular orthogonal geometry over a finite field, with $dim(V) \geq 2$ and $char(F) \neq 2$. Then the bilinear form of V is universal.

Proof. Suppose first that V contains a null vector $\mathbf{u}$. Since V is nonsingular, there must exist a vector $\mathbf{v}$ in V for which $\{\mathbf{u},\mathbf{v}\}$ is linearly independent, and $\langle \mathbf{u},\mathbf{v}\rangle \neq 0$. Let $\mathbf{w} = \alpha\mathbf{u} + \beta\mathbf{v}$. For any $c \neq 0$, we want to determine α and β so that

$$c = \langle \mathbf{w},\mathbf{w}\rangle = 2\alpha\langle \mathbf{u},\mathbf{v}\rangle + \beta^2\langle \mathbf{v},\mathbf{v}\rangle \tag{11.6}$$

But, setting $\beta = 1$, this is easily solved for α. Hence, in this case, $\langle,\rangle$ is universal.

Now suppose that V has no null vectors, and that $\{\mathbf{u},\mathbf{v}\}$ are linearly independent, with

$$\langle \mathbf{u},\mathbf{u}\rangle = a \neq 0, \quad \langle \mathbf{v},\mathbf{v}\rangle = b \neq 0, \quad \langle \mathbf{u},\mathbf{v}\rangle = 0$$

Let $\mathbf{w} = \alpha\mathbf{u} + \beta\mathbf{v}$. We want to find α and β for which

$$c = \langle \mathbf{w},\mathbf{w}\rangle = a\alpha^2 + b\beta^2$$

Replacing a by ac and b by bc and dividing by $c \neq 0$, our goal is to show that, in any finite field of characteristic different from 2, the equation

$$a\alpha^2 + b\beta^2 = 1 \tag{11.7}$$

always has a solution (α,β).

If a is a square, then we may set $\beta = 0$, to get

$$\alpha^2 = a^{-1}, \quad \text{or} \quad \alpha = \sqrt{a^{-1}}$$

Similarly, if b is a square, we may set $\alpha = 0$ and solve for β. So let us assume that a and b are nonsquares.

Observe that -1 is the sum of squares in F_q, since if $q = p^n$, the characteristic of F is p, and so

$$-1 = 1^2 + \cdots + 1^2$$

where there are $p-1$ summands on the right. Hence, any number $c \in F$ is the sum of squares, since

$$4c = (1+c)^2 + (-1)(1-c)^2$$

From this, we deduce that the sum of two squares cannot always be a square, for then all elements of F_q would be squares, contradicting Theorem 11.18. Hence, there exist nonzero squares r and s in F_q for which $r^2 + s^2$ is a nonsquare.

Thus, a, b and $r^2 + s^2$ are all nonsquares in F_q. Since Theorem 11.18 implies that the product of any two nonsquares is a square, we deduce that

$$b = u^2a \quad \text{and} \quad r^2 + s^2 = v^2a$$

for some $u,v \in F$. Setting $\alpha = r/av$ and $\beta = s/uva$ gives

$$a\alpha^2 + b\beta^2 = \frac{ar^2}{a^2v^2} + \frac{bs^2}{u^2v^2a^2} = \frac{ar^2u^2 + bs^2}{u^2v^2a^2} = \frac{b(r^2+s^2)}{u^2v^2a^2} = \frac{b}{u^2a} = 1$$

This completes the proof. ∎

Now we can proceed with the business at hand. Let us settle the case $\mathrm{char}(F) = 2$ first.

Theorem 11.20 Let V be an orthogonal geometry over a finite field F, with $\mathrm{char}(F) = 2$. If V is not symplectic, then V has an ordered orthogonal basis $\mathcal{B} = (\mathbf{u}_1, \ldots, \mathbf{u}_k, \mathbf{z}_1, \ldots, \mathbf{z}_m)$ for which $\langle \mathbf{u}_i, \mathbf{u}_i \rangle = 1$ and $\langle \mathbf{z}_i, \mathbf{z}_i \rangle = 0$, and so $M_{\mathcal{B}}$ has the diagonal form

$$M_{\mathcal{B}} = Z_{k,m} = \begin{bmatrix} 1 & & & & \\ & \ddots & & & \\ & & 1 & & \\ & & & 0 & \\ & & & & \ddots \\ & & & & & 0 \end{bmatrix}$$

with k ones and m zeros on the diagonal. Furthermore, the number k is the rank of the bilinear form, and is uniquely determined by V. In particular, if V is nonsingular, then V has an orthonormal basis.

Proof. Referring to (11.5), since every element of F has a square root, we may take $r_i = \sqrt{a_i}$. ∎

The case $\mathrm{char}(F) \neq 2$ is a bit more difficult.

Theorem 11.21 Let V be an orthogonal geometry over a finite field F, with $\mathrm{char}(F) \neq 2$. Then there exists a nonzero number d, and an orthogonal basis $\mathcal{B} = (\mathbf{u}_1, \ldots, \mathbf{u}_k, \mathbf{z}_1, \ldots, \mathbf{z}_m)$ for which

$$\langle \mathbf{u}_i, \mathbf{u}_i \rangle = 1 \text{ for } 1 \le i \le k-1, \quad \langle \mathbf{u}_k, \mathbf{u}_k \rangle = d, \quad \langle \mathbf{z}_i, \mathbf{z}_i \rangle = 0$$

Hence, the matrix of the form in this basis is

$$M_{\mathcal{B}} = \begin{bmatrix} 1 & & & & & \\ & \ddots & & & & \\ & & 1 & & & \\ & & & d & & \\ & & & & 0 & \\ & & & & & \ddots \\ & & & & & & 0 \end{bmatrix}$$

The rank k of this matrix is uniquely determined by V. The number d is uniquely determined, up to multiplication by a square in F, by V. Moreover, the set $\{r^2d \mid 0 \neq r \in F\}$, which is the discriminant of the form when V is nonsingular, is uniquely determined by V.

Proof. We know that there is an ordered orthogonal basis $\mathcal{B} = (u_1,\ldots,u_k,z_1,\ldots,z_m)$ for which $\langle u_i,u_i\rangle = a_i \neq 0$ and $\langle z_i,z_i\rangle = 0$. Hence, $M_{\mathcal{B}}$ has the diagonal form

$$(11.8) \qquad M_{\mathcal{B}} = \begin{bmatrix} a_1 & & & & & \\ & \ddots & & & & \\ & & a_k & & & \\ & & & 0 & & \\ & & & & \ddots & \\ & & & & & 0 \end{bmatrix}$$

Now, consider the orthogonal geometry $V_1 = span\{u_1,u_2\}$. Then V_1 is nonsingular, since $\langle u_1,u_1\rangle \neq 0$, and so the form $\langle,\rangle$, restricted to V_1 is universal. Hence, there exists a $v_1 \in V_1$ for which $\langle v_1,v_1\rangle = 1$.

Since $\{u_1,u_2\}$ is a basis for V_1, we have $v_1 = ru_1 + su_2$. If $s = 0$, then we form the ordered basis $\mathcal{B}_1 = (v_1,u_2,\ldots,u_k,z_1,\ldots,z_m)$. The matrix of the form with respect to this basis is the same as (11.8), except that it has a 1 in the upper left corner (in place of a_1). If $s \neq 0$, then we form the ordered basis $\mathcal{B}_1 = (v_1,u_1,u_3,\ldots,u_k,z_1,\ldots,z_m)$, which will have the effect of replacing a_1 by 1 and a_2 by a_1.

We now repeat the process with the subspace V_2 generated by the second and third vectors in the new ordered basis $\mathcal{B}_1$. Continuing in this way, we can replace each a_i by a 1, for $1 \leq i \leq k-1$. We leave the remainder of the proof to the reader. ∎

Isometries

We now turn to a discussion of isometries on metric vector spaces.

Definition Let V and W be metric vector spaces. We use the same notation $\langle,\rangle$ for the bilinear form for each space. A *bijective* linear map $\tau:V\to W$ is called an **isometry** if

$$\langle \tau u,\tau v\rangle = \langle u,v\rangle$$

for all vectors u and v in V. If an isometry exists from V to W, we say that V and W are **isometric**, and write $V \approx W$. It is evident that the set of all isometries from V to V forms a group under composition, called the **group** of V.

If V is a *nonsingular* orthogonal geometry, an isometry from V to V is called an **orthogonal transformation.** If V is a *nonsingular* symplectic geometry, an isometry from V to V is called a **symplectic transformation.** □

Note that an isometry $\tau \in \mathcal{L}(V)$ is always injective if V is nonsingular. (It is customary in the theory of metric vector spaces to require an isometry to be bijective, unlike the special case of real or complex inner product spaces.) Here are a few of the basic properties of isometries.

Theorem 11.22 Let $\tau \in \mathcal{L}(V,W)$ be a linear transformation between finite dimensional metric vector spaces V and W.

1) Let $\mathcal{B} = \{\mathbf{v}_1,\ldots,\mathbf{v}_n\}$ be a basis for V. Then τ is an isometry if and only if τ is bijective, and
$$\langle \tau\mathbf{v}_i,\tau\mathbf{v}_j\rangle = \langle \mathbf{v}_i,\mathbf{v}_j\rangle$$
for all i,j.
2) If $\mathrm{char}(F) \neq 2$, then τ is an isometry if and only if it is bijective and
$$\langle \tau(\mathbf{v}),\tau(\mathbf{v})\rangle = \langle \mathbf{v},\mathbf{v}\rangle$$
for all $\mathbf{v} \in V$. ∎

Theorem 11.23 Let $\tau \in \mathcal{L}(V)$ be a linear operator on a finite dimensional metric vector space V. Let $\mathcal{B} = (\mathbf{v}_1,\ldots,\mathbf{v}_n)$ be an ordered basis for V, and let $M_{\mathcal{B}}$ be the matrix of the form relative to $\mathcal{B}$. Then τ is an isometry if and only if

$$[\tau]_{\mathcal{B}}^{\mathsf{T}} M_{\mathcal{B}}[\tau]_{\mathcal{B}} = M_{\mathcal{B}} \tag{11.9}$$

Proof. Dropping the subscript $\mathcal{B}$ for readability, we have
$$\langle \mathbf{x},\mathbf{y}\rangle = [\mathbf{x}]^{\mathsf{T}} M_{\mathcal{B}}[\mathbf{y}]$$
and
$$\langle \tau(\mathbf{x}),\tau(\mathbf{y})\rangle = [\tau(\mathbf{x})]^{\mathsf{T}} M_{\mathcal{B}}[\tau(\mathbf{y})] = [\mathbf{x}]^{\mathsf{T}}[\tau]^{\mathsf{T}} M_{\mathcal{B}}[\tau][\mathbf{y}]$$
Hence
$$\langle \mathbf{x},\mathbf{y}\rangle = \langle \tau(\mathbf{x}),\tau(\mathbf{y})\rangle$$
for all $\mathbf{x},\mathbf{y} \in V$ if and only if
$$[\mathbf{x}]^{\mathsf{T}} M_{\mathcal{B}}[\mathbf{y}] = [\mathbf{x}]^{\mathsf{T}}[\tau]^{\mathsf{T}} M_{\mathcal{B}}[\tau][\mathbf{y}]$$
for all $\mathbf{x},\mathbf{y} \in V$, which holds if and only if
$$M_{\mathcal{B}} = [\tau]^{\mathsf{T}} M_{\mathcal{B}}[\tau]$$
∎

If τ is an isometry, then (11.9) holds, and we may take determinants to get
$$\det(M_{\mathcal{B}}) = \det([\tau]_{\mathcal{B}})^2\det(M_{\mathcal{B}})$$

Therefore, if V is nonsingular, then $\det(M_{\mathcal{B}}) \neq 0$, and so

$$\det([\tau]_{\mathcal{B}}) = \pm 1$$

Since the determinant is an invariant under similarity, we can make the following definition.

Definition Let $\tau \in \mathcal{L}(V)$ be an orthogonal transformation. The **determinant** of τ is the determinant of any matrix $[\tau]_{\mathcal{B}}$ representing τ. If $\det(\tau) = 1$, then τ is called a **rotation**, and if $\det(\tau) = -1$, then τ is called a **reflection.** ∎

Because the Riesz representation theorem is valid in any nonsingular metric vector space, we can define the adjoint τ^* of a linear map τ exactly as we did in Chapter 9, that is, by the condition

$$\langle \tau(\mathbf{v}), \mathbf{w} \rangle = \langle \mathbf{v}, \tau^*(\mathbf{w}) \rangle$$

Theorem 11.24 Let $\tau \in \mathcal{L}(V)$ be a linear operator on a finite dimensional nonsingular metric vector space V.

1) τ is an isometry (orthogonal transformation) if and only if it is unitary, that is, if and only if τ is bijective and $\tau\tau^* = \iota$.
2) Let τ be an isometry. If $V = S \oplus S^{\perp}$ and S is invariant under τ, then so is $S^{\perp}$.

Proof. We prove part (2). Since S is invariant under τ, we have $\tau(S) \subset S$. But $\mathit{dim}(\tau(S)) = \mathit{dim}(S)$, and so $\tau(S) = S$, and $S = \tau^{-1}(S)$. Now, suppose that $\mathbf{v} \in S^{\perp}$. Then, for any $\mathbf{s} \in S$, in view of part (1), and the fact that $\tau^{-1}(\mathbf{s}) \in S$, we have

$$\langle \tau(\mathbf{v}), \mathbf{s} \rangle = \langle \mathbf{v}, \tau^{-1}(\mathbf{s}) \rangle = 0$$

and so $\tau(\mathbf{v}) \in S^{\perp}$. ∎

Symmetries

Suppose that V is a nonsingular metric vector space over F, where $\mathrm{char}(F) \neq 2$, and let $\mathbf{u} \in V$ be nonnull. Consider the linear map

$$\sigma_{\mathbf{u}}(\mathbf{v}) = \mathbf{v} - \frac{2\langle \mathbf{v}, \mathbf{u} \rangle}{\langle \mathbf{u}, \mathbf{u} \rangle}\mathbf{u}$$

It is not hard to verify that $\sigma_{\mathbf{u}}$ has the following properties.

1) $\sigma_{\mathbf{u}}$ is an isometry
2) $\sigma_{\mathbf{u}}(\mathbf{u}) = -\mathbf{u}$
3) $\sigma_{\mathbf{u}}(\mathbf{x}) = \mathbf{x}$ for all $\mathbf{x} \in (\mathit{span}\{\mathbf{u}\})^{\perp}$

In view of these properties, we refer to $\sigma_{\mathbf{u}}$ as the **symmetry** determined by $\mathbf{u}$. Note that properties (2) and (3) uniquely determine the linear map $\sigma_{\mathbf{u}}$, since $V = span\{\mathbf{u}\} \oplus (span\{\mathbf{u}\})^{\perp}$. Note also that

$$\sigma_{\mathbf{u}} = -\iota \mid_{\langle \mathbf{u} \rangle} \oplus \iota \mid_{\langle \mathbf{u} \rangle^{\perp}}$$

where ι is the identity map, and $\langle \mathbf{u} \rangle$ is the subspace spanned by the vector $\mathbf{u}$.

We will require the following property of symmetries.

Theorem 11.25 Let V be a nonsingular orthogonal geometry over a field F, with char(F) $\neq 2$. If $\mathbf{u}$ and $\mathbf{v}$ are nonzero vectors in V, with $\langle \mathbf{u},\mathbf{u} \rangle = \langle \mathbf{v},\mathbf{v} \rangle \neq 0$, then there exists a symmetry σ for which $\sigma(\mathbf{u}) = \mathbf{v}$ or $\sigma(\mathbf{u}) = -\mathbf{v}$.

Proof. Suppose first that $\mathbf{u}+\mathbf{v}$ is nonnull. Then the symmetry $\sigma_{\mathbf{u}+\mathbf{v}}$ is defined, and

$$\sigma_{\mathbf{u}+\mathbf{v}}(\mathbf{u}+\mathbf{v}) = -(\mathbf{u}+\mathbf{v})$$

Further, since $\langle \mathbf{u}-\mathbf{v},\mathbf{u}+\mathbf{v} \rangle = 0$, we have

$$\sigma_{\mathbf{u}+\mathbf{v}}(\mathbf{u}-\mathbf{v}) = \mathbf{u}-\mathbf{v}$$

Combining these two shows that $\sigma_{\mathbf{u}+\mathbf{v}}(\mathbf{u}) = -\mathbf{v}$, as desired.

If $\mathbf{u}+\mathbf{v}$ is null, then $\mathbf{u}-\mathbf{v}$ must be nonnull, for otherwise $\mathbf{u}$ would be null. Hence, the symmetry $\sigma_{\mathbf{u}-\mathbf{v}}$ is defined. Moreover,

$$\sigma_{\mathbf{u}-\mathbf{v}}(\mathbf{u}-\mathbf{v}) = -(\mathbf{u}-\mathbf{v})$$

and

$$\sigma_{\mathbf{u}-\mathbf{v}}(\mathbf{u}+\mathbf{v}) = \mathbf{u}+\mathbf{v}$$

These equations show that $\sigma_{\mathbf{u}-\mathbf{v}}(\mathbf{u}) = \mathbf{v}$, as desired. ∎

The next result indicates the importance of symmetries.

Theorem 11.26 Let V be a nonsingular orthogonal geometry over a field F with char(F) $\neq 2$. Then any orthogonal transformation $\tau: V \to V$ is the product of symmetries on V.

Proof. We proceed by induction on $d = dim(V)$, leaving the case $d = 1$ to the reader. Assume the theorem is true for $d-1$, and let $dim(V) = d$. Choose a nonnull vector $\mathbf{u} \in V$. Since $\langle \tau(\mathbf{u}),\tau(\mathbf{u}) \rangle = \langle \mathbf{u},\mathbf{u} \rangle$, we may apply Theorem 11.25 to deduce the existence of a symmetry σ on V for which $\sigma(\tau(\mathbf{u})) = \epsilon\mathbf{u}$, where $\epsilon = \pm 1$. Since σ is an isometry, if $\mathbf{x} \in \langle \mathbf{u} \rangle^{\perp}$, then

$$\langle \sigma\tau(\mathbf{x}),\mathbf{u} \rangle = \langle \sigma\tau(\mathbf{x}),\epsilon\sigma\tau(\mathbf{u}) \rangle = \epsilon\langle \mathbf{x},\mathbf{u} \rangle = 0$$

and so $\sigma\tau(\langle \mathbf{u} \rangle^{\perp}) \subset \langle \mathbf{u} \rangle^{\perp}$. Hence, $\langle \mathbf{u} \rangle$ and $\langle \mathbf{u} \rangle^{\perp}$ are both invariant under $\sigma\tau$. By the induction hypothesis, applied to the (d–1)-dimensional space $\langle \mathbf{u} \rangle^{\perp}$,

(11.10) $$\sigma\tau\,|_{\langle \mathbf{u}\rangle^{\perp}} = \sigma_{\mathbf{w}_1}\cdots\sigma_{\mathbf{w}_k}$$

where $\mathbf{w}_i \in \langle \mathbf{u}\rangle^{\perp}$ and $\sigma_{\mathbf{w}_i}$ is the symmetry on $\langle \mathbf{u}\rangle^{\perp}$ defined by

$$\sigma_{\mathbf{w}_i}(\mathbf{v}) = \mathbf{v} - \frac{2\langle \mathbf{v},\mathbf{w}_i\rangle}{\langle \mathbf{w}_i,\mathbf{w}_i\rangle}\mathbf{w}_i$$

But this is also a symmetry on V, where $\sigma_{\mathbf{w}_i}(\mathbf{u}) = \mathbf{u}$. Thus, thinking of $\sigma_{\mathbf{w}_i}$ as defined on all of V, we have

(11.11) $$\sigma_{\mathbf{w}_1}\cdots\sigma_{\mathbf{w}_k}(\mathbf{u}) = \mathbf{u} = \epsilon\sigma\tau(\mathbf{u})$$

Now we distinguish two cases. If $\epsilon = 1$, then (11.10) and (11.11) show that

$$\sigma\tau = \sigma_{\mathbf{w}_1}\cdots\sigma_{\mathbf{w}_k}$$

on both $\langle \mathbf{u}\rangle$ and $\langle \mathbf{u}\rangle^{\perp}$, which implies that $\sigma\tau = \sigma_{\mathbf{w}_1}\cdots\sigma_{\mathbf{w}_k}$ on V. Finally, since a symmetry is its own inverse, we have

$$\tau = \sigma\sigma_{\mathbf{w}_1}\cdots\sigma_{\mathbf{w}_k}$$

On the other hand, if $\epsilon = -1$, then (11.11) gives

$$\sigma_{\mathbf{u}}\sigma_{\mathbf{w}_1}\cdots\sigma_{\mathbf{w}_k}(\mathbf{u}) = \sigma_{\mathbf{u}}(\mathbf{u}) = -\mathbf{u} = \sigma\tau(\mathbf{u})$$

and since $\sigma_{\mathbf{u}}$ fixes all vectors in $\langle \mathbf{u}\rangle^{\perp}$, and $\langle \mathbf{u}\rangle^{\perp}$ is invariant under $\sigma\tau$, (11.10) gives, for $\mathbf{x} \in \langle \mathbf{u}\rangle^{\perp}$,

$$\sigma_{\mathbf{u}}\sigma_{\mathbf{w}_1}\cdots\sigma_{\mathbf{w}_k}(\mathbf{x}) = \sigma_{\mathbf{u}}\sigma\tau\,|_{\langle \mathbf{u}\rangle^{\perp}}(\mathbf{x}) = \sigma\tau\,|_{\langle \mathbf{u}\rangle^{\perp}}(\mathbf{x})$$

Thus, in this case,

$$\sigma\tau = \sigma_{\mathbf{u}}\sigma_{\mathbf{w}_1}\cdots\sigma_{\mathbf{w}_k}$$

and so

$$\tau = \sigma\sigma_{\mathbf{u}}\sigma_{\mathbf{w}_1}\cdots\sigma_{\mathbf{w}_k}$$ ∎

Witt's Cancellation Theorem

We now come to one of the major results of orthogonal geometry, due to Witt. To wit:

Theorem 11.27 (Witt's cancellation theorem) Let V be a nonsingular orthogonal geometry over a field F, with $\mathrm{char}(F) \neq 2$. Suppose that

$$V = S \oplus S^{\perp} = T \oplus T^{\perp}$$

where S and T are nonsingular. Then

$$S \approx T \;\Rightarrow\; S^{\perp} \approx T^{\perp}$$

Proof. Let $\tau: S \to T$ be an isometry. We proceed by induction on $\mathit{dim}(S)$. Suppose first that $\mathit{dim}(S) = 1$, and that $S = \mathit{span}\{\mathbf{s}\}$. Then

$T = span\{\tau(s)\}$ and $\langle \tau(s),\tau(s)\rangle = \langle s,s\rangle$. According to Theorem 11.25, there is a symmetry σ for which $\sigma(s) = \epsilon\tau(s)$, where $\epsilon = \pm 1$. Hence, σ is an isometry of V for which $\sigma(S) = T$. It follows that

$$\mathbf{x} \in S^{\perp} \Leftrightarrow \langle \mathbf{x},\mathbf{s}\rangle = 0 \Leftrightarrow \langle \sigma(\mathbf{x}),\sigma(\mathbf{s})\rangle = 0 \Leftrightarrow \langle \sigma(\mathbf{x}),\tau(\mathbf{s})\rangle = 0 \Leftrightarrow \sigma(\mathbf{x}) \in T^{\perp}$$

and so the restriction $\sigma|_{S^{\perp}}$ is an isometry from $S^{\perp}$ to $T^{\perp}$, which shows that $S^{\perp} \approx T^{\perp}$.

Now suppose the theorem is true for $dim(S) < k$, and let $dim(S) = k$. Let $\tau:S \to T$ be an isometry. Since S is nonsingular, we can choose a nonnull vector $\mathbf{s} \in S$, and write

$$S = span\{\mathbf{s}\} \oplus U$$

where U is nonsingular. Moreover,

$$T = span\{\tau(\mathbf{s})\} \oplus \tau(U)$$

Thus,

$$V = span\{\mathbf{s}\} \oplus U \oplus S^{\perp}$$

and

$$V = span\{\tau(\mathbf{s})\} \oplus \tau(U) \oplus T^{\perp}$$

Now we may apply Witt's Theorem for the one-dimensional case to deduce that

$$U \oplus S^{\perp} \approx \tau(U) \oplus T^{\perp}$$

Suppose that $\sigma:U \oplus S^{\perp} \to \tau(U) \oplus T^{\perp}$ is an isometry. Then, since $\sigma(U \oplus S^{\perp}) = \sigma(U) \oplus \sigma(S^{\perp})$, we have

$$\sigma(U) \oplus \sigma(S^{\perp}) = \tau(U) \oplus T^{\perp}$$

But $\sigma(U) \approx \tau(U)$, and since $dim(\sigma(U)) = dim(U) < dim(S)$, the induction hypothesis implies that $S^{\perp} \approx \sigma(S^{\perp}) \approx T^{\perp}$. ∎

Witt's Extension Theorem

Suppose that V and V′ are nonsingular orthogonal geometries, and that $\sigma:V \to V'$ is an isometry. Suppose also that U is a *nonsingular* subspace of V, and $\tau:U \to \tau(U) \subset V'$ is an isometry. We want to show that τ can be extended to an isometry on V.

Since U is nonsingular, so is $\tau(U)$. We deduce that

$$V' = \tau(U) \oplus \tau(U)^{\perp} = \sigma(U) \oplus \sigma(U^{\perp})$$

Since $\tau(U) \approx U \approx \sigma(U)$, the Witt cancellation theorem implies that $\tau(U)^{\perp} \approx \sigma(U^{\perp})$. If $\nu:\sigma(U^{\perp}) \to \tau(U)^{\perp}$ is an isometry, then the product $\nu\sigma:U^{\perp} \to \tau(U)^{\perp}$ is also an isometry, and so

$$\bar{\tau} = \tau \oplus \nu\sigma:U \oplus U^{\perp} \to V'$$

is an isometry with the property that $\bar{\tau}|_U = \tau$. This is the extension of τ that we have been seeking.

We now propose to show that the assumption that U is nonsingular is not needed. The plan is to show first that any subspace U of V can be embedded in a nonsingular subspace, and that any isometry on U can be extended to this nonsingular subspace. Then we may appeal to the nonsingular case, as described earlier.

Theorem 11.28 Let V be a nonsingular orthogonal geometry over F, with $char(F) \neq 2$. Let U be a subspace of V. Write $U = Rad(U) \oplus W$ and W is nonsingular (which we can do by Theorem 11.9). Suppose that $\mathcal{B} = \{b_1,\ldots,b_k\}$ is a basis for $Rad(U)$. Then there exist vectors $\{z_1,\ldots,z_k\}$ for which

1) the pairs (b_i,z_i) are hyperbolic pairs, and so the spaces $H_i = span\{b_i,z_i\}$ are hyperbolic planes, and
2) U is contained in the *nonsingular* space $H_1 \oplus \cdots \oplus H_k \oplus W$.

Moreover, if $\tau:U \to \tau(U) \subset V'$ is an isometry, where V′ is nonsingular, then there exists an isometry $\bar{\tau}:V \to V'$ for which $\bar{\tau}|_U = \tau$.

Proof. We prove (1) and (2) by induction on $k = dim(Rad(U))$. For $k = 0$, there is nothing to prove. To get a feel for the procedure, let $k = 1$. Thus, $\mathcal{B} = \{b_1\}$ is a basis for $Rad(U)$.

We want to find a $z_1 \in V$ for which (b_1,z_1) is a hyperbolic pair, that is,

$$(11.12) \qquad \langle b_1,b_1\rangle = \langle z_1,z_1\rangle = 0 \quad \text{and} \quad \langle b_1,z_1\rangle = 1$$

and for which, letting $H_1 = span\{b_1,z_1\}$, we have

$$H_1 \cap W = \{0\} \quad \text{and} \quad H_1 \perp W$$

Suppose we find a $z_1 \in W^{\perp}$ for which (11.12) holds. Then if $x = rb_1 + sz_1 \in H_1 \cap W$, we get

$$0 = \langle rb_1 + sz_1,z_1\rangle = r$$

and so $x = sz_1 \in W \cap W^{\perp} = \{0\}$, since W is nonsingular. Hence,

$$H_1 \cap W = \{0\}$$

Thus, since $b_1,z_1 \in W^{\perp}$ imply $H_1 \perp W$, we need only find a $z_1 \in W^{\perp}$ for which (11.12) holds. Since $b_1 \in W^{\perp}$, and $Rad(W^{\perp}) = Rad(W) = \{0\}$, there must exist a vector $x \in W^{\perp}$ such that $\langle b_1,x\rangle \neq 0$. Let us set $z_1 = rb_1 + sx$, and show there exists an r and s for which (11.12) holds, that is, for which

$$1 = \langle b_1,z_1\rangle = \langle b_1,rb_1 + sx\rangle = s\langle b_1,x\rangle$$

and

$$0 = \langle z_1,z_1\rangle = \langle rb_1 + sx,rb_1 + sx\rangle = 2rs\langle b_1,x\rangle + s^2\langle x,x\rangle$$

Since $\langle \mathbf{b}_1,\mathbf{x}\rangle \neq 0$, the first of these equations can be solved for s, and the second can then be solved for r. Thus, the desired vector $\mathbf{z}_1$ exists for which $(\mathbf{b}_1,\mathbf{z}_1)$ is a hyperbolic pair, and (1) and (2) hold for the case $k = 1$.

Assume for the purposes of induction that (1) and (2) are true for $dim(Rad(U)) < k$, and let $dim(Rad(U)) = k$. Then

$$U = Rad(U) \oplus W = span\{\mathbf{b}_k\} \oplus span\{\mathbf{b}_1,\ldots,\mathbf{b}_{k-1}\} \oplus W$$

and if we let $U_0 = span\{\mathbf{b}_1,\ldots,\mathbf{b}_{k-1}\} \oplus W$, then $Rad(U_0) = span\{\mathbf{b}_1,\ldots,\mathbf{b}_{k-1}\}$. Now, since $\mathbf{b}_k \in U_0^{\perp}$, and since

$$Rad(U_0^{\perp}) = Rad(U_0) = span\{\mathbf{b}_1,\ldots,\mathbf{b}_{k-1}\}$$

does not contain $\mathbf{b}_k$, we deduce as before the existence of a vector $\mathbf{x} \in U_0^{\perp}$ for which $\langle \mathbf{b}_k,\mathbf{x}\rangle \neq 0$. Again as before, we deduce the existence of a vector $\mathbf{z}_k \in U_0^{\perp}$ for which $(\mathbf{b}_k,\mathbf{z}_k)$ is a hyperbolic pair. Let $H_k = span\{\mathbf{b}_k,\mathbf{z}_k\}$.

Since $H_k \subset U_0^{\perp}$, we have $U_0 \subset H_k^{\perp}$, and since $H_k^{\perp}$ is nonsingular, and $dim(Rad(U_0)) = k-1$, we may apply the induction hypothesis to U_0, as a subspace of $H_k^{\perp}$. Hence, there exists hyperbolic planes $H_1,\ldots,H_{k-1}$ in $H_k^{\perp}$, for which

$$U_0 \subset \overline{U}_0 = H_1 \oplus \cdots \oplus H_{k-1} \oplus W$$

and since $\overline{U}_0 \subset H_k^{\perp}$, we have $H_k \perp \overline{U}_0$ and

$$U \subset H_1 \oplus \cdots \oplus H_k \oplus W$$

To prove the final statement of the theorem, suppose that $\tau{:}U \to \tau(U) \subset V'$ is an isometry, where V' is nonsingular. We know that

$$U = \langle \mathbf{b}_1\rangle \oplus \cdots \oplus \langle \mathbf{b}_k\rangle \oplus W \subset H_1 \oplus \cdots \oplus H_k \oplus W$$

where $\langle \mathbf{b}_i\rangle$ is the subspace spanned by $\mathbf{b}_i$, and $H_i = span\{\mathbf{b}_i,\mathbf{z}_i\}$. Since τ is an isometry, we have

$$\tau(U) = \langle \tau(\mathbf{b}_1)\rangle \oplus \cdots \oplus \langle \tau(\mathbf{b}_k)\rangle \oplus \tau(W)$$

Now, let $\overline{\tau}(\mathbf{b}_i) = \tau(\mathbf{b}_i)$ and $\overline{\tau}(\mathbf{w}) = \tau(\mathbf{w})$ for all $\mathbf{w} \in W$. We need only choose $\overline{\tau}(\mathbf{z}_i)$ so as to make $\overline{\tau}$ an isometry.

To this end, let us apply the first part of the theorem to the nonsingular space V', with subspace $\tau(U)$. Since $\{\tau(\mathbf{b}_1),\ldots,\tau(\mathbf{b}_k)\}$ is a basis for $Rad\{\tau(U)\}$, we deduce the existence of vectors $\mathbf{w}_i \in V'$ for which

$$\tau(U) \subset K_1 \oplus \cdots \oplus K_k \oplus W'$$

where the $K_i = span\{\tau(\mathbf{b}_i),\mathbf{w}_i\}$ are hyperbolic planes in V'. Hence, if we let $\overline{\tau}(\mathbf{z}_i) = \mathbf{w}_i$, it follows easily that $\overline{\tau}$ is an isometry. ∎

Theorem 11.28, together with the discussion preceding the theorem, gives the following.

Theorem 11.29 (Witt's extension theorem) Let V and V′ be isometric nonsingular orthogonal geometries over a field F, with $char(F) \neq 2$. Suppose that U is a nonsingular subspace of V, and $\tau: U \to U' \subset V'$ is an isometry. Then τ can be extended to all of V, that is, there is an isometry $\overline{\tau}: V \to V'$ for which $\overline{\tau}|_U = \tau$. ∎

We consider an application of Witt's extension theorem. Let V be a nonsingular orthogonal geometry over a field F, with $char(F) \neq 2$. Suppose that U and U′ are *maximal* null subspaces of V. (That is, U and U′ are not properly contained in any null subspaces of V.) We propose to show that $dim(U) = dim(U')$.

If $dim(U) \leq dim(U')$, then there is a vector space isomorphism $\tau: U \to \tau(U) \subset U'$, which is also an isometry, since U and U′ are null. Thus, Witt's extension theorem implies the existence of an isometry $\overline{\tau}: V \to V$ that extends τ. In particular, $\overline{\tau}^{-1}(U')$ is a null space that contains U, and so $\overline{\tau}^{-1}(U') = U$, which shows that $dim(U) = dim(U')$. We now have the following.

Theorem 11.30 Let V be a nonsingular orthogonal geometry over a field F, with $char(F) \neq 2$. Then all maximal null subspaces of V have the same dimension, which is called the **Witt index** of V, and is denoted by $w(V)$. ∎

Maximal Hyperbolic Subspaces

Since a hyperbolic space is completely determined (up to isometry) by its dimension, it is of interest to know something about maximal hyperbolic subspaces of a nonsingular orthogonal geometry. (In the symplectic case, if V is nonsingular, then V is hyperbolic.) We will denote a hyperbolic space by $\mathcal{H}$, and a hyperbolic space of dimension 2k by $\mathcal{H}_{2k}$, thus

$$\mathcal{H}_{2k} = H_1 \oplus \cdots \oplus H_k$$

where each H_i is a hyperbolic plane.

Note that a two-dimensional space is a hyperbolic plane if and only if it is nonsingular and contains a null vector. (We assume that $char(F) \neq 2$.)

Suppose that V is isotropic, that is, V contains a null vector. If U_k is a nonempty null subspace of V of dimension k, then $Rad(U_k) = U_k$, and so we may apply Theorem 11.28, to deduce that

$$\mathrm{U_k} \subset \mathcal{H}_{2k} = \mathrm{H}_1 \oplus \cdots \oplus \mathrm{H}_k$$

where $\mathrm{H_i}$ is generated by a hyperbolic pair $(\mathbf{x}_i,\mathbf{y}_i)$. Thus, any null subspace $\mathrm{U_k}$ is contained in a hyperbolic space $\mathcal{H}_{2k}$ with $dim(\mathcal{H}_{2k}) = 2\,dim(\mathrm{U_k})$. This implies that the Witt index of V is at most $dim(\mathrm{V})/2$.

On the other hand, suppose that

$$\mathcal{H}_{2k} = \mathrm{H}_1 \oplus \cdots \oplus \mathrm{H}_k$$

is a hyperbolic space in V, and that $\mathrm{H_i}$ is generated by the hyperbolic pair $(\mathbf{x}_i,\mathbf{y}_i)$. Then the set $\mathcal{B} = \{\mathbf{x}_1,\ldots,\mathbf{x}_k\}$ is independent, for if

$$r_1\mathbf{x}_1 + \cdots + r_k\mathbf{x}_k = 0$$

it follows that

$$0 = \langle r_1\mathbf{x}_1 + \cdots + r_k\mathbf{x}_k, \mathbf{y}_j\rangle = r_j\langle \mathbf{x}_j,\mathbf{y}_j\rangle = r_j$$

for all j. Moreover, since $\langle \mathbf{x}_i,\mathbf{x}_j\rangle = 0$ for all i,j, the subspace $\mathrm{U_k} = span\{\mathcal{B}\}$ is a k-dimensional null space. Thus, any hyperbolic space $\mathcal{H}_{2k}$ in V contains a null space $\mathrm{U_k}$. This implies that if $\mathcal{H}_{2m}$ is a maximal hyperbolic subspace of V, then $m \le w(\mathrm{V})$. Furthermore, since V must contain a null space $\mathrm{U}_{w(\mathrm{V})}$, it must also contain a hyperbolic subspace of dimension $2w(\mathrm{V})$. In other words, the maximum dimension of a hyperbolic subspace of V is $2w(\mathrm{V})$.

Now, suppose that

$$\mathcal{H} = \mathrm{H}_1 \oplus \cdots \oplus \mathrm{H}_k$$

and

$$\mathcal{K} = \mathrm{K}_1 \oplus \cdots \oplus \mathrm{K}_m$$

are *maximal* hyperbolic subspaces of V, and that $\mathrm{H_i}$ is spanned by the hyperbolic pair $(\mathbf{u}_i,\mathbf{v}_i)$, and $\mathrm{K_i}$ is spanned by the hyperbolic pair $(\mathbf{x}_i,\mathbf{y}_i)$. We wish to show that $dim(\mathcal{H}) = dim(\mathcal{K})$.

Suppose that $dim(\mathcal{H}) \le dim(\mathcal{K})$, and consider the vector space monomorphism $\tau:\mathcal{H}\to\tau(\mathcal{H}) \subset \mathcal{K}$ defined by the conditions

$$\tau(\mathbf{u}_i) = \mathbf{x}_i, \quad \tau(\mathbf{v}_i) = \mathbf{y}_i$$

According to Theorem 11.22, τ is an isometry, and so $\mathcal{H} \approx \tau(\mathcal{H})$. Thus, Witt's extension theorem implies the existence of an isometry $\overline{\tau}:\mathrm{V}\to\mathrm{V}$ that extends τ. In particular, $\overline{\tau}^{-1}(\mathcal{K})$ is a hyperbolic space that contains $\mathcal{H}$, and so $\overline{\tau}^{-1}(\mathcal{K}) = \mathcal{H}$, which shows that $dim(\mathcal{K}) = dim(\mathcal{H})$. We have shown that all maximal hyperbolic subspaces of V have the same dimension, namely, $2w(\mathrm{V})$.

Now suppose that $\mathcal{H}$ is a maximal hyperbolic subspace of V. Since hyperbolic spaces are nonsingular, we have $\mathrm{V} = \mathcal{H} \oplus \mathcal{H}^{\perp}$. Then $\mathcal{H}^{\perp}$ is anisotropic, that is, it contains no null vectors. To see this, suppose to the contrary that $\mathbf{x} \in \mathcal{H}^{\perp}$ is a null vector. Since there is a null subspace $\mathrm{U} \subset \mathcal{H}$ for which $dim(\mathrm{U}) = dim(\mathcal{H})/2 = w(\mathrm{V})$, the null space $U' = span\{U,\mathbf{x}\}$ has dimension $w(V)+1$, which contradicts the meaning of the Witt index. Hence, $H^{\perp}$ contains no null vectors.

Conversely, suppose that $\mathcal{H}_{2k}$ is hyperbolic, that $V = \mathcal{H}_{2k} \oplus \mathcal{H}_{2k}^{\perp}$ and that $\mathcal{H}_{2k}^{\perp}$ is anisotropic. Then $\mathcal{H}_{2k}$ is maximal hyperbolic. For if not, then $\mathcal{H}_{2k}$ is properly contained in a hyperbolic subspace $\mathcal{H}_{2m}$, and we can write

$$\mathcal{H}_{2m} = \mathcal{H}_{2k} \oplus \mathcal{K}$$

where $\mathcal{K}$ is the orthogonal complement of $\mathcal{H}$ in $\mathcal{H}_{2m}$. Now, we claim that $\mathcal{K}$ is a hyperbolic space of dimension $2(m-k)$. For we do have

$$(11.13) \qquad \mathcal{H}_{2m} = \mathcal{H}_{2k} \oplus \mathcal{H}_{2(m-k)}$$

for some hyperbolic space $\mathcal{H}_{2(m-k)}$, and so by Witt's cancellation theorem, $\mathcal{K} \approx \mathcal{H}_{2(m-k)}$.

Since $\mathcal{H}_{2(m-k)}$ contains a null vector, (11.13) implies that there is a null vector $\mathbf{x} \in \mathcal{H}_{2k}^{\perp}$, contrary to assumption. Hence, $\mathcal{H}_{2k}$ is maximal.

Our discussion of hyperbolic subspaces has established the following key result.

Theorem 11.31 Let V be a nonsingular orthogonal geometry over F with $\text{char}(F) \neq 2$. Then all maximal hyperbolic subspaces of V have dimension $2w(V)$, where $w(V)$ is the Witt index of V. Moreover,

$$V = \mathcal{H} \oplus S$$

where $\mathcal{H}$ is a maximal hyperbolic subspace of V, or $\mathcal{H} = \{\mathbf{0}\}$ if V has no null vectors, and S is an *anisotropic* subspace of V, that is, S contains no null vectors. ∎

According to Theorems 11.9 and 11.31, any orthogonal geometry V over a field F, with $\text{char}(F) \neq 2$, can be written in the form

$$Rad(V) \oplus \mathcal{H} \oplus S$$

where $Rad(V)$ is a null space, $\mathcal{H}$ is a hyperbolic space, and S is anisotropic.

EXERCISES

1. Prove that a form is symmetric if and only if the matrix $M_{\mathcal{B}}$ of the form is symmetric.
2. Prove that a form is alternate if and only if its matrix $M_{\mathcal{B}} = (a_{i,j})$ is alternate, that is,

$$a_{i,i} = 0, \qquad a_{i,j} = -a_{j,i}\ (i \neq j)$$

3. Show that a metric vector space V is nonsingular if and only if the matrix $M_{\mathfrak{B}}$ of the form is nonsingular, for any ordered basis $\mathfrak{B}$.
4. Does Minkowski space contain any null vectors? If so, find them.
5. Is Minkowski space isometric to Euclidean space $\mathbb{R}^4$?
6. If $\langle , \rangle$ is a symmetric bilinear form on V, show that $Q(\mathbf{x}) = \langle \mathbf{x},\mathbf{x}\rangle/2$ is a quadratic form.
7. Show that τ is an isometry if and only if $Q(\tau(\mathbf{v})) = Q(\mathbf{v})$ where Q is the quadratic form associated with the bilinear form on V. (Here $char(F) \neq 2$.)
8. Show that a bijective map $\tau:V\to W$ is an isometry if and only if for any basis $\{\mathbf{v}_1,\ldots,\mathbf{v}_n\}$ of V, we have
$$\langle \tau\mathbf{v}_i,\tau\mathbf{v}_j\rangle = \langle \mathbf{v}_i,\mathbf{v}_j\rangle$$
9. Show that if V is a nonsingular orthogonal geometry over a field F, with $char(F) \neq 2$, then any totally isotropic subspace of V is also a null space.
10. Find a metric vector space V for which $Rad(V)^{\perp}$ is singular. Is $V = Rad(V) \oplus Rad(V)^{\perp}$?
11. Prove that if x is any nonsquare in a finite field F_q, then all nonsquares have the form r^2x, for some $r \in F$. Hence, the product of any two nonsquares in F_q is a square.
12. Formulate Sylvester's law of inertia in terms of quadratic forms on V.
13. Let V be any orthogonal geometry over the real field $\mathbb{R}$. Prove that V can be written as a direct sum $V = \mathcal{P} \oplus \mathcal{N} \oplus \mathcal{Z}$, where the bilinear form on V is positive definite on $\mathcal{P}$, negative definite on $\mathcal{N}$ and zero on $\mathcal{Z}$. Moreover, the dimensions of $\mathcal{P}$, $\mathcal{N}$ and $\mathcal{Z}$ are uniquely determined by V.
14. Prove that two one-dimensional metric vector spaces are isometric if and only if they have the same discriminant.
15. a) Let U be a subspace of V. Show that the inner product $\langle \mathbf{x}+U,\mathbf{y}+U\rangle = \langle \mathbf{x},\mathbf{y}\rangle$ is well-defined if and only if $U \subset Rad(V)$.
    b) If $U \subset Rad(V)$, when is V/U is nonsingular?
16. Let $V = N \oplus S$, where N is a null space.
    a) Prove that $N = Rad(V)$ if and only if S is nonsingular.
    b) If S is nonsingular, prove that $S \approx V/Rad(V)$.
17. Let $dim(V) = dim(W)$. Prove that $V/Rad(V) \approx W/Rad(W)$ implies $V \approx W$.
18. Let $V = S \oplus T$.
    a) Prove that $Rad(V) = Rad(S) \oplus Rad(T)$
    b) $V/Rad(V) \approx S/Rad(S) \oplus T/Rad(T)$

c) $dim(Rad(V)) = dim(Rad(S)) + dim(Rad(T))$
d) V is nonsingular if and only if S and T are both nonsingular.

19. Verify in detail that the adjoint is well-defined and linear.
20. Prove that $\tau \in \mathcal{L}(V)$, where V is nonsingular, is an isometry if and only if it is bijective and unitary.
21. If $char(F) \neq 2$, prove that a $\tau \in \mathcal{L}(V,W)$ is an isometry if and only if it is bijective and $\langle \tau(\mathbf{v}), \tau(\mathbf{v}) \rangle = \langle \mathbf{v}, \mathbf{v} \rangle$ for all $\mathbf{v} \in V$.
22. Let $\mathcal{B} = \{\mathbf{v}_1, \ldots, \mathbf{v}_n\}$ be a basis for V. Prove that $\tau \in \mathcal{L}(V,W)$ is an isometry if and only if it is bijective and $\langle \tau\mathbf{v}_i, \tau\mathbf{v}_j \rangle = \langle \mathbf{v}_i, \mathbf{v}_j \rangle$ for all i,j.
23. Let V be a nonsingular orthogonal geometry, and let $\tau \in \mathcal{L}(V)$ be an isometry.
    a) Show that $dim(ker(\iota - \tau)) = dim(im(\iota - \tau)^{\perp})$.
    b) Show that $ker(\iota - \tau) = im(\iota - \tau)^{\perp}$. How would you describe $ker(\iota - \tau)$ in words?
    c) If τ is a symmetry, what is $dim(ker(\iota - \tau))$?
    d) Can you characterize symmetries by means of $dim(ker(\iota - \tau))$?
24. A linear transformation $\tau \in \mathcal{L}(V)$ is called **unipotent** if $\tau - \iota$ is nilpotent. Suppose that V is an anisotropic metric vector space, and that τ is unipotent and isometric. Show that $\tau = \iota$.
25. Let V be a hyperbolic space of dimension 2m, and let U be a hyperbolic subspace of V of dimension 2k. Show that for each $k \leq j \leq m$, there is a hyperbolic subspace $\mathcal{H}_{2j}$ of V for which $U \subset \mathcal{H}_{2j} \subset V$.
26. Let V be a symplectic geometry or an orthogonal geometry with $char(F) \neq 2$. Prove that a subspace S of V is a hyperbolic plane if and only if S is nonsingular, has dimension 2 and contains a null vector.

CHAPTER 12

Metric Spaces

Contents: *The Definition. Open and Closed Sets. Convergence in a Metric Space. The Closure of a Set. Dense Subsets. Continuity. Completeness. Isometries. The Completion of a Metric Space. Exercises.*

The Definition

In Chapter 9, we studied the basic properties of real and complex inner product spaces. Much of what we did does not depend on whether the space in question is finite or infinite dimensional. However, as we discussed in Chapter 9, the presence of an inner product, and hence a metric, on a vector space, raises a host of new issues related to convergence. In this chapter, we discuss briefly the concept of a metric space. This will enable us to study the convergence properties of real and complex inner product spaces.

A metric space is not an algebraic structure. Rather it is designed to model the abstract properties of distance.

Definition A **metric space** is a pair (M,d), where M is a nonempty set and $d:M\times M\to\mathbb{R}$ is a real-valued function, called a **metric** on M, with the following properties. The expression $d(x,y)$ is read "the distance from x to y."

1) (**Positive definiteness**) For all $x,y\in M$,

$$d(x,y) \geq 0$$

2) **(Symmetry)** For all $x,y \in M$,

$$d(x,y) = d(y,x)$$

3) **(Triangle inequality)** For all $x,y,z \in M$,

$$d(x,y) \leq d(x,z) + d(z,y)$$

□

As is customary, when there is no cause for confusion, we simply say "let M be a metric space."

Example 12.1 Any nonempty set M is a metric space under the **discrete metric**, defined by

$$d(x,y) = \begin{cases} 0 & \text{if } x = y \\ 1 & \text{if } x \neq y \end{cases}$$

□

Example 12.2

1) The set $\mathbb{R}^n$ is a metric space, under the metric defined, for $\mathbf{x} = (x_1,\ldots,x_n)$ and $\mathbf{y} = (y_1,\ldots,y_n)$ by

$$d(\mathbf{x},\mathbf{y}) = \sqrt{(x_1 - y_1)^2 + \cdots + (x_n - y_n)^2}$$

This is called the **Euclidean metric** on $\mathbb{R}^n$. We note that $\mathbb{R}^n$ is also a metric space under the metric

$$d_1(\mathbf{x},\mathbf{y}) = |x_1 - y_1| + \cdots + |x_n - y_n|$$

Of course, $(\mathbb{R}^n, d)$ and $(\mathbb{R}^n, d_1)$ are *different* metric spaces.

2) The set $\mathbb{C}^n$ is a metric space under the **unitary metric**

$$d(\mathbf{x},\mathbf{y}) = \sqrt{|x_1 - y_1|^2 + \cdots + |x_n - y_n|^2}$$

where $\mathbf{x} = (x_1,\ldots,x_n)$ and $\mathbf{y} = (y_1,\ldots,y_n)$ are in $\mathbb{C}^n$. □

Example 12.3

1) The set C[a,b] of all real-valued (or complex-valued) continuous functions on [a,b] is a metric space, under the metric

$$d(f,g) = \sup_{x \in [a,b]} |f(x) - g(x)|$$

We refer to this metric as the **sup metric.**

2) The set C[a,b] of all real-valued (or complex-valued) continuous functions on [a,b] is a metric space, under the metric

$$d_1(f(x),g(x)) = \int_a^b |f(x) - g(x)| \, dx$$

□

Example 12.4 Many important *sequence spaces* are metric spaces. We will often use boldface Roman letters to denote sequences, as in $\mathbf{x} = (x_n)$ and $\mathbf{y} = (y_n)$.

1) The set $\ell^\infty_{\mathbb{R}}$ of all bounded sequences of real numbers is a metric space under the metric defined by
$$d(\mathbf{x},\mathbf{y}) = \sup_n |x_n - y_n|$$
The set $\ell^\infty_{\mathbb{C}}$ of all bounded complex sequences, with the same metric, is also a metric space. As is customary, we will usually denote both of these spaces by ℓ^∞.

2) For $p \geq 1$, let ℓ^p be the set of all sequences $\mathbf{x} = (x_n)$ of real (or complex) numbers for which
$$\sum_{n=1}^{\infty} |x_n|^p < \infty$$
We define the **p-norm** of $\mathbf{x}$ by
$$\|\mathbf{x}\|_p = \left(\sum_{n=1}^{\infty} |x_n|^p\right)^{1/p}$$
Then ℓ^p is a metric space, under the metric
$$d(\mathbf{x},\mathbf{y}) = \|\mathbf{x}-\mathbf{y}\|_p = \left(\sum_{n=1}^{\infty} |x_n - y_n|^p\right)^{1/p}$$
The fact that ℓ^p is a metric follows from some rather famous results about sequences of real or complex numbers, whose proofs we leave as (well-hinted) exercises.

Hölder's inequality Let $p,q \geq 1$ and $p+q = pq$. If $\mathbf{x} \in \ell^p$ and $\mathbf{y} \in \ell^q$, then $\mathbf{xy} = (x_n y_n) \in \ell^1$ and
$$\|\mathbf{xy}\|_1 \leq \|\mathbf{x}\|_p \|\mathbf{y}\|_q$$
that is,
$$\sum_{n=1}^{\infty} |x_n y_n| \leq \left(\sum_{n=1}^{\infty} |x_n|^p\right)^{1/p} \left(\sum_{n=1}^{\infty} |y_n|^q\right)^{1/q}$$

A special case of this (with $p = q = 2$) is the **Cauchy-Schwarz inequality**
$$\sum_{n=1}^{\infty} |x_n y_n| \leq \sqrt{\sum_{n=1}^{\infty} |x_n|^2}\sqrt{\sum_{n=1}^{\infty} |y_n|^2}$$

Minkowski's inequality For $p \geq 1$, if $\mathbf{x},\mathbf{y} \in \ell^p$, then $\mathbf{x}+\mathbf{y} = (x_n + y_n) \in \ell^p$ and
$$\|\mathbf{x}+\mathbf{y}\|_p \leq \|\mathbf{x}\|_p + \|\mathbf{y}\|_p$$
that is,

$$\left(\sum_{n=1}^{\infty} |x_n + y_n|^p\right)^{1/p} \leq \left(\sum_{n=1}^{\infty} |x_n|^p\right)^{1/p} + \left(\sum_{n=1}^{\infty} |y_n|^p\right)^{1/p} \quad \square$$

Definition If M is a metric space under d, then any nonempty subset S of M is also a metric under the restriction of d to $S \times S$. The metric space S thus obtained is called a **subspace** of M.

Open and Closed Sets

Definition Let M be a metric space. Let $x_0 \in M$ and let r be a *positive* real number.

1) The **open ball** centered at x_0, with radius r, is

$$B(x_0,r) = \{x \in M \mid d(x,x_0) < r\}$$

2) The **closed ball** centered at x_0, with radius r, is

$$\overline{B}(x_0,r) = \{x \in M \mid d(x,x_0) \leq r\}$$

3) The **sphere** centered at x_0, with radius r, is

$$S(x_0,r) = \{x \in M \mid d(x,x_0) = r\} \quad \square$$

Definition A subset S of a metric space M is said to be **open** if each point of S is the center of an open ball that is contained *completely* in S. More specifically, S is open if for all $x \in S$, there exists an $r > 0$ such that $B(x,r) \subset S$. Note that the empty set is open. A set $T \subset M$ is **closed** if its complement $T^c = M - T$ is open. $\square$

It is easy to show that an open ball is an open set and a closed ball is a closed set. If $x \in M$, we refer to any open set S containing x as an **open neighborhood** of x. It is also easy to see that a set is open if and only if it contains an open neighborhood of each of its points.

The next example shows that it is possible for a set to be both open and closed, or neither open nor closed.

Example 12.5 In the metric space $\mathbb{R}$, the open balls are just the open intervals

$$B(x_0,r) = (x_0 - r, x_0 + r)$$

and the closed balls are the closed intervals

$$\overline{B}(x_0,r) = [x_0 - r, x_0 + r]$$

Consider the half-open interval $S = (a,b]$, for $a < b$. This set is not open, since it contains no open ball centered at $b \in S$, and it is not closed, since its complement $S^c = (-\infty,a] \cup (b,\infty)$ is not open (it contains no open ball about a).

Observe also that the empty set is both open and closed, as is the entire space $\mathbb{R}$. (Although we will not do so, it is possible to show that these are the *only* two sets that are both open and closed in $\mathbb{R}$.) □

It is not our intention to enter into a detailed discussion of open and closed sets, the subject of which belongs to a branch of mathematics known as *topology*. In order to put these concepts in perspective, however, we have the following result, whose proof is left to the reader.

Theorem 12.1 The collection $\mathcal{O}$ of all open subsets of a metric space M has the following properties

1) $\emptyset \in \mathcal{O}$, $M \in \mathcal{O}$
2) If $S, T \in \mathcal{O}$, then $S \cap T \in \mathcal{O}$
3) If $\{S_i \mid i \in K\}$ is any collection of open sets, then $\bigcup_{i \in K} S_i \in \mathcal{O}$. ∎

These three properties form the basis for an axiom system that is designed to generalize notions such as convergence and continuity, and lead to the following definition.

Definition Let X be a nonempty set. A collection $\mathcal{O}$ of subsets of X is called a **topology** for X if it has the following properties

1) $\emptyset \in \mathcal{O}$, $X \in \mathcal{O}$
2) If $S, T \in \mathcal{O}$, then $S \cap T \in \mathcal{O}$
3) If $\{S_i \mid i \in K\}$ is any collection of sets in $\mathcal{O}$, then $\bigcup_{i \in K} S_i \in \mathcal{O}$.

We refer to subsets in $\mathcal{O}$ as **open sets**, and the pair $(X,\mathcal{O})$ as a **topological space.** □

According to Theorem 12.1, the open sets (as we defined them earlier) in a metric space M form a topology for M, called the topology **induced** by the metric.

Topological spaces are the most general setting in which we can define concepts such as convergence and continuity, which is why these concepts are called *topological concepts*. However, since the topologies with which we will be dealing are induced by a metric, we will generally phrase the definitions of the topological properties that we will need directly in terms of the metric.

Convergence in a Metric Space

Convergence of sequences in a metric space is defined as follows.

Definition A sequence (x_n) in a metric space M **converges** to $x \in M$, written $(x_n) \to x$, if

$$\lim_{n\to\infty} d(x_n,x) = 0$$

Equivalently, $(x_n) \to x$ if for any $\epsilon > 0$, there exists an $N > 0$ such that

$$n > N \Rightarrow d(x_n,x) < \epsilon$$

or, equivalently

$$n > N \Rightarrow x_n \in B(x,\epsilon)$$

In this case, x is called the **limit** of the sequence (x_n). ∎

If M is a metric space, and S is a subset of M, by a *sequence in* S, we mean a sequence whose terms all lie in S. We next characterize closed sets, and therefore also open sets, using convergence.

Theorem 12.2 Let M be a metric space. A subset $S \subset M$ is closed if and only if whenever (x_n) is a sequence in S, and $(x_n) \to x$, then $x \in S$. In loose terms, a subset S is closed if it is closed under the taking of sequential limits.

Proof. Suppose that S is closed, and let $(x_n) \to x$, where $x_n \in S$ for all n. Suppose that $x \notin S$. Then since $x \in S^c$ and S^c is open, there exists an $\epsilon > 0$ for which $x \in B(x,\epsilon) \subset S^c$. But this implies that

$$B(x,\epsilon) \cap \{x_n\} = \emptyset$$

which contradicts the fact that $(x_n) \to x$. Hence, $x \in S$.

Conversely, suppose that S is closed under the taking of limits. We show that S^c is open. Let $x \in S^c$, and suppose to the contrary that no open ball about x is contained in S^c. Consider the open balls $B(x,1/n)$, for $n = 1,2,\ldots$. Since none of these balls is contained in S^c, for each n, there is an $x_n \in S \cap B(x,1/n)$. It is clear that $(x_n) \to x$, and so $x \in S$. But x cannot be in both S and S^c, and so some ball about x is in S^c, which implies that S^c is open. Thus, S is closed. ∎

The Closure of a Set

Definition Let S be any subset of a metric space M. The **closure** of S, denoted by $cl(S)$, is the smallest closed set containing S. □

We should hasten to add that, since the entire space M is closed, and since the intersection of any collection of closed sets is closed (exercise), the closure of any set S does exist, and is, in fact, the intersection of all closed sets containing S. The following definition will allow us to characterize the closure in another way.

Definition Let S be a nonempty subset of a metric space M. An element $x \in M$ is said to be a **limit point**, or **accumulation point** of S if every open ball centered at x meets S at a point other than x itself. Let us denote the set of all limit points of S by $l(S)$. □

Here are some key facts concerning limit points and closures.

Theorem 12.3 Let S be a nonempty subset of a metric space M.

1) An element $x \in M$ is a limit point of S if and only if there is a sequence (x_n) in S for which $x_n \neq x$ for all n, and $(x_n) \to x$.
2) S is closed if and only if $l(S) \subset S$. In words, S is closed if and only if it contains all of its limit points.
3) $cl(S) = S \cup l(S)$.
4) An element x is in $cl(S)$ if and only if there is a sequence (x_n) in M for which $(x_n) \to x$.

Proof.

1) Assume first that $x \in l(S)$. For each n, there exists a point $x_n \neq x$ such that $x_n \in B(x,1/n) \cap S$. Thus, we have
$$d(x_n,x) < 1/n$$
and so $(x_n) \to x$. For the converse, suppose that $(x_n) \to x$, where $x \neq x_n \in S$. If $B(x,r)$ is any ball centered at x, then there is some N such that $n > N$ implies $x_n \in B(x,r)$. Hence, for any ball $B(x,r)$ centered at x, there is a point $x_n \neq x$, such that $x_n \in S \cap B(x,r)$. Thus, x is a limit point of S.
2) As for part (2), if S is closed, then by part (1), any $x \in l(S)$ is the limit of a sequence (x_n) in S, and so must be in S. Hence, $l(S) \subset S$. Conversely, if $l(S) \subset S$, then S is closed. For if (x_n) is any sequence in S, and $(x_n) \to x$, then there are two possibilities. First, we might have $x_n = x$ for some n, in which case $x = x_n \in S$. Second, we might have $x_n \neq x$ for all n, in which case $(x_n) \to x$ implies that $x \in l(S) \subset S$. In either case, $x \in S$ and so S is closed under the taking of limits, which implies that S is closed.
3) Clearly, $S \subset T = S \cup l(S)$. To show that T is closed, we show that it contains all of its limit points. So let $x \in l(T)$. Hence, there is a sequence $(x_n) \in T$ for which $x_n \neq x$ and $(x_n) \to x$. Of course, each x_n is either in S, or is a limit point of S. We must show that $x \in T$, that is, that x is either in S or is a limit point of S.

Suppose for the purposes of contradiction that $x \notin S$ and $x \notin l(S)$. Then there is a ball $B(x,r)$ for which $B(x,r) \cap S \neq \emptyset$. However, since $(x_n) \to x$, there must be an $x_n \in B(x,r)$. Since x_n cannot be in S, it must be a limit point of S. Referring to Figure 12.1, if $d(x_n,x) = d < r$, then consider the ball $B(x_n,\frac{r-d}{2})$. This ball is completely contained in $B(x,r)$ and must contain an element y of S, since its center x_n is a limit point of S. But then $y \in S \cap B(x,r)$, a contradiction. Hence, $x \in S$ or $x \in l(S)$. In either case, $x \in T = S \cup l(S)$ and so T is closed.

Thus, T is closed and contains S, and so $cl(S) \subset T$. On the other hand, $T = S \cap l(S) \subset cl(S)$, and so $cl(S) = T$.

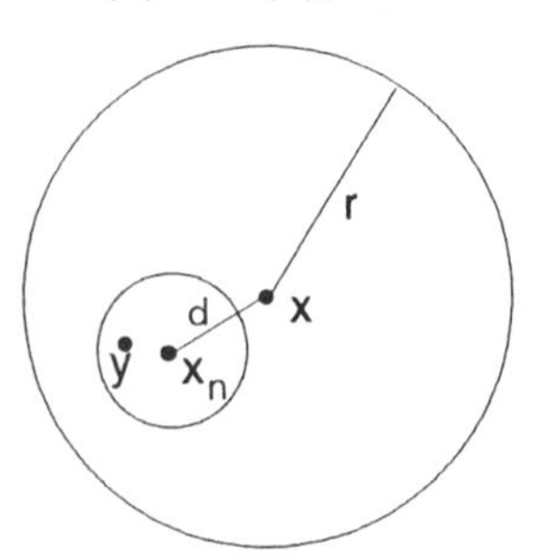

Figure 12.1

4) If $x \in cl(S)$, then there are two possibilities. If $x \in S$, then the constant sequence (x_n), with $x_n = x$ for all x, is a sequence in S that converges to x. If $x \notin S$, then $x \in l(S)$, and so there is a sequence (x_n) in S for which $x_n \neq x$ and $(x_n) \to x$. In either case, there is a sequence in S converging to x. Conversely, if there is a sequence (x_n) in S for which $(x_n) \to x$, then either $x_n = x$ for some n, in which case $x \in S \subset cl(S)$, or else $x_n \neq x$ for all n, in which case $x \in l(S) \subset cl(S)$. ∎

Dense Subsets

The following concept is meant to convey the idea of a subset $S \subset M$ being "arbitrarily close" to every point in M.

Definition A subset S of a metric space M is **dense** in M if $cl(S) = M$. A metric space is said to be **separable** if it contains a *countable* dense subset. □

Thus, a subset S of M is dense if every open ball about any point $x \in M$ contains at least one point of S.

Certainly, any metric space contains a dense subset, namely, the space itself. However, as the next examples show, not every metric space contains a countable dense subset.

Example 12.6

1) The real line $\mathbb{R}$ is separable, since the rational numbers $\mathbb{Q}$ form a countable dense subset. Similarly, $\mathbb{R}^n$ is separable, since the set $\mathbb{Q}^n$ is countable and dense.
2) The complex plane $\mathbb{C}$ is separable, as is $\mathbb{C}^n$ for all n.
3) A discrete metric space is separable if and only if it is countable. We leave proof of this as an exercise. □

Example 12.7 The space ℓ^∞ is *not* separable. Recall that ℓ^∞ is the set of all bounded sequences of real numbers (or complex numbers), with metric

$$d(\mathbf{x},\mathbf{y}) = \sup_n |x_n - y_n|$$

To see that this space is not separable, consider the set S of all binary sequences

$$S = \{(x_n) \mid x_i = 0 \text{ or } 1 \text{ for all } i\}$$

This set is in one-to-one correspondence with the set of all subsets of $\mathbb{N}$, and so is uncountable. (It has cardinality $2^{\aleph_0} > \aleph_0$.) Now, each sequence in S is certainly bounded and so lies in ℓ^∞. Moreover, if $\mathbf{x} \neq \mathbf{y} \in \ell^\infty$, then the two sequences must differ in at least one position, and so $d(\mathbf{x},\mathbf{y}) = 1$.

In other words, we have a subset S of ℓ^∞ that is uncountable, and for which the distance between any two distinct elements is 1. This implies that the uncountable collection of balls $\{B(\mathbf{s},1/3) \mid \mathbf{s} \in S\}$ is mutually disjoint. Hence, no countable set can meet every ball, which implies that no countable set can be dense in ℓ^∞. □

Example 12.8 The metric spaces ℓ^p are separable, for $p \geq 1$. The set S of all sequences of the form

$$\mathbf{s} = (q_1,\ldots,q_n,0,\ldots)$$

for all $n > 0$, where the q_i's are rational, is a countable set. Let us show that it is dense in ℓ^p. Any $\mathbf{x} \in \ell^p$ satisfies

$$\sum_{n=1}^{\infty} |x_n|^p < \infty$$

Hence, for any $\epsilon > 0$, there exists an N such that

$$\sum_{n=N+1}^{\infty} |x_n|^p < \frac{\epsilon}{2}$$

Since the rational numbers are dense in $\mathbb{R}$, we can find rational numbers q_i for which

$$|x_i - q_i|^p < \frac{\epsilon}{2N}$$

for all $i = 1,\ldots,N$. Hence, if $\mathbf{s} = (q_1,\ldots,q_N,0,\ldots)$, then

$$d(\mathbf{x},\mathbf{s})^p = \sum_{n=1}^{N} |x_n - q_n|^p + \sum_{n=N+1}^{\infty} |x_n|^p < \frac{\epsilon}{2} + \frac{\epsilon}{2} = \epsilon$$

which shows that there is an element of S arbitrarily close to any element of ℓ^p. Thus, S is dense in ℓ^p, and so ℓ^p is separable. □

Continuity

Continuity plays a central role in the study of linear operators on infinite dimensional inner product spaces.

Definition Let $f:M \to M'$ be a function from the metric space (M,d) to the metric space (M',d'). We say that f is **continuous at** $x_0 \in M$ if for any $\epsilon > 0$, there exists a $\delta > 0$ such that

$$d(x,x_0) < \delta \;\Rightarrow\; d'(f(x),f(x_0)) < \epsilon$$

or, equivalently,

$$f\Big(B(x_0,\delta)\Big) \subset B(f(x_0),\epsilon)$$

(See Figure 12.2.) A function is **continuous** if it is continuous at every $x_0 \in M$. □

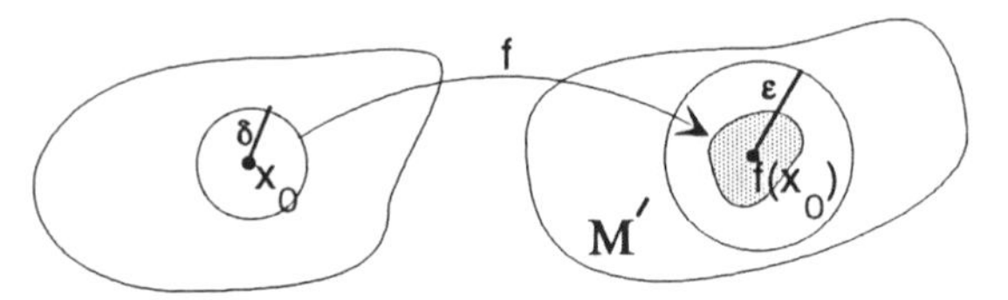

Figure 12.2

We can use the notion of convergence to characterize continuity for functions between metric spaces.

Theorem 12.4 A function $f:M \to M'$ is continuous if and only if whenever (x_n) is a sequence in M that converges to $x_0 \in M$, then the sequence $(f(x_n))$ converges to $f(x_0)$, in short,

$$(x_n) \to x_0 \;\Rightarrow\; (f(x_n)) \to f(x_0)$$

Proof. Suppose first that f is continuous at x_0, and let $(x_n) \to x_0$. Then, given $\epsilon > 0$, the continuity of f implies the existence of a $\delta > 0$ such that

$$f(B(x_0,\delta)) \subset B(f(x_0),\epsilon)$$

Since $(x_n) \to x$, there exists an $N > 0$ such that $x_n \in B(x_0, \delta)$ for $n > N$, and so

$$n > N \;\Rightarrow\; f(x_n) \in B(f(x_0), \epsilon)$$

Thus, $f(x_n) \to f(x_0)$.

Conversely, suppose that $(x_n) \to x_0$ implies $(f(x_n)) \to f(x_0)$. Suppose, for the purposes of contradiction, that f is not continuous at x_0. Then there exists an $\epsilon > 0$ such that, for all $\delta > 0$

$$f\Big(B(x_0, \delta)\Big) \subsetneqq B(f(x_0), \epsilon)$$

Thus, for all $n > 0$,

$$f\Big(B(x_0, \tfrac{1}{n})\Big) \subsetneqq B(f(x_0), \epsilon)$$

and so we may construct a sequence (x_n) by choosing each term x_n with the property that

$$x_n \in B(x_0, \tfrac{1}{n}), \quad \text{but} \quad f(x_n) \notin B(f(x_0), \epsilon)$$

Hence, $(x_n) \to x_0$, but $f(x_n)$ does not converge to $f(x_0)$. This contradiction implies that f must be continuous at x_0. ∎

The next theorem says that the distance function is a continuous function in both variables.

Theorem 12.5 Let (M, d) be a metric space. If $(x_n) \to x$ and $(y_n) \to y$, then $d(x_n, y_n) \to d(x, y)$.

Proof. According to Exercise 2,

$$| \, d(x_n, y_n) - d(x, y) \, | \le d(x_n, x) + d(y_n, y)$$

But the right side tends to 0 as $n \to \infty$, and so $d(x_n, y_n) \to d(x, y)$. ∎

Completeness

The reader who has studied analysis will recognize the following definitions.

Definition A sequence (x_n) in a metric space M is a **Cauchy sequence** if, for any $\epsilon > 0$, there exists an $N > 0$ for which

$$n, m > N \;\Rightarrow\; d(x_n, x_m) < \epsilon$$ □

We leave it to the reader to show that any convergent sequence is a Cauchy sequence. When the converse holds, the space is said to be *complete*.

Definition Let M be a metric space.

1) M is said to be **complete** if every Cauchy sequence in M converges in M.
2) A subspace S of M is **complete** if it is complete as a metric space. Thus, S is complete if every Cauchy sequence (s_n) in S converges *to an element in* S. □

Before considering examples, we prove a very useful result about completeness of subspaces.

Theorem 12.6 Let M be a metric space.

1) Any complete subspace of M is closed.
2) If M is complete then a subspace S of M is complete if and only if it is closed.

Proof. To prove (1), assume that S is a complete subspace of M. Let (x_n) be a sequence in S for which $(x_n) \to x \in M$. Then (x_n) is a Cauchy sequence in S, and since S is complete, (x_n) must converge to an element of S. Since limits of sequences are unique, we have $x \in S$. Hence, S is closed.

To prove part (2), first assume that S is complete. Then part (1) shows that S is closed. Conversely, suppose that S is closed, and let (x_n) be a Cauchy sequence in S. Since (x_n) is also a Cauchy sequence in the complete space M, it must converge to some $x \in M$. But since S is closed, we have $(x_n) \to x \in S$. Hence, S is complete. ∎

Now let us consider some examples of complete (and incomplete) metric spaces.

Example 12.9 It is well-known that the metric space $\mathbb{R}$ is complete. (However, a proof of this fact would lead us outside the scope of this book.) Similarly, the complex numbers $\mathbb{C}$ are complete. □

Example 12.10 The Euclidean space $\mathbb{R}^n$ and the unitary space $\mathbb{C}^n$ are complete. Let us prove this for $\mathbb{R}^n$. Suppose that $(\mathbf{x}_k)$ is a Cauchy sequence in $\mathbb{R}^n$, where

$$\mathbf{x}_k = (x_{k,1}, \ldots, x_{k,n})$$

Thus,

$$d(\mathbf{x}_k, \mathbf{x}_m)^2 = \sum_{i=1}^{n} (x_{k,i} - x_{m,i})^2 \to 0 \quad \text{as} \quad k,m \to \infty$$

and so, for each coordinate position i,

$$(x_{k,i} - x_{m,i})^2 \le d(\mathbf{x}_k, \mathbf{x}_m)^2 \to 0$$

which shows that the sequence $(x_{k,i})_{k=1,2,\dots}$ of ith coordinates is a Cauchy sequence in $\mathbb{R}$. Since $\mathbb{R}$ is complete, we must have

$$(x_{k,i}) \to y_i \quad \text{as} \quad k \to \infty$$

If $\mathbf{y} = (y_1, \dots, y_n)$, then

$$d(\mathbf{x}_k, \mathbf{y})^2 = \sum_{i=1}^{n} (x_{k,i} - y_i)^2 \to 0 \quad \text{as} \quad k \to \infty$$

and so $(\mathbf{x}_n) \to \mathbf{y} \in \mathbb{R}^n$. This proves that $\mathbb{R}^n$ is complete. ▯

Example 12.11 The metric space $(C[a,b], d)$ of all real-valued (or complex-valued) continuous functions on $[a,b]$, with metric

$$d(f,g) = \sup_{x \in [a,b]} | f(x) - g(x) |$$

is complete. To see this, we first observe that the limit with respect to d is the uniform limit on $[a,b]$, that is $d(f_n, f) \to 0$ if and only if for any $\epsilon > 0$, there is an $N > 0$ for which

$$n > N \Rightarrow | f_n(x) - f(x) | \leq \epsilon \quad \text{for } all \quad x \in [a,b]$$

Now, let (f_n) be a Cauchy sequence in $(C[a,b], d)$. Thus, for any $\epsilon > 0$, there is an N for which

$$\text{(12.1)} \qquad m,n > N \Rightarrow | f_n(x) - f_m(x) | \leq \epsilon \quad \text{for } all \quad x \in [a,b]$$

This implies that, for each $x \in [a,b]$, the sequence $(f_n(x))$ is a Cauchy sequence of real (or complex) numbers, and so it converges. We can therefore define a function f on $[a,b]$ by

$$f(x) = \lim_{n \to \infty} f_n(x)$$

Letting $m \to \infty$ in (12.1), we get

$$n > N \Rightarrow | f_n(x) - f(x) | \leq \epsilon \quad \text{for } all \quad x \in [a,b]$$

Thus, $f_n(x)$ converges to $f(x)$ *uniformly*. It is well-known that the uniform limit of continuous functions is continuous, and so $f(x) \in C[a,b]$. Thus, $(f_n(x)) \to f(x) \in C[a,b]$, and so $(C[a,b], d)$ is complete. ▯

Example 12.12 The metric space $(C[a,b], d_1)$ of all real-valued (or complex-valued) continuous functions on $[a,b]$, with metric

$$d_1(f(x), g(x)) = \int_a^b | f(x) - g(x) | \, dx$$

is *not* complete. For convenience, we take $[a,b] = [0,1]$ and leave the general case for the reader. Consider the sequence of functions $f_n(x)$

whose graphs are shown in Figure 12.3. (The definition of $f_n(x)$ should be clear from the graph.)

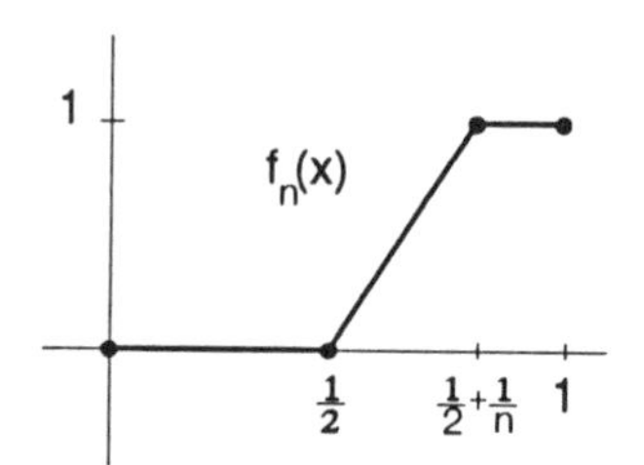

Figure 12.3

We leave it to the reader to show that the sequence $(f_n(x))$ is Cauchy, but does not converge in $(C[0,1],d_1)$. □

Example 12.13 The metric space ℓ^∞ is complete. To see this, suppose that $(\mathbf{x}_n)$ is a Cauchy sequence in ℓ^∞, where

$$\mathbf{x}_n = (x_{n,1}, x_{n,2}, \ldots)$$

Then, for each coordinate position i, we have

$$(12.2) \qquad |x_{n,i} - x_{m,i}| \le \sup_j |x_{n,j} - x_{m,j}| \to 0 \text{ as } n,m \to \infty$$

Hence, for each i, the sequence $(x_{n,i})_{n=1,2,\ldots}$ of ith coordinates is a Cauchy sequence in $\mathbb{R}$ (or $\mathbb{C}$). Since $\mathbb{R}$ (or $\mathbb{C}$) is complete, we have

$$(x_{n,i}) \to y_i \text{ as } n \to \infty$$

for each coordinate position $i = 1, 2, \ldots$. We want to show that $\mathbf{y} = (y_i) \in \ell^\infty$ and that $(\mathbf{x}_n) \to \mathbf{y}$.

Letting $m \to \infty$ in (12.2) gives

$$(12.3) \qquad \sup_j |x_{n,j} - y_j| \to 0 \text{ as } n \to \infty$$

and so, for some n,

$$|x_{n,j} - y_j| < 1 \text{ for all } j$$

and so

$$|y_j| < 1 + |x_{n,j}| \text{ for all } j$$

But since $\mathbf{x}_n \in \ell^\infty$, it is a bounded sequence, and therefore so is (y_j). That is, $\mathbf{y} = (y_j) \in \ell^\infty$. Since (12.3) implies that $(\mathbf{x}_n) \to \mathbf{y}$, we see that ℓ^∞ is complete. □

Example 12.14 The metric space ℓ^p is complete. To prove this, let $(\mathbf{x}_n)$ be a Cauchy sequence in ℓ^p, where

$$\mathbf{x}_n = (x_{n,1}, x_{n,2}, \ldots)$$

Then, for each coordinate position i,

$$|x_{n,i}-x_{m,i}|^p \le \sum_{j=1}^{\infty} |x_{n,j}-x_{m,j}|^p = d(\mathbf{x}_n,\mathbf{x}_m)^p \to 0$$

which shows that the sequence $(x_{n,i})_{n=1,2,...}$ of ith coordinates is a Cauchy sequence in $\mathbb{R}$ (or $\mathbb{C}$). Since $\mathbb{R}$ (or $\mathbb{C}$) is complete, we have

$$(x_{n,i}) \to y_i \quad \text{as} \quad n \to \infty$$

We want to show that $\mathbf{y} = (y_i) \in \ell^p$ and that $(\mathbf{x}_n) \to \mathbf{y}$.

To this end, observe that for any $\epsilon > 0$, there is an N for which

$$n,m > N \Rightarrow \sum_{i=1}^{r} |x_{n,i}-x_{m,i}|^p \le \epsilon$$

for all $r > 0$. Now, we let $m \to \infty$, to get

$$n > N \Rightarrow \sum_{i=1}^{r} |x_{n,i}-y_i|^p \le \epsilon$$

for all $r > 0$. Letting $r \to \infty$, we get, for any $n > N$,

$$\sum_{i=1}^{\infty} |x_{n,i}-y_i|^p < \epsilon$$

which implies that $(\mathbf{x}_n) - \mathbf{y} \in \ell^p$, and so $\mathbf{y} = \mathbf{y} - (\mathbf{x}_n) + (\mathbf{x}_n) \in \ell^p$, and in addition, $(\mathbf{x}_n) \to \mathbf{y}$. □

As we will see in the next chapter, the property of completeness plays a major role in the theory of inner product spaces. Inner product spaces for which the induced metric space is complete are called **Hilbert spaces.**

Isometries

A function between two metric spaces that preserves distance is called an *isometry.* Here is the formal definition.

Definition Let (M,d) and (M',d') be metric spaces. A function $f:M \to M'$ is called an **isometry** if

$$d'(f(x),f(y)) = d(x,y)$$

for all $x,y \in M$. If $f:M \to M'$ is a *bijective* isometry from M to M′, we say that M and M′ are **isometric** and write $M \approx M'$. ∎

Theorem 12.7 Let $f:(M,d) \to (M',d')$ be an isometry. Then

1) f is injective
2) f is continuous

3) $f^{-1}:f(M) \to M$ is also an isometry, and hence also continuous.

Proof. To prove (1), we observe that

$$f(x) = f(y) \Leftrightarrow d'(f(x),f(y)) = 0 \Leftrightarrow d(x,y) = 0 \Leftrightarrow x = y$$

To prove (2), let $(x_n) \to x$ in M, then

$$d'(f(x_n),f(x)) = d(x_n,x) \to 0 \quad \text{as } n \to \infty$$

and so $(f(x_n)) \to f(x)$, which proves that f is continuous. Finally, we have

$$d(f^{-1}(f(x)),f^{-1}(f(y))) = d(x,y) = d'(f(x),f(y))$$

and so $f^{-1}:f(M) \to M$ is an isometry. ∎

The Completion of a Metric Space

While not all metric spaces are complete, any metric space can be embedded in a complete metric space. To be more specific, we have the following important theorem.

Theorem 12.8 Let (M,d) be any metric space. Then there is a *complete* metric space (M',d') and an isometry $\tau:M \to \tau(M) \subset M'$ for which $\tau(M)$ is dense in M'. The metric space (M',d') is called a **completion** of (M,d). Moreover, (M',d') is unique, up to bijective isometry.

Proof. The proof is a bit lengthy, so we divide it into various parts. We can simplify the notation considerably by thinking of sequences (x_n) in M as functions $f:\mathbb{N} \to M$, where $f(n) = x_n$.

Cauchy Sequences in M

The basic idea is to let the elements of M' be equivalence classes of Cauchy sequences in M. So let CS(M) denote the set of all Cauchy sequences in M. If $f,g \in CS(M)$ then, intuitively speaking, the terms f(n) get closer together as $n \to \infty$, and so do the terms g(n). Therefore, it seems reasonable that $d(f(n),g(n))$ should approach a finite limit as $n \to \infty$. Indeed, according to Exercise 2,

$$| d(f(n),g(n)) - d(f(m),g(m)) | \leq d(f(n),f(m)) + d(g(n),g(m)) \to 0$$

as $n,m \to \infty$, and so $d(f(n),g(n))$ is a Cauchy sequence of real numbers, which implies that

$$\lim_{n \to \infty} d(f(n),g(n)) < \infty \qquad (12.4)$$

(That is, the limit exists and is finite.)

Equivalence Classes of Cauchy Sequences in M

We would *like* to define a metric d' on the set $\mathrm{CS(M)}$ by

$$d'(f,g)=\lim_{n\to\infty} d(f(n),g(n))$$

However, it is possible that

$$\lim_{n\to\infty} d(f(n),g(n))=0$$

for *distinct* sequences f and g, so this does not define a metric. Thus, we are lead to define an equivalence relation on $\mathrm{CS(M)}$ by

$$f\sim g \Leftrightarrow \lim_{n\to\infty} d(f(n),g(n))=0$$

Let $\overline{\mathrm{CS(M)}}$ be the set of all equivalence classes of Cauchy sequences, and define, for $\bar{f}, \bar{g}\in\overline{\mathrm{CS(M)}}$

$$d'(\bar{f},\bar{g})=\lim_{n\to\infty} d(f(n),g(n)) \tag{12.5}$$

where $f\in\bar{f}$ and $g\in\bar{g}$.

To see that d' is well-defined, suppose that $f'\in\bar{f}$ and $g'\in\bar{g}$. Then since $f'\sim f$ and $g'\sim g$, we have

$$|\, d(f'(n),g'(n))-d(f(n),g(n))\,| \le d(f'(n),f(n))+d(g'(n),g(n))\to 0$$

as $n\to\infty$. Thus,

$$f'\sim f \text{ and } g'\sim g \Rightarrow \lim_{n\to\infty} d(f'(n),g'(n))=\lim_{n\to\infty} d(f(n),g(n))$$

$$\Rightarrow\ d'(f',g')=d'(f,g)$$

which shows that d' is well-defined. To see that d' is a metric, we verify the triangle inequality, leaving the rest to the reader. If f,g and h are Cauchy sequences, then

$$d(f(n),g(n))\le d(f(n),h(n))+d(h(n),g(n))$$

Taking limits gives

$$\lim_{n\to\infty} d(f(n),g(n))\le\lim_{n\to\infty} d(f(n),h(n))+\lim_{n\to\infty} d(h(n),g(n))$$

and so

$$d'(\bar{f},\bar{g})\le d'(\bar{f},\bar{h})+d'(\bar{h},\bar{g})$$

Embedding (M,d) in (M',d')

For each $x\in M$, consider the constant Cauchy sequence $[x]$, where $[x](n)=x$ for all n. The map $\tau:M\to M'$ defined by

$$\tau(x)=\overline{[x]}$$

is an isometry, since

$$d'(\tau(x),\tau(y)) = d'(\overline{[x]},\overline{[y]}) = \lim_{n\to\infty} d([x](n),[y](n)) = d(x,y)$$

Moreover, $\tau(M)$ is dense in M'. This follows from the fact that we can approximate any Cauchy sequence in M by a constant sequence. In particular, let $\bar{f} \in M'$. Since $f \in \bar{f}$ is a Cauchy sequence, for any $\epsilon > 0$, there exists an N such that

$$n,m \geq N \Rightarrow d(f(n),f(m)) < \epsilon$$

Now, for the constant sequence $[f(N)]$ we have

$$d'(\overline{[f(N)]},\bar{f}) = \lim_{n\to\infty} d(f(N),f(n)) \leq \epsilon$$

and so $\tau(M)$ is dense in M'.

(M',d') is Complete

Suppose that

$$\overline{f_1}, \overline{f_2}, \overline{f_3}, \ldots$$

is a Cauchy sequence in M'. We wish to find a Cauchy sequence g in M for which

$$d'(\overline{f_k},\bar{g}) = \lim_{n\to\infty} d(f_k(n),g(n)) \to 0 \text{ as } k\to\infty$$

Since $\overline{f_k} \in M'$, and since $\tau(M)$ is dense in M', there is a constant sequence $[c_k]$ for which

$$d'(\overline{f_k},\overline{[c_k]}) = \lim_{n\to\infty} d(f_k(n),c_k) < \frac{1}{k}$$

Let g be the sequence defined by

$$g(k) = c_k$$

This is a Cauchy sequence in M, since

$$\begin{aligned} d(c_k,c_j) &= d'(\overline{[c_k]},\overline{[c_j]}) \\ &\leq d'(\overline{[c_k]},\overline{f_k}) + d'(\overline{f_k},\overline{f_j}) + d'(\overline{f_j},\overline{[c_j]}) \leq \frac{1}{k} + d'(\overline{f_k},\overline{f_j}) + \frac{1}{j} \to 0 \end{aligned}$$

as $k,j\to\infty$. To see that $\overline{f_k}$ converges to $\bar{g}$, observe that

$$\begin{aligned} d'(\overline{f_k},\bar{g}) &\leq d'(\overline{f_k},\overline{[c_k]}) + d'(\overline{[c_k]},\bar{g}) < \frac{1}{k} + \lim_{n\to\infty} d(c_k,g(n)) \\ &= \frac{1}{k} + \lim_{n\to\infty} d(c_k,c_n) \end{aligned}$$

Now, since g is a Cauchy sequence, for any $\epsilon > 0$, there is an N such that

$$k,n \geq N \Rightarrow d(c_k,c_n) < \epsilon$$

In particular,

$$k \geq N \Rightarrow \lim_{n\to\infty} d(c_k,c_n) \leq \epsilon$$

and so

$$k \geq N \Rightarrow d'(\overline{f_k},\overline{g}) \leq \frac{1}{k} + \epsilon$$

which implies that $\overline{f_k} \to g$, as desired.

Uniqueness

Finally, we must show that if (M',d') and (M'',d'') are both completions of (M,d), then $M' \approx M''$. Note that we have bijective isometries $\tau:M\to\tau(M) \subset M'$ and $\sigma:M\to\sigma(M) \subset M''$. Hence, the map $\rho = \sigma\tau^{-1}:\tau(M)\to\sigma(M)$ is a bijective isometry from $\tau(M)$ onto $\sigma(M)$, where $\tau(M)$ is dense in M'. (See Figure 12.4.)

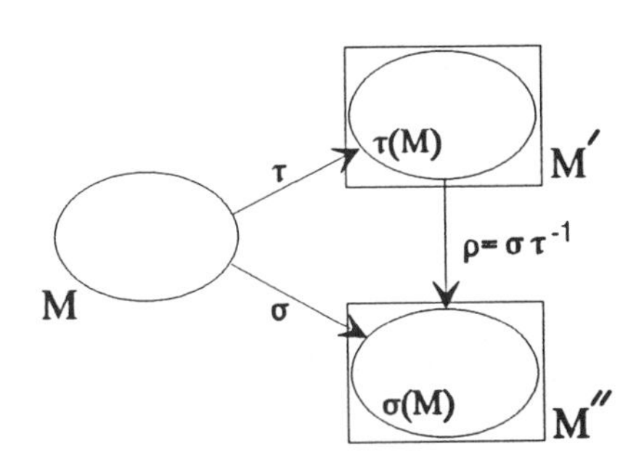

Figure 12.4

Our goal is to show that ρ can be extended to a bijective isometry $\overline{\rho}$ from M' to M''.

Let $x \in M'$. Then there is a sequence (a_n) in $\tau(M)$ for which $(a_n)\to x$. Since (a_n) is a Cauchy sequence in $\tau(M)$, $(\rho(a_n))$ is a Cauchy sequence in $\sigma(M) \subset M''$, and since M'' is complete, we have $(\rho(a_n))\to y$ for some $y \in M''$. Let us define $\overline{\rho}(x) = y$.

To see that $\overline{\rho}$ is well-defined, suppose that $(a_n)\to x$ and $(b_n)\to x$, where both sequences lie in $\tau(M)$. Then

$$d''(\rho(a_n),\rho(b_n)) = d'(a_n,b_n) \to 0 \text{ as } n\to\infty$$

and so $(\rho(a_n))$ and $(\rho(b_n))$ converge to the same element of M'', which implies that $\overline{\rho}(x)$ does not depend on the choice of sequence in $\tau(M)$ converging to x. Thus, $\overline{\rho}$ is well-defined. Moreover, if $a \in \tau(M)$, then the constant sequence $[a]$ converges to a, and so $\overline{\rho}(a) = \lim \rho(a) = \rho(a)$, which shows that $\overline{\rho}$ is an extension of ρ.

To see that $\overline{\rho}$ is an isometry, suppose that $(a_n)\to x$ and

$(b_n) \to y$. Then $(\rho(a_n)) \to \overline{\rho}(x)$ and $(\rho(b_n)) \to \overline{\rho}(y)$, and since d'' is continuous, we have

$$d''(\overline{\rho}(x),\overline{\rho}(y)) = \lim_{n\to\infty} d''(\rho(a_n),\rho(b_n)) = \lim_{n\to\infty} d'(a_n,b_n) = d'(x,y)$$

Thus, we need only show that $\overline{\rho}$ is surjective. Note first that $\sigma(M) = im(\rho) \subset im(\overline{\rho})$. Thus, if $im(\overline{\rho})$ is closed, we can deduce from the fact that $\sigma(M)$ is dense in M'' that $im(\overline{\rho}) = M''$. So, suppose that $(\overline{\rho}(x_n))$ is a sequence in $im(\overline{\rho})$, and $(\overline{\rho}(x_n)) \to z$. Then $(\overline{\rho}(x_n))$ is a Cauchy sequence, and therefore so is (x_n). Thus, $(x_n) \to x \in M'$. But $\overline{\rho}$ is continuous, and so $(\overline{\rho}(x_n)) \to \overline{\rho}(x)$, which implies that $\overline{\rho}(x) = z$, and so $z \in im(\overline{\rho})$. Hence, $\overline{\rho}$ is surjective, and $M' \approx M''$. ∎

EXERCISES

1. Prove the generalized triangle inequality

$$d(x_1,x_n) \leq d(x_1,x_2) + d(x_2,x_3) + \cdots + d(x_{n-1},x_n)$$

2. a) Use the triangle inequality to prove that

$$| d(x,y) - d(a,b) | \leq d(x,a) + d(y,b)$$

   b) Prove that

$$| d(x,z) - d(y,z) | \leq d(x,y)$$

3. Let $S \subset \ell^\infty$ be the subspace of all binary sequences (sequences of 0s and 1s). Describe the metric on S.
4. Let $M = \{0,1\}^n$ be the set of all binary n-tuples. Define a function $h: S \times S \to \mathbb{R}$ by letting $h(\mathbf{x},\mathbf{y})$ be the number of positions in which $\mathbf{x}$ and $\mathbf{y}$ differ. For example, $h[(11010),(01001)] = 3$. Prove that h is a metric. (It is called the **Hamming distance function** and plays an important role in coding theory.)
5. Let $1 \leq p < \infty$.
   a) If $\mathbf{x} = (x_n) \in \ell^p$ show that $x_n \to 0$
   b) Find a sequence that converges to 0 but is not an element of any ℓ^p for $1 \leq p < \infty$.
6. a) Show that if $\mathbf{x} = (x_n) \in \ell^p$, then $\mathbf{x} \in \ell^q$ for all $q > p$.
   b) Find a sequence $\mathbf{x} = (x_n)$ that is in ℓ^p for $p > 1$, but is not in ℓ^1.
7. Show that a subset S of a metric space M is open if and only if S contains an open neighborhood of each of its points.
8. Show that the intersection of any collection of closed sets in a metric space is closed.
9. Let (M,d) be a metric space. The **diameter** of a nonempty

subset $S \subset M$ is

$$\delta(S) = \sup_{x,y \in S} d(x,y)$$

A set S is **bounded** if $\delta(S) < \infty$.

a) Prove that S is bounded if and only if there is some $x \in M$ and $r \in \mathbb{R}$ for which $S \subset B(x,r)$.

b) Prove that $\delta(S) = 0$ if and only if S consists of a single point.

c) Prove that $S \subset T$ implies $\delta(S) \leq \delta(T)$.

d) If S and T is bounded, show that $S \cup T$ is also bounded.

10. Let (M,d) be a metric space. Let d' be the function defined by

$$d'(x,y) = \frac{d(x,y)}{1 + d(x,y)}$$

a) Show that $(M.d')$ is a metric space, and that M is bounded under this metric, even if it is not bounded under the metric d.

b) Show that the metric spaces (M,d) and (M,d') have the same open sets.

11. If S and T are subsets of a metric space (M,d), we define the **distance** between S and T by

$$\rho(S,T) = \inf_{x \in S, t \in T} d(x,y)$$

a) Is it true that $\rho(S,T) = 0$ if and only if $S = T$? Is ρ a metric?

b) Show that $x \in cl(S)$ if and only if $\rho(\{x\},S) = 0$.

12. Prove that $x \in M$ is a limit point of $S \subset M$ if and only if every neighborhood of x meets S in a point other than x itself.

13. Prove that $x \in M$ is a limit point of $S \subset M$ if and only if every open ball $B(x,r)$ contains infinitely many points of S.

14. Prove that limits are unique, that is, $(x_n) \to x$, $(x_n) \to y$ implies that $x = y$.

15. Let S be a subset of a metric space M. Prove that $x \in cl(S)$ if and only if there exists a sequence (x_n) in S that converges to x.

16. Prove that the closure has the following properties.

a) $S \subset cl(S)$ b) $cl(cl(S)) = S$

c) $cl(S \cup T) = cl(S) \cup cl(T)$ d) $cl(S \cap T) \subset cl(S) \cap cl(T)$

Can the last part be strengthened to equality?

17. a) Prove that the closed ball $\overline{B}(x,r)$ is always a closed subset.

b) Find an example of a metric space in which the closure of an open ball $B(x,r)$ is not equal to the closed ball $\overline{B}(x,r)$.

18. Provide the details to show that $\mathbb{R}^n$ is separable.
19. Prove that $\mathbb{C}^n$ is separable.
20. Prove that a discrete metric space is separable if and only if it is countable.
21. Prove that the metric space $\mathcal{B}[a,b]$ of all bounded functions on $[a,b]$, with metric
$$d(f,g) = \sup_{x \in [a,b]} |f(x) - g(x)|$$
is not separable.
22. Show that a function $f:(M,d)\to(M',d')$ is continuous if and only if the inverse image of any open set is open, that is, if and only if $f^{-1}(U) = \{x \in M \mid f(x) \in U\}$ is open in M whenever U is an open set in M'.
23. Repeat the previous exercise, replacing the word *open* by the word *closed.*
24. Give an example to show that if $f:(M,d)\to(M',d')$ is a continuous function and U is an open set in M, it need not be the case that $f(U)$ is open in M'.
25. Show that any convergent sequence is a Cauchy sequence.
26. If $(x_n)\to x$ in a metric space M, show that any subsequence (x_{n_k}) of (x_n) also converges to x.
27. Suppose that (x_n) is a Cauchy sequence in a metric space M, and that some subsequence (x_{n_k}) of (x_n) converges. Prove that (x_n) converges to the same limit as the subsequence.
28. Prove that if (x_n) is a Cauchy sequence, then the set $\{x_n\}$ is bounded. What about the converse? Is a bounded sequence necessarily a Cauchy sequence?
29. Let (x_n) and (y_n) be Cauchy sequences in a metric space M. Prove that the sequence $d_n = d(x_n,y_n)$ converges.
30. Show that the space of all convergent sequences of real numbers (or complex numbers) is complete as a subspace of ℓ^∞.
31. Let $\mathcal{P}$ denote the metric space of all polynomials over $\mathbb{C}$, with metric $d(p,q) = \sup_{[a,b]} |p(x) - q(x)|$. Is $\mathcal{P}$ complete?
32. Let $S \subset \ell^\infty$ be the subspace of all sequences with finite support (that is, with a finite number of nonzero terms). Is S complete?
33. Prove that the metric space $\mathbb{Z}$ of all integers, with metric $d(n,m) = |n - m|$, is complete.
34. Show that the subspace S of the metric space $C[a,b]$ (under the sup metric) consisting of all functions $f \in C[a,b]$ for which $f(a) = f(b)$ is complete.
35. If $M \approx M'$ and M is complete, show that M' is also complete.
36. Show that the metric spaces $C[a,b]$ and $C[c,d]$, under the sup metric, are isometric.

37. (**Hölder's inequality**) Prove Hölder's inequality

$$\sum_{n=1}^{\infty} |x_n y_n| \le \left(\sum_{n=1}^{\infty} |x_n|^p\right)^{1/p} \left(\sum_{n=1}^{\infty} |y_n|^q\right)^{1/q}$$

as follows.

a) Show that $s = t^{p-1} \Rightarrow t = s^{q-1}$

b) Let u and v be positive real numbers, and consider the rectangle R in $\mathbb{R}^2$ with corners $(0,0)$, $(u,0)$, $(0,v)$ and (u,v), with area uv. Argue geometrically (i.e., draw a picture) to show that

$$uv \le \int_0^u t^{p-1} dt + \int_0^v s^{q-1} ds$$

and so

$$uv \le \frac{u^p}{p} + \frac{v^q}{q}$$

c) Now let $X = \Sigma |x_n|^p < \infty$ and $Y = \Sigma |y_n|^q < \infty$. Apply the results of part (b), to

$$u = \frac{|x_n|}{X^{1/p}}, \quad v = \frac{|y_n|}{Y^{1/q}}$$

and then sum on n to deduce Hölder's inequality.

38. (**Minkowski's inequality**) Prove Minkowski's inequality

$$\left(\sum_{n=1}^{\infty} |x_n + y_n|^p\right)^{1/p} \le \left(\sum_{n=1}^{\infty} |x_n|^p\right)^{1/p} + \left(\sum_{n=1}^{\infty} |y_n|^p\right)^{1/p}$$

as follows.

a) Prove it for $p = 1$ first.

b) Assume $p > 1$. Show that

$$|x_n + y_n|^p \le |x_n|\,|x_n + y_n|^{p-1} + |y_n|\,|x_n + y_n|^{p-1}$$

c) Sum this from $n = 1$ to k, and apply Hölder's inequality to each sum on the right, to get

$$\sum_{n=1}^{k} |x_n + y_n|^p$$

$$\le \left\{\left(\sum_{n=1}^{k} |x_n|^p\right)^{1/p} + \left(\sum_{n=1}^{k} |y_n|^p\right)^{1/p}\right\}\left(\sum_{n=1}^{k} |x_n + y_n|^p\right)^{1/q}$$

Divide both sides of this by the last factor on the right, and let $n \to \infty$ to deduce Minkowski's inequality.

39. Prove that ℓ^p is a metric space.

CHAPTER 13

Hilbert Spaces

Contents: *A Brief Review. Hilbert Spaces. Infinite Series. An Approximation Problem. Hilbert Bases. Fourier Expansions. A Characterization of Hilbert Bases. Hilbert Dimension. A Characterization of Hilbert Spaces. The Riesz Representation Theorem. Exercises.*

Now that we have the necessary background on the topological properties of metric spaces, we can resume our study of inner product spaces without qualification as to dimension. As in Chapter 9, *we restrict attention to real and complex inner product spaces.* Hence F will denote either $\mathbb{R}$ or $\mathbb{C}$.

A Brief Review

Let us begin by reviewing some of the results from Chapter 9. Recall that an inner product space V over F is a vector space V, together with an inner product $\langle,\rangle:V\times V\to F$. If $F=\mathbb{R}$, then the inner product is bilinear, and if $F=\mathbb{C}$, the inner product is sesquilinear.

An inner product induces a norm on V, defined by

$$\|\mathbf{v}\| = \sqrt{\langle \mathbf{v},\mathbf{v}\rangle}$$

We recall in particular the following properties of the norm.

Theorem 13.1

1) (**The Cauchy-Schwarz inequality**) For all $\mathbf{u},\mathbf{v} \in V$,

$$|\langle \mathbf{u},\mathbf{v}\rangle| \le \|\mathbf{u}\| \, \|\mathbf{v}\|$$

with equality if and only if $\mathbf{u} = r\mathbf{v}$ for some $r \in F$.

2) (**The triangle inequality**) For all $\mathbf{u},\mathbf{v} \in V$,

$$\|\mathbf{u}+\mathbf{v}\| \le \|\mathbf{u}\| + \|\mathbf{v}\|$$

with equality if and only if $\mathbf{u} = r\mathbf{v}$ for some $r \in F$.

3) (**The parallelogram law**)

$$\|\mathbf{u}+\mathbf{v}\|^2 + \|\mathbf{u}-\mathbf{v}\|^2 = 2\|\mathbf{u}\|^2 + 2\|\mathbf{v}\|^2$$

∎

We have seen that the inner product can be recovered from the norm, as follows.

Theorem 13.2

1) If V is a real inner product space, then

$$\langle \mathbf{u},\mathbf{v}\rangle = \tfrac{1}{4}(\|\mathbf{u}+\mathbf{v}\|^2 - \|\mathbf{u}-\mathbf{v}\|^2)$$

2) If V is a complex inner product space, then

$$\langle \mathbf{u},\mathbf{v}\rangle = \tfrac{1}{4}(\|\mathbf{u}+\mathbf{v}\|^2 - \|\mathbf{u}-\mathbf{v}\|^2) + \tfrac{1}{4}i(\|\mathbf{u}+i\mathbf{v}\|^2 - \|\mathbf{u}-i\mathbf{v}\|^2)$$

The inner product also induces a metric on V defined by

$$d(\mathbf{u},\mathbf{v}) = \|\mathbf{u}-\mathbf{v}\|$$

Thus, any inner product space is a metric space.

Definition Let V and W be inner product spaces, and let $\tau \in \mathcal{L}(V,W)$.

1) τ is an **isometry** if it preserves the inner product, that is, if

$$\langle \tau(\mathbf{u}),\tau(\mathbf{v})\rangle = \langle \mathbf{u},\mathbf{v}\rangle$$

for all $\mathbf{u},\mathbf{v} \in V$.

2) A bijective isometry is called an **isometric isomorphism.** When $\tau:V\to W$ is an isometric isomorphism, we say that V and W are **isometrically isomorphic.** □

It is easy to see that an isometry is always injective but need not be surjective, even if $V = W$. (See Example 10.3.)

Theorem 13.3 A linear transformation $\tau \in \mathcal{L}(V,W)$ is an isometry if and only if it preserves the norm, that is,

$$\| \tau(\mathbf{v}) \| = \| \mathbf{v} \|$$

for all $\mathbf{v} \in V$. ∎

The following result points out one of the main differences between real and complex inner product spaces.

Theorem 13.4 Let V be an inner product space, and let $\tau \in \mathcal{L}(V)$.
1) If $\langle \tau(\mathbf{v}), \mathbf{w} \rangle = 0$ for all $\mathbf{v}, \mathbf{w} \in V$, then $\tau = 0$.
2) If V is a *complex* inner product space, and $\langle \tau(\mathbf{v}), \mathbf{v} \rangle = 0$ for all $\mathbf{v} \in V$, then $\tau = 0$.
3) Part (2) does *not* hold in general for real inner product spaces. ∎

Hilbert Spaces

Since an inner product space is a metric space, all that we learned about metric spaces applies to inner product spaces. In particular, if $(\mathbf{x}_n)$ is a sequence of vectors in an inner product space V, then

$$(\mathbf{x}_n) \rightarrow \mathbf{x} \text{ if and only if } \| \mathbf{x}_n - \mathbf{x} \| \rightarrow 0 \text{ as } n \rightarrow \infty$$

The fact that the inner product is continuous as a function of either of its coordinates is extremely useful.

Theorem 13.5 Let V be an inner product space. Then
1) $(\mathbf{x}_n) \rightarrow \mathbf{x}, (\mathbf{y}_n) \rightarrow \mathbf{y} \Rightarrow \langle \mathbf{x}_n, \mathbf{y}_n \rangle \rightarrow \langle \mathbf{x}, \mathbf{y} \rangle$
2) $(\mathbf{x}_n) \rightarrow \mathbf{x} \Rightarrow \| \mathbf{x}_n \| \rightarrow \| \mathbf{x} \|$ ∎

Complete inner product spaces play an especially important role in both theory and practice.

Definition An inner product space that is complete under the metric induced by the inner product is said to be a **Hilbert space**. □

Example 13.1 One of the most important examples of a Hilbert space is the space ℓ^2 of Example 10.2. Recall that the inner product is defined by

$$\langle \mathbf{x}, \mathbf{y} \rangle = \sum_{n=1}^{\infty} x_n \overline{y}_n$$

(In the real case, the conjugate is unnecessary.) The metric induced by this inner product is

$$d(\mathbf{x},\mathbf{y}) = \|\mathbf{x}-\mathbf{y}\|_2 = \left(\sum_{n=1}^{\infty} |x_n - y_n|^2\right)^{1/2}$$

which agrees with the definition of the metric space ℓ^2 given in Chapter 12. In other words, the metric in Chapter 12 is induced by this inner product. As we saw in Chapter 12, this inner product space is complete, and so it is a Hilbert space. (In fact, it is the prototype of all Hilbert spaces, introduced by David Hilbert in 1912, even before the axiomatic definition of Hilbert space was given by Johnny von Neumann in 1927.) □

The previous example raises the question of whether or not the other metric spaces ℓ^p $(p \neq 2)$, with distance given by

$$(13.1) \qquad d(\mathbf{x},\mathbf{y}) = \|\mathbf{x}-\mathbf{y}\|_p = \left(\sum_{n=1}^{\infty} |x_n - y_n|^p\right)^{1/p}$$

are complete inner product spaces. The fact is that they are not even inner product spaces! More specifically, there is no inner product whose induced metric is given by (13.1). To see this, observe that, according to Theorem 13.1, any norm that comes from an inner product must satisfy the parallelogram law

$$\|\mathbf{x}+\mathbf{y}\|^2 + \|\mathbf{x}-\mathbf{y}\|^2 = 2\|\mathbf{x}\|^2 + 2\|\mathbf{y}\|^2$$

But the norm in (13.1) does not satisfy this law. To see this, take $\mathbf{x} = (1,1,0\ldots)$ and $\mathbf{y} = (1,-1,0\ldots)$. Then

$$\|\mathbf{x}+\mathbf{y}\|_p = 2, \quad \|\mathbf{x}-\mathbf{y}\|_p = 2$$

and

$$\|\mathbf{x}\|_p = 2^{1/p}, \quad \|\mathbf{y}\|_p = 2^{1/p}$$

Thus, the left side of the parallelogram law is 8, and the right side is $4 \cdot 2^{2/p}$, which equals 8 if and only if $p = 2$.

Just as any metric space has a completion, so does any inner product space.

Theorem 13.6 Let V be an inner product space. Then there exists a Hilbert space H and an isometry $\tau: V \to H$ for which $\tau(V)$ is dense in H. Moreover, H is unique up to isometric isomorphism.

Proof. We know that the *metric space* (V,d), where d is induced by the inner product, has a unique completion (V',d'), which consists of equivalence classes of Cauchy sequences in V. If $(\mathbf{x}_n) \in \overline{(\mathbf{x}_n)} \in V'$ and $(\mathbf{y}_n) \in \overline{(\mathbf{y}_n)} \in V'$, then we set

$$\overline{(\mathbf{x}_n)} + \overline{(\mathbf{y}_n)} = \overline{(\mathbf{x}_n + \mathbf{y}_n)}, \qquad r\overline{(\mathbf{x}_n)} = \overline{(r\mathbf{x}_n)}$$

and

$$\langle \overline{(\mathbf{x}_n)}, \overline{(\mathbf{y}_n)} \rangle = \lim_{n\to\infty} \langle \mathbf{x}_n, \mathbf{y}_n \rangle$$

It is easy to see that, since $(\mathbf{x}_n)$ and $(\mathbf{y}_n)$ are Cauchy sequences, so are $(\mathbf{x}_n + \mathbf{y}_n)$ and $(r\mathbf{x}_n)$. In addition, these definitions are well-defined, that is, they are independent of the choice of representative from each equivalence class. For instance, if $(\hat{\mathbf{x}}_n) \in \overline{(\mathbf{x}_n)}$ then

$$\lim_{n\to\infty} \| \mathbf{x}_n - \hat{\mathbf{x}}_n \| = 0$$

and so

$$| \langle \mathbf{x}_n, \mathbf{y}_n \rangle - \langle \hat{\mathbf{x}}_n, \mathbf{y}_n \rangle | = | \langle \mathbf{x}_n - \hat{\mathbf{x}}_n, \mathbf{y}_n \rangle | \le \| \mathbf{x}_n - \hat{\mathbf{x}}_n \| \, \| \mathbf{y}_n \| \to 0$$

(The Cauchy sequence $(\mathbf{y}_n)$ is bounded.) Hence,

$$\langle \overline{(\mathbf{x}_n)}, \overline{(\mathbf{y}_n)} \rangle = \lim_{n\to\infty} \langle \mathbf{x}_n, \mathbf{y}_n \rangle = \lim_{n\to\infty} \langle \hat{\mathbf{x}}_n, \mathbf{y}_n \rangle = \langle \overline{(\hat{\mathbf{x}}_n)}, \overline{(\mathbf{y}_n)} \rangle$$

We leave it to the reader to show that V' is an inner product space under these operations.

Moreover, the inner product on V' induces the metric d', since

$$\begin{aligned}
\langle \overline{(\mathbf{x}_n - \mathbf{y}_n)}, \overline{(\mathbf{x}_n - \mathbf{y}_n)} \rangle &= \lim_{n\to\infty} \langle \mathbf{x}_n - \mathbf{y}_n, \mathbf{x}_n - \mathbf{y}_n \rangle \\
&= \lim_{n\to\infty} d(\mathbf{x}_n, \mathbf{y}_n)^2 \\
&= d'((\mathbf{x}_n), (\mathbf{y}_n))^2
\end{aligned}$$

Hence, the metric space isometry $\tau: V \to V'$ is an isometry of inner product spaces, since

$$\langle \tau(\mathbf{x}), \tau(\mathbf{y}) \rangle = d'(\tau(\mathbf{x}), \tau(\mathbf{y}))^2 = d(\mathbf{x}, \mathbf{y})^2 = \langle \mathbf{x}, \mathbf{y} \rangle$$

Thus, V' is a complete inner product space, and $\tau(V)$ is a dense subspace of V' that is isometrically isomorphic to V. We leave the issue of uniqueness to the reader. ∎

The next result concerns subspaces of inner product spaces.

Theorem 13.7

1) Any complete subspace of an inner product space is closed.
2) A subspace of a Hilbert space is a Hilbert space if and only if it is closed.
3) Any finite dimensional subspace of an inner product space is closed and complete.

Proof. Parts (1) and (2) follow from Theorem 12.6. Let us prove that a finite dimensional subspace S of an inner product space V is closed. Suppose that $(\mathbf{x}_n)$ is a sequence in S, $(\mathbf{x}_n) \to \mathbf{x}$, and $\mathbf{x} \notin S$.

Let $\mathfrak{B} = \{\mathbf{b}_1, \ldots, \mathbf{b}_m\}$ be an orthonormal Hamel basis for S. The Fourier expansion

$$\mathbf{s} = \sum_{i=1}^{m} \langle \mathbf{x}, \mathbf{b}_i \rangle \mathbf{b}_i$$

in S has the property that $\mathbf{x} - \mathbf{s} \neq \mathbf{0}$ but

$$\langle \mathbf{x} - \mathbf{s}, \mathbf{b}_j \rangle = \langle \mathbf{x}, \mathbf{b}_j \rangle - \langle \mathbf{s}, \mathbf{b}_j \rangle = 0$$

Thus, if we write $\mathbf{y} = \mathbf{x} - \mathbf{s}$ and $\mathbf{y}_n = \mathbf{x}_n - \mathbf{s} \in S$, the sequence $(\mathbf{y}_n)$, which is in S, converges to a vector $\mathbf{y}$ that is orthogonal to S. But this is impossible, because $\mathbf{y}_n \perp \mathbf{y}$ implies that

$$\| \mathbf{y}_n - \mathbf{y} \|^2 = \| \mathbf{y}_n \|^2 + \| \mathbf{y} \|^2 \geq \| \mathbf{y} \|^2 \not\to 0$$

This proves that S is closed.

To see that any finite dimensional subspace S of an inner product space is complete, let us embed S (as an inner product space in its own right) in its completion S′. Then S (or rather an isometric copy of S) is a finite dimensional subspace of a complete inner product space S′, and as such it is closed. However, S is dense in S′ and so S = S′, which shows that S is complete. ∎

Infinite Series

Since an inner product space allows both addition of vectors and convergence of sequences, we can define the concept of infinite sums, or infinite series.

Definition Let V be an inner product space, and let $(\mathbf{x}_k)$ be a sequence in V. The **nth partial sum** of the sequence is $\mathbf{s}_n = \mathbf{x}_1 + \cdots + \mathbf{x}_n$. If the sequence (s_n) of partial sums converges to a vector $\mathbf{s} \in V$, that is, if

$$\| \mathbf{s}_n - \mathbf{s} \| \to 0 \text{ as } n \to \infty$$

then we say that the *series* Σx_n **converges** to $\mathbf{s}$, and write

$$\sum_{n=1}^{\infty} x_n = \mathbf{s}$$ □

We can also define absolute convergence.

Definition A series $\Sigma \mathbf{x}_k$ is said to be **absolutely convergent** if the series

$$\sum_{n=1}^{\infty} \| \mathbf{x}_k \|$$

converges. □

The key relationship between convergence and absolute convergence is given in the next theorem. Note that completeness is required to guarantee that absolute convergence implies convergence.

Theorem 13.8 Let V be an inner product space. Then V is complete if and only if absolute convergence of a series implies convergence.

Proof. Suppose that V is complete, and that $\Sigma \| \mathbf{x}_k \| < \infty$. Then the sequence $\mathbf{s}_n$ of partial sums is a Cauchy sequence, for if $n > m$, we have

$$\| \mathbf{s}_n - \mathbf{s}_m \| = \| \sum_{k=m+1}^{n} \mathbf{x}_k \| \leq \sum_{k=m+1}^{n} \| \mathbf{x}_k \| \to 0$$

Hence, the sequence $(\mathbf{s}_n)$ converges, that is, the series $\Sigma \mathbf{x}_k$ converges.

Conversely, suppose that absolute convergence implies convergence, and let $(\mathbf{x}_n)$ be a Cauchy sequence in V. We wish to show that this sequence converges. Since $(\mathbf{x}_n)$ is a Cauchy sequence, for each $k > 0$, there exists an N_k with the property that

$$i,j \geq N_k \;\Rightarrow\; \| \mathbf{x}_i - \mathbf{x}_j \| < \frac{1}{2^k}$$

Clearly, we can choose $N_1 < N_2 < \cdots$, in which case

$$\| \mathbf{x}_{N_{k+1}} - \mathbf{x}_{N_k} \| < \frac{1}{2^k}$$

and so

$$\sum_{k=1}^{\infty} \| \mathbf{x}_{N_{k+1}} - \mathbf{x}_{N_k} \| \leq \sum_{k=1}^{\infty} \frac{1}{2^k} < \infty$$

Thus, according to hypothesis, the series

$$\sum_{k=1}^{\infty} (\mathbf{x}_{N_{k+1}} - \mathbf{x}_{N_k})$$

converges. But this is a telescoping series, whose nth partial sum is

$$\mathbf{x}_{N_{n+1}} - \mathbf{x}_{N_1}$$

and so the *subsequence* $(\mathbf{x}_{N_k})$ converges. Since any Cauchy sequence that has a convergent subsequence must itself converge, the sequence $(\mathbf{x}_k)$ converges, and so V is complete. ▮

An Approximation Problem

Suppose that V is an inner product space, and that S is a subset of V. It is of considerable interest to be able to find, for any $\mathbf{x} \in V$, a vector in S that is closest to $\mathbf{x}$ in the metric induced by the inner product, should such a vector exist. This is the *approximation problem* for V.

Suppose that $\mathbf{x} \in V$, and let

$$\delta = \inf_{\mathbf{s} \in S} \| \mathbf{x} - \mathbf{s} \|$$

Then there is a sequence $\mathbf{s}_n$ for which

$$\delta_n = \| \mathbf{x} - \mathbf{s}_n \| \to \delta$$

as shown in Figure 13.1.

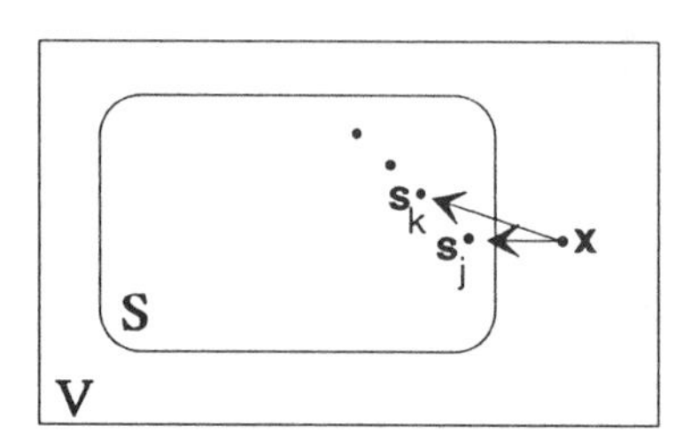

Figure 13.1

Let us see what we can learn about this sequence. First, if we let $\mathbf{y}_k = \mathbf{x} - \mathbf{s}_k$, then according to the parallelogram law

$$\| \mathbf{y}_k + \mathbf{y}_j \|^2 + \| \mathbf{y}_k - \mathbf{y}_j \|^2 = 2(\| \mathbf{y}_k \|^2 + \| \mathbf{y}_j \|^2)$$

or

$$(13.2) \qquad \| \mathbf{y}_k - \mathbf{y}_j \|^2 = 2(\| \mathbf{y}_k \|^2 + \| \mathbf{y}_j \|^2) - 4 \| \frac{\mathbf{y}_k + \mathbf{y}_j}{2} \|^2$$

Now, if the set S is **convex**, that is, if

$$\mathbf{x},\mathbf{y} \in S \Rightarrow r\mathbf{x} + (1-r)\mathbf{y} \in S \text{ for all } 0 \le r \le 1$$

(in words S contains the line segment between any two of its points) then $(\mathbf{s}_k + \mathbf{s}_j)/2 \in S$ and so

$$\| \frac{\mathbf{y}_k + \mathbf{y}_j}{2} \| = \| \mathbf{x} - \frac{\mathbf{s}_k + \mathbf{s}_j}{2} \| \ge \delta$$

Thus, (13.2) gives

$$\| \mathbf{y}_k - \mathbf{y}_j \|^2 \le 2(\| \mathbf{y}_k \|^2 + \| \mathbf{y}_j \|^2) - 4\delta^2 \to 0$$

as $k,j \to \infty$. Hence, if S is convex, then $(\mathbf{y}_n) = (\mathbf{x} - \mathbf{s}_n)$ is a Cauchy sequence, and therefore so is $(\mathbf{s}_n)$.

If we also require that S be complete, then the Cauchy sequence $(\mathbf{s}_n)$ converges to a vector $\widehat{\mathbf{s}} \in S$, and by the continuity of the norm, we must have $\| \mathbf{x} - \widehat{\mathbf{s}} \| = \delta$. Let us summarize and add a remark about uniqueness.

Theorem 13.9 Let V be an inner product space, and let S be a complete convex subset of V. Then for any $\mathbf{x} \in V$, there exists a *unique* $\hat{\mathbf{s}} \in S$ for which

$$\| \mathbf{x} - \hat{\mathbf{s}} \| = \inf_{\mathbf{s} \in S} \| \mathbf{x} - \mathbf{s} \|$$

The vector $\hat{\mathbf{s}}$ is called the **best approximation** to $\mathbf{x}$ in S.

Proof. Only the uniqueness remains to be established. Suppose that

$$\| \mathbf{x} - \hat{\mathbf{s}} \| = \delta = \| \mathbf{x} - \mathbf{s}' \|$$

Then, by the parallelogram law,

$$\begin{aligned} \| \hat{\mathbf{s}} - \mathbf{s}' \|^2 &= \| (\mathbf{x} - \mathbf{s}') - (\mathbf{x} - \hat{\mathbf{s}}) \|^2 \\ &\le 2 \| \mathbf{x} - \hat{\mathbf{s}} \|^2 + 2 \| \mathbf{x} - \mathbf{s}' \|^2 - \| 2\mathbf{x} - \hat{\mathbf{s}} - \mathbf{s}' \|^2 \\ &= 2 \| \mathbf{x} - \hat{\mathbf{s}} \|^2 + 2 \| \mathbf{x} - \mathbf{s}' \|^2 - 4 \| \mathbf{x} - \frac{\hat{\mathbf{s}} + \mathbf{s}'}{2} \|^2 \\ &\le 2\delta^2 + 2\delta^2 - 4\delta^2 = 0 \end{aligned}$$

and so $\hat{\mathbf{s}} = \mathbf{s}'$. ∎

Since any subspace S of an inner product space V is convex, Theorem 13.9 applies to complete subspaces. However, in this case, we can say more.

Theorem 13.10 Let V be an inner product space, and let S be a complete *subspace* of V. Then for any $\mathbf{x} \in V$, the best approximation to $\mathbf{x}$ in S is the unique vector $\mathbf{s}' \in S$ for which $\mathbf{x} - \mathbf{s}' \perp S$.

Proof. Suppose that $\mathbf{x} - \mathbf{s}' \perp S$, where $\mathbf{s}' \in S$. Then for any $\mathbf{s} \in S$, we have $\mathbf{x} - \mathbf{s}' \perp \mathbf{s} - \mathbf{s}'$ and so

$$\| \mathbf{x} - \mathbf{s} \|^2 = \| \mathbf{x} - \mathbf{s}' \|^2 + \| \mathbf{s}' - \mathbf{s} \|^2 \ge \| \mathbf{x} - \mathbf{s}' \|^2$$

Hence $\mathbf{s}' = \hat{\mathbf{s}}$ is the best approximation to $\mathbf{x}$ in S. Now we need only show that $\mathbf{x} - \hat{\mathbf{s}} \perp S$, where $\hat{\mathbf{s}}$ is the best approximation to $\mathbf{x}$ in S. For any $\mathbf{s} \in S$, a little computation reminiscent of completing the square gives

$$\begin{aligned} \| \mathbf{x} - r\mathbf{s} \|^2 &= \langle \mathbf{x} - r\mathbf{s}, \mathbf{x} - r\mathbf{s} \rangle \\ &= \| \mathbf{x} \|^2 - \bar{r}\langle \mathbf{x}, \mathbf{s} \rangle - r\langle \mathbf{s}, \mathbf{x} \rangle + r\bar{r} \| \mathbf{s} \|^2 \\ &= \| \mathbf{x} \|^2 + \| \mathbf{s} \|^2 \left(r\bar{r} - \bar{r}\frac{\langle \mathbf{x}, \mathbf{s} \rangle}{\| \mathbf{s} \|^2} - r\frac{\overline{\langle \mathbf{x}, \mathbf{s} \rangle}}{\| \mathbf{s} \|^2} \right) \end{aligned}$$

$$= \|\mathbf{x}\|^2 + \|\mathbf{s}\|^2\left(r - \frac{\langle \mathbf{x},\mathbf{s}\rangle}{\|\mathbf{s}\|^2}\right)\left(\bar{r} - \frac{\overline{\langle \mathbf{x},\mathbf{s}\rangle}}{\|\mathbf{s}\|^2}\right) - \frac{|\langle \mathbf{x},\mathbf{s}\rangle|^2}{\|\mathbf{s}\|}$$

$$= \|\mathbf{x}\|^2 + \|\mathbf{s}\|^2\left|r - \frac{\langle \mathbf{x},\mathbf{s}\rangle}{\|\mathbf{s}\|^2}\right|^2 - \frac{|\langle \mathbf{x},\mathbf{s}\rangle|^2}{\|\mathbf{s}\|}$$

Now, the last expression is smallest when

$$r = r_0 = \frac{\langle \mathbf{x},\mathbf{s}\rangle}{\|\mathbf{s}\|^2}$$

in which case

$$\|\mathbf{x} - r_0\mathbf{s}\|^2 \leq \|\mathbf{x}\|^2 - \frac{|\langle \mathbf{x},\mathbf{s}\rangle|^2}{\|\mathbf{s}\|}$$

Replacing $\mathbf{x}$ by $\mathbf{x} - \widehat{\mathbf{s}}$ gives

$$\|\mathbf{x} - \widehat{\mathbf{s}} - r_0\mathbf{s}\|^2 \leq \|\mathbf{x} - \widehat{\mathbf{s}}\|^2 - \frac{|\langle \mathbf{x} - \widehat{\mathbf{s}},\mathbf{s}\rangle|^2}{\|\mathbf{s}\|} \leq \delta - \frac{|\langle \mathbf{x} - \widehat{\mathbf{s}},\mathbf{s}\rangle|^2}{\|\mathbf{s}\|}$$

But $\widehat{\mathbf{s}} + r_0\mathbf{s} \in S$, and so the left side must be at least δ, implying that

$$\frac{|\langle \mathbf{x} - \widehat{\mathbf{s}},\mathbf{s}\rangle|^2}{\|\mathbf{s}\|} = 0$$

or, equivalently,

$$\langle \mathbf{x} - \widehat{\mathbf{s}},\mathbf{s}\rangle = 0$$

Hence, $\mathbf{x} - \widehat{\mathbf{s}} \perp S$. ∎

According to Theorem 13.10, if S is a complete subspace of an inner product space V, then for any $\mathbf{x} \in V$, we may write

$$\mathbf{x} = \widehat{\mathbf{s}} + (\mathbf{x} - \widehat{\mathbf{s}})$$

where $\widehat{\mathbf{s}} \in S$ and $\mathbf{x} - \widehat{\mathbf{s}} \in S^{\perp}$. Hence, $V = S + S^{\perp}$, and since $S \cap S^{\perp} = \{\mathbf{0}\}$, we also have $V = S \oplus S^{\perp}$. This is the projection theorem for arbitrary inner product spaces.

Theorem 13.11 (The projection theorem) If S is a complete subspace of an inner product space V, then

$$V = S \oplus S^{\perp}$$

In particular, if S is a closed subspace of a Hilbert space H, then

$$H = S \oplus S^{\perp}$$ ∎

Theorem 13.12 Let S, T and T' be subspaces of an inner product space V.

1) If $V = S \odot T$, then $T = S^{\perp}$.
2) If $S \oplus T = S \oplus T'$, then $T = T'$.

Proof. If $V = S \odot T$ then $T \subset S^{\perp}$ by definition of orthogonal direct sum. On the other hand, if $\mathbf{z} \in S^{\perp}$, then $\mathbf{z} = \mathbf{s} + \mathbf{t}$, for some $\mathbf{s} \in S$ and $\mathbf{t} \in T$. Hence,

$$0 = \langle \mathbf{z},\mathbf{s} \rangle = \langle \mathbf{s},\mathbf{s} \rangle + \langle \mathbf{t},\mathbf{s} \rangle = \langle \mathbf{s},\mathbf{s} \rangle$$

and so $\mathbf{s} = \mathbf{0}$, implying that $\mathbf{z} = \mathbf{t} \in T$. Thus, $S^{\perp} \subset T$. Part (2) follows from part (1). ∎

Let us denote the closure of the span of a set S of vectors by *cspan*(S).

Theorem 13.13 Let H be a Hilbert space.
1) If A is a *subset* of H, then

$$cspan(A) = A^{\perp\perp}$$

2) If S is a *subspace* of H, then

$$cl(S) = S^{\perp\perp}$$

3) If K is a closed subspace of H, then

$$K = K^{\perp\perp}$$

Proof. We leave it as an exercise to show that $[cspan(A)]^{\perp} = A^{\perp}$. Hence

$$H = cspan(A) \odot [cspan(A)]^{\perp} = cspan(A) \odot A^{\perp}$$

But since $A^{\perp}$ is closed, we also have

$$H = A^{\perp} \odot A^{\perp\perp}$$

and so by Theorem 13.12, $cspan(A) = A^{\perp\perp}$. The rest follows easily from part (1). ∎

In the exercises, we provide an example of a closed subspace K of an inner product space V for which $K \neq K^{\perp\perp}$. Hence, we cannot drop the requirement that H be a Hilbert space in Theorem 13.13.

Corollary 13.14 If A is a *subset* of a Hilbert space H then *span*(A) is dense in H if and only if $A^{\perp} = \{\mathbf{0}\}$.

Proof. As in the previous proof,

$$H = cspan(A) \odot A^{\perp}$$

and so $A^{\perp} = \{\mathbf{0}\}$ if and only if $H = cspan(A)$. ∎

Hilbert Bases

We recall the following definition from Chapter 9.

Definition A maximal orthonormal set in a Hilbert space H is called a **Hilbert basis** for H. □

Zorn's lemma can be used to show that any nontrivial Hilbert space has a Hilbert basis. Again, we should mention that the concepts of Hilbert basis and Hamel basis (a maximal linearly independent set) are quite different. We will show later in this chapter that any two Hilbert bases for a Hilbert space have the same dimension.

Since an orthonormal set $\mathcal{O}$ is maximal if and only if $\mathcal{O}^{\perp} = \{\mathbf{0}\}$, Corollary 13.14 gives the following characterization of Hilbert bases.

Theorem 13.15 Let $\mathcal{O}$ be an orthonormal subset of a Hilbert space H. The following are equivalent.

1) $\mathcal{O}$ is a Hilbert basis
2) $\mathcal{O}^{\perp} = \{\mathbf{0}\}$
3) $\mathcal{O}$ is a **total subset** of H, that is, $cspan(\mathcal{O}) = H$. ∎

Part (3) of this theorem says that a subset of a Hilbert space is a Hilbert basis if and only if it is a *total orthonormal set.*

Fourier Expansions

We now want to take a closer look at best approximations. Our goal is to find an explicit expression for the best approximation to any vector $\mathbf{x}$ from within a closed subspace S of a Hilbert space H. We will find it convenient to consider three cases, depending on whether S has finite, countably infinite, or uncountable dimension.

The Finite Dimensional Case

Suppose that $\mathcal{O} = \{\mathbf{u}_1, \ldots, \mathbf{u}_n\}$ is an orthonormal set in a Hilbert space H. Recall that the Fourier expansion of any $\mathbf{x} \in H$, with respect to $\mathcal{O}$, is given by

$$\widehat{\mathbf{x}} = \sum_{k=1}^{n} \langle \mathbf{x}, \mathbf{u}_k \rangle \mathbf{u}_k$$

where $\langle \mathbf{x}, \mathbf{u}_k \rangle$ is the Fourier coefficient of $\mathbf{x}$ with respect to $\mathbf{u}_k$. Observe that

$$\langle \mathbf{x} - \widehat{\mathbf{x}}, \mathbf{u}_k \rangle = \langle \mathbf{x}, \mathbf{u}_k \rangle - \langle \widehat{\mathbf{x}}, \mathbf{u}_k \rangle = 0$$

and so $\mathbf{x} - \widehat{\mathbf{x}} \perp span(\mathcal{O})$. Thus, according to Theorem 13.10, the Fourier expansion $\widehat{\mathbf{x}}$ is the best approximation to $\mathbf{x}$ in $span(\mathcal{O})$.

Moreover, since $\mathbf{x}-\widehat{\mathbf{x}} \perp \widehat{\mathbf{x}}$, we have

$$\|\widehat{\mathbf{x}}\|^2 = \|\mathbf{x}\|^2 - \|\mathbf{x}-\widehat{\mathbf{x}}\|^2 \le \|\mathbf{x}\|^2$$

and so

$$\|\widehat{\mathbf{x}}\| \le \|\mathbf{x}\|$$

with equality if and only if $\mathbf{x}=\widehat{\mathbf{x}}$, which happens if and only if $\mathbf{x} \in span(\mathcal{O})$. Let us summarize.

Theorem 13.16 Let $\mathcal{O} = \{\mathbf{u}_1,\ldots,\mathbf{u}_n\}$ be a finite orthonormal set in a Hilbert space H. For any $\mathbf{x} \in H$, the Fourier expansion $\widehat{\mathbf{x}}$ of $\mathbf{x}$ is the best approximation to $\mathbf{x}$ in $span(\mathcal{O})$. We also have **Bessel's inequality**

$$\|\widehat{\mathbf{x}}\| \le \|\mathbf{x}\|$$

or, equivalently

$$\sum_{k=1}^{n} |\langle \mathbf{x},\mathbf{u}_k\rangle|^2 \le \|\mathbf{x}\|^2 \tag{13.3}$$

with equality if and only if $\mathbf{x} \in span(\mathcal{O})$. ∎

The Countably Infinite Dimensional Case

In the countably infinite case, we will be dealing with infinite sums, and so questions of convergence will arise. Thus, we begin with the following.

Theorem 13.17 Let $\mathcal{O} = \{\mathbf{u}_1,\mathbf{u}_2,\ldots\}$ be a countably infinite orthonormal set in a Hilbert space H. The series

$$\sum_{k=1}^{\infty} r_k \mathbf{u}_k \tag{13.4}$$

converges in H if and only if the series

$$\sum_{k=1}^{\infty} |r_k|^2 \tag{13.5}$$

converges in $\mathbb{R}$. If these series converge, then they converge unconditionally (that is, any series formed by rearranging the order of the terms also converges). Finally, if the series (13.4) converges then

$$\left\| \sum_{k=1}^{\infty} r_k \mathbf{u}_k \right\|^2 = \sum_{k=1}^{\infty} |r_k|^2$$

Proof. Denote the partial sums of the first series by $\mathbf{s}_n$ and the partial sums of the second series by $\mathbf{p}_n$. Then for $m \le n$

$$\|s_n - s_m\|^2 = \left\| \sum_{k=m+1}^{n} r_k u_k \right\|^2 = \sum_{k=m+1}^{n} |r_k|^2 = |p_n - p_m|$$

Hence (s_n) is a Cauchy sequence in H if and only if (p_n) is a Cauchy sequence in $\mathbb{R}$. Since both H and $\mathbb{R}$ are complete, (s_n) converges if and only if (p_n) converges.

If the series (13.5) converges, then it converges absolutely, and hence unconditionally. (A *real* series converges unconditionally if and only if it converges absolutely.) But if (13.5) converges unconditionally, then so does (13.4). The last part of the theorem follows from the continuity of the norm. ∎

Now let $\mathcal{O} = \{u_1, u_2, \ldots\}$ be a countably infinite orthonormal set in H. The **Fourier expansion** of a vector $x \in H$ is defined to be the sum

$$\widehat{x} = \sum_{k=1}^{\infty} \langle x, u_k \rangle u_k \tag{13.6}$$

To see that this sum converges, observe that, for any $n > 0$, (13.3) gives

$$\sum_{k=1}^{n} |\langle x, u_k \rangle|^2 \leq \|x\|^2$$

and so

$$\sum_{k=1}^{\infty} |\langle x, u_k \rangle|^2 \leq \|x\|^2$$

which shows that the series on the left converges. Hence, according to Theorem 13.17, the Fourier expansion (13.6) converges unconditionally.

Moreover, since the inner product is continuous,

$$\langle x - \widehat{x}, u_k \rangle = \langle x, u_k \rangle - \langle \widehat{x}, u_k \rangle = 0$$

and so $x - \widehat{x} \in [span(\mathcal{O})]^{\perp} = [cspan(\mathcal{O})]^{\perp}$. Hence, $\widehat{x}$ is the best approximation to x in $cspan(\mathcal{O})$. Finally, since $x - \widehat{x} \perp \widehat{x}$, we again have

$$\|\widehat{x}\|^2 = \|x\|^2 - \|x - \widehat{x}\|^2 \leq \|x\|^2$$

and so

$$\|\widehat{x}\| \leq \|x\|$$

with equality if and only if $x = \widehat{x}$, which happens if and only if $x \in cspan(\mathcal{O})$. Thus, the following analog of Theorem 13.16 holds.

Theorem 13.18 Let $\mathcal{O} = \{u_1, u_2, \ldots\}$ be a countably infinite orthonormal set in a Hilbert space H. For any $x \in H$, the Fourier expansion

$$\widehat{x} = \sum_{k=1}^{\infty} \langle x, u_k \rangle u_k$$

of $\mathbf{x}$ converges unconditionally and is the best approximation to $\mathbf{x}$ in $cspan(\mathcal{O})$. We also have **Bessel's inequality**

$$\|\hat{\mathbf{x}}\| \leq \|\mathbf{x}\|$$

or, equivalently

$$\sum_{k=1}^{\infty} |\langle \mathbf{x},\mathbf{u}_k \rangle|^2 \leq \|\mathbf{x}\|^2$$

with equality if and only if $\mathbf{x} \in cspan(\mathcal{O})$. ∎

The Arbitrary Case

To discuss the case of an arbitrary orthonormal set $\mathcal{O} = \{\mathbf{u}_k \mid k \in K\}$, let us first define and discuss the concept of the sum of an arbitrary number of terms. (This is a bit of a digression, since we could proceed without all of the coming details – but they are interesting.)

Definition Let $\mathcal{X} = \{\mathbf{x}_k \mid k \in K\}$ be an arbitrary family of vectors in an inner product space V. The sum $\sum_{k \in K} \mathbf{x}_k$ is said to **converge** to a vector $\mathbf{x} \in V$, and we write

$$\mathbf{x} = \sum_{k \in K} \mathbf{x}_k \tag{13.7}$$

if for any $\epsilon > 0$, there exists a *finite* set $S \subset K$ for which

$$T \supset S,\ T \text{ finite} \Rightarrow \left\| \sum_{k \in T} \mathbf{x}_k - \mathbf{x} \right\| \leq \epsilon$$ ∎

For those readers familiar with the language of convergence of nets, the set $\mathcal{P}_0(K)$ of all finite subsets of K is a directed set under inclusion, and the function

$$S \rightarrow \sum_{k \in S} \mathbf{x}_k$$

is a net in H. Convergence of (13.7) is convergence of this net. In any case, we will refer to the preceding definition as the *net* definition of convergence.

It is not hard to verify the following basic properties of net convergence for arbitrary sums.

Theorem 13.19 Let $\mathcal{X} = \{\mathbf{x}_k \mid k \in K\}$ be an arbitrary family of vectors in an inner product space V. If

$$\sum_{k \in K} \mathbf{x}_k = \mathbf{x} \quad \text{and} \quad \sum_{k \in K} \mathbf{y}_k = \mathbf{y}$$

then

1) $\sum_{k \in K} rx_k = rx$ for any $r \in F$

2) $\sum_{k \in K} (x_k + y_k) = x + y$

3) $\sum_K \langle x_k, y \rangle = \langle x, y \rangle$ and $\sum_K \langle y, x_k \rangle = \langle y, x \rangle$ ∎

The next result gives a useful description of convergence, which does not require explicit mention of the sum.

Theorem 13.20 Let $\mathcal{K} = \{x_k \mid k \in K\}$ be an arbitrary family of vectors in an inner product space V.

1) If the sum

$$\sum_{k \in K} x_k$$

converges, then for any $\epsilon > 0$, there exists a *finite* set $I \subset K$ such that

$$J \cap I = \emptyset, \text{ J finite } \Rightarrow \left\| \sum_{k \in J} x_k \right\| \leq \epsilon$$

2) If V is a Hilbert space, then the converse of (1) also holds.

Proof. For part (1), given $\epsilon > 0$, let $S \subset K$, S finite, be such that

$$T \supset S, \text{ T finite } \Rightarrow \left\| \sum_{k \in T} x_k - x \right\| \leq \frac{\epsilon}{2}$$

If $J \cap S = \emptyset$, J finite, then

$$\begin{aligned} \| \sum_J x_k \| &= \| (\sum_J x_k + \sum_S x_k - x) - (\sum_S x_k - x) \| \\ &\leq \| \sum_{J \cup S} x_k - x \| + \| \sum_S x_k - x \| \leq \frac{\epsilon}{2} + \frac{\epsilon}{2} = \epsilon \end{aligned}$$

As for part (2), for each $n > 0$, let $I_n \subset K$ be a finite set for which

$$J \cap I_n = \emptyset, \text{ J finite } \Rightarrow \left\| \sum_{j \in J} x_j \right\| \leq \frac{1}{n}$$

and let

$$y_n = \sum_{k \in I_n} x_k$$

Then (y_n) is a Cauchy sequence, since

$$\begin{aligned} \| y_n - y_m \| &= \| \sum_{I_n} x_k - \sum_{I_m} x_k \| = \| \sum_{I_n - I_m} x_k - \sum_{I_m - I_n} x_k \| \\ &\leq \| \sum_{I_n - I_m} x_k \| + \| \sum_{I_m - I_n} x_k \| \leq \frac{1}{m} + \frac{1}{n} \to 0 \end{aligned}$$

Since V is assumed complete, we have $(\mathbf{y}_n) \to \mathbf{y}$.

Now, given $\epsilon > 0$, there exists an N such that

$$n \geq N \Rightarrow \|\mathbf{y}_n - \mathbf{y}\| = \|\sum_{I_n} \mathbf{x}_k - \mathbf{y}\| \leq \frac{\epsilon}{2}$$

Setting $n = \max\{N, 2/\epsilon\}$ gives

$T \supset I_n$, T finite $\Rightarrow$

$$\begin{aligned}\|\sum_{T} \mathbf{x}_k - \mathbf{y}\| &= \|\sum_{I_n} \mathbf{x}_k - \mathbf{y} + \sum_{T-I_n} \mathbf{x}_k\| \\ &\leq \|\sum_{I_n} \mathbf{x}_k - \mathbf{y}\| + \|\sum_{T-I_n} \mathbf{x}_k\| \leq \frac{\epsilon}{2} + \frac{1}{n} \leq \epsilon\end{aligned}$$

and so $\sum_{k \in K} \mathbf{x}_k$ converges to $\mathbf{y}$. ∎

The following theorem tells us that convergence of an arbitrary sum implies something very special about the terms.

Theorem 13.21 Let $\mathcal{K} = \{\mathbf{x}_k \mid k \in K\}$ be an arbitrary family of vectors in an inner product space V. If the sum

$$\sum_{k \in K} \mathbf{x}_k$$

converges, then at most a countable number of terms $\mathbf{x}_k$ can be nonzero.

Proof. According to Theorem 13.20, for each $n > 0$, we can let $I_n \subset K$, I_n finite, be such that

$$J \cap I_n = \emptyset,\ J \text{ finite} \Rightarrow \left\|\sum_{j \in J} \mathbf{x}_j\right\| \leq \frac{1}{n}$$

Let $I = \bigcup_n I_n$. Then I is countable, and

$$k \notin I \Rightarrow \{k\} \cap I_n = \emptyset \text{ for all } n \Rightarrow \|\mathbf{x}_k\| \leq \frac{1}{n} \text{ for all } n \Rightarrow \mathbf{x}_k = \mathbf{0} \quad ∎$$

Here is the analog of Theorem 13.17.

Theorem 13.22 Let $\mathcal{O} = \{\mathbf{u}_k \mid k \in K\}$ be an arbitrary orthonormal family of vectors in a Hilbert space H. The two series

$$\sum_{k \in K} r_k \mathbf{u}_k \quad \text{and} \quad \sum_{k \in K} |r_k|^2$$

converge or diverge together. If these series converge then

$$\left\|\sum_{k \in K} r_k \mathbf{u}_k\right\|^2 = \sum_{k \in K} |r_k|^2$$

Proof. The first series converges if and only if for any $\epsilon > 0$, there exists a finite set $I \subset K$ such that

$$J \cap I = \emptyset, \ J \text{ finite} \Rightarrow \left\| \sum_{k \in J} r_k u_k \right\|^2 \leq \epsilon^2$$

or, equivalently

$$J \cap I = \emptyset, \ J \text{ finite} \Rightarrow \sum_{k \in J} |r_k|^2 \leq \epsilon^2$$

and this is precisely what it means for the second series to converge. We leave proof of the remaining statement to the reader. ∎

The following is a useful characterization of arbitrary sums of nonnegative real terms.

Theorem 13.23 Let $\{r_k \mid k \in K\}$ be a collection of nonnegative real numbers. Then

$$\sum_{k \in K} r_k = \sup_{\substack{J \text{ finite} \\ J \subset K}} \sum_{k \in J} r_k \tag{13.8}$$

provided that either of the preceding expressions are finite.

Proof. Suppose that

$$\sup_{\substack{J \text{ finite} \\ J \subset K}} \sum_{k \in J} r_k = R < \infty$$

Then, for any $\epsilon > 0$, there exists a finite set $S \subset K$ such that

$$R \geq \sum_{k \in S} r_k \geq R - \epsilon$$

Hence, if $T \subset K$ is a finite set for which $T \supset S$, then since $r_k \geq 0$,

$$R \geq \sum_{k \in T} r_k \geq \sum_{k \in S} r_k \geq R - \epsilon$$

and so

$$\left\| R - \sum_{k \in T} r_k \right\| \leq \epsilon$$

which shows that $\sum r_k$ converges to R. Finally, if the sum on the left of (13.8) converges, then the supremum on the right is finite, and so (13.8) holds. ∎

The reader may have noticed that we have two definitions of convergence for countably infinite series — the net version and the traditional version involving the limit of partial sums. Let us write

$$\sum_{k \in \mathbb{N}^+} x_k \quad \text{and} \quad \sum_{k=1}^{\infty} x_k$$

for the net version and the partial sum version, respectively. Here is the relationship between these two definitions.

Theorem 13.24 Let H be a Hilbert space. If $\mathbf{x}_k \in H$ for all k, then the following are equivalent.

1) $\sum_{k \in \mathbb{N}^+} \mathbf{x}_k$ converges (net version) to $\mathbf{x}$

2) $\sum_{k=1}^{\infty} \mathbf{x}_k$ converges unconditionally to $\mathbf{x}$

Proof. Assume that (1) holds. Suppose that π is any permutation of $\mathbb{N}^+$. Given any $\epsilon > 0$, there is a finite set $S \subset \mathbb{N}^+$ for which

$$T \supset S, \ T \text{ finite} \Rightarrow \left\| \sum_{k \in T} \mathbf{x}_k - \mathbf{x} \right\| \leq \epsilon$$

Let us denote the set of integers $\{1,\ldots,n\}$ by I_n, and choose a positive integer n so that $\pi(I_n) \supset S$. Then

$$m \geq n \Rightarrow \pi(I_m) \supset \pi(I_n) \supset S$$
$$\Rightarrow \left\| \sum_{k=1}^{m} \mathbf{x}_{\pi(k)} - \mathbf{x} \right\| = \left\| \sum_{k \in \pi(I_m)} \mathbf{x}_k - \mathbf{x} \right\| \leq \epsilon$$

and so (2) holds.

Next, assume that (2) holds, but that the series in (1) does not converge. Then there exists an $\epsilon > 0$ such that, for any finite subset $I \subset \mathbb{N}^+$, there exists a finite subset J with $J \cap I = \emptyset$ for which

$$\left\| \sum_{k \in J} \mathbf{x}_k \right\| > \epsilon$$

From this, we deduce the existence of a countably infinite sequence J_n of mutually disjoint finite subsets of $\mathbb{N}^+$ with the property that

$$\max(J_n) = M_n < m_{n+1} = \min(J_{n+1})$$

and

$$\left\| \sum_{k \in J_n} \mathbf{x}_k \right\| > \epsilon$$

Now, we choose any permutation $\pi:\mathbb{N}^+ \to \mathbb{N}^+$ with the following properties

1) $\pi([m_n,M_n]) \subset [m_n,M_n]$

2) if $J_n = \{j_{n,1},\ldots,j_{n,u_n}\}$ then
$$\pi(m_n) = j_{n,1},\ \pi(m_n+1) = j_{n,2},\ldots,\pi(m_n+u_n-1) = j_{n,u_n}$$

The intention in property (2) is that, for each n, π takes a set of *consecutive* integers to the integers in J_n.

For any such permutation π, we have

$$\left\| \sum_{k=m_n}^{m_n+u_n-1} \mathbf{x}_{\pi(k)} \right\| = \left\| \sum_{k\in J_n} \mathbf{x}_k \right\| > \epsilon$$

which shows that the sequence of partial sums of the series

$$\sum_{k=1}^{\infty} \mathbf{x}_{\pi(k)}$$

is not Cauchy, and so this series does not converge. This contradicts (2), and shows that (2) implies at least that (1) converges. But if (1) converges to $\mathbf{y} \in H$, then since (1) implies (2), and since unconditional limits are unique, we have $\mathbf{y} = \mathbf{x}$. Hence, (2) implies (1). ∎

Now we can return to a discussion of Fourier expansions. Let $\mathcal{O} = \{\mathbf{u}_k \mid k \in K\}$ be an arbitrary orthonormal set in a Hilbert space H. Given any $\mathbf{x} \in H$, we may apply Theorem 13.16 to all finite subsets of $\mathcal{O}$, to deduce that

$$\sup_{\substack{J \text{ finite} \\ J \subset K}} \sum_{k\in J} |\langle \mathbf{x},\mathbf{u}_k\rangle|^2 \leq \|\mathbf{x}\|^2$$

and so Theorem 13.23 tells us that the sum

$$\sum_{k\in K} |\langle \mathbf{x},\mathbf{u}_k\rangle|^2$$

converges. Hence, according to Theorem 13.22, the **Fourier expansion**

$$\hat{\mathbf{x}} = \sum_{k\in K} \langle \mathbf{x},\mathbf{u}_k\rangle \mathbf{u}_k$$

of $\mathbf{x}$ also converges, and

$$\|\hat{\mathbf{x}}\|^2 = \sum_{k\in K} |\langle \mathbf{x},\mathbf{u}_k\rangle|^2$$

Note that, according to Theorem 13.21, $\hat{\mathbf{x}}$ is a countably infinite sum of terms of the form $\langle \mathbf{x},\mathbf{u}_k\rangle\mathbf{u}_k$, and so is in $cspan(\mathcal{O})$.

In view of part (3) of Theorem 13.19, we have

$$\langle \mathbf{x}-\hat{\mathbf{x}},\mathbf{u}_k\rangle = \langle \mathbf{x},\mathbf{u}_k\rangle - \langle \hat{\mathbf{x}},\mathbf{u}_k\rangle = 0$$

and so $\mathbf{x}-\hat{\mathbf{x}} \in [span(\mathcal{O})]^{\perp} = [cspan(\mathcal{O})]^{\perp}$. Hence, $\hat{\mathbf{x}}$ is the best approximation to $\mathbf{x}$ in $cspan(\mathcal{O})$. Finally, since $\mathbf{x}-\hat{\mathbf{x}} \perp \hat{\mathbf{x}}$, we again have

$$\|\hat{\mathbf{x}}\|^2 = \|\mathbf{x}\|^2 - \|\mathbf{x}-\hat{\mathbf{x}}\|^2 \leq \|\mathbf{x}\|^2$$

and so

$$\|\hat{\mathbf{x}}\| \leq \|\mathbf{x}\|$$

with equality if and only if $\mathbf{x} = \hat{\mathbf{x}}$, which happens if and only if $\mathbf{x} \in cspan(\mathcal{O})$. Thus, we arrive at the most general form of a key theorem about Hilbert spaces.

Theorem 13.25 Let $\mathcal{O} = \{\mathbf{u}_k \mid k \in K\}$ be an orthonormal family of vectors in a Hilbert space H. For any $\mathbf{x} \in H$, the Fourier expansion

$$\hat{\mathbf{x}} = \sum_{k \in K} \langle \mathbf{x}, \mathbf{u}_k \rangle \mathbf{u}_k$$

of $\mathbf{x}$ converges in H, and is the unique best approximation to $\mathbf{x}$ in $cspan(\mathcal{O})$. Moreover, we have **Bessel's inequality**

$$\|\hat{\mathbf{x}}\| \leq \|\mathbf{x}\|$$

or, equivalently

$$\sum_{k \in K} |\langle \mathbf{x}, \mathbf{u}_k \rangle|^2 \leq \|\mathbf{x}\|^2$$

with equality if and only if $\mathbf{x} \in cspan(\mathcal{O})$. ∎

A Characterization of Hilbert Bases

Recall from Theorem 13.15 that an orthonormal set $\mathcal{O} = \{\mathbf{u}_k \mid k \in K\}$ in a Hilbert space H is a Hilbert basis if and only if

$$cspan(\mathcal{O}) = H$$

Theorem 13.25 then leads to the following characterization of Hilbert bases.

Theorem 13.26 Let $\mathcal{O} = \{\mathbf{u}_k \mid k \in K\}$ be an orthonormal family in a Hilbert space H. The following are equivalent.

1) $\mathcal{O}$ is a Hilbert basis (a maximal orthonormal set)
2) $\mathcal{O}^\perp = \{\mathbf{0}\}$
3) $\mathcal{O}$ is total (that is, $cspan(\mathcal{O}) = H$)
4) $\mathbf{x} = \hat{\mathbf{x}}$ for all $\mathbf{x} \in H$
5) Equality holds in Bessel's inequality for all $\mathbf{x} \in H$, that is,

$$\|\mathbf{x}\| = \|\hat{\mathbf{x}}\|$$

for all $\mathbf{x} \in H$.

6) **Parseval's identity**

$$\langle \mathbf{x}, \mathbf{y} \rangle = \langle \hat{\mathbf{x}}, \hat{\mathbf{y}} \rangle$$

holds for all $\mathbf{x}, \mathbf{y} \in H$, that is,

$$\langle \mathbf{x}, \mathbf{y} \rangle = \sum_{k \in K} \langle \mathbf{x}, \mathbf{u}_k \rangle \overline{\langle \mathbf{y}, \mathbf{u}_k \rangle}$$

Proof. Parts (1), (2) and (3) are equivalent by Theorem 13.15. Part

(4) implies part (3), since $\hat{\mathbf{x}} \in cspan(\mathcal{O})$, and (3) implies (4) since the unique best approximation of any $\mathbf{x} \in cspan(\mathcal{O})$ is itself, and so $\mathbf{x} = \hat{\mathbf{x}}$. Parts (3) and (5) are equivalent by Theorem 13.25. Parseval's identity follows from part (4) by part (3) of Theorem 13.19. Finally, Parseval's identity for $\mathbf{y} = \mathbf{x}$ implies that equality holds in Bessel's inequality. ∎

Hilbert Dimension

We now wish to show that all Hilbert bases for a Hilbert space H have the same cardinality, and so we can define the Hilbert dimension of H to be that cardinality.

Theorem 13.27 All Hilbert bases for a Hilbert space H have the same cardinality. This cardinality is called the **Hilbert dimension** of H. We will denote the Hilbert dimension of H by $hdim(\mathrm{H})$.

Proof. If H has a finite Hilbert basis, then that set is also a Hamel basis, and so all Hilbert bases have size $dim(\mathrm{H})$. Suppose next that $\mathcal{B} = \{\mathbf{b}_k \mid k \in K\}$ and $\mathcal{C} = \{\mathbf{c}_j \mid j \in J\}$ are infinite Hilbert bases for H. Then for each $\mathbf{b}_k$, we have

$$\mathbf{b}_k = \sum_{j \in J_k} \langle \mathbf{b}_k, \mathbf{c}_j \rangle \mathbf{c}_j$$

where J_k is the *countable* set $\{j \mid \langle \mathbf{b}_k, \mathbf{c}_j \rangle \neq 0\}$. Moreover, since no $\mathbf{c}_j$ can be orthogonal to every $\mathbf{b}_k$, we have $\bigcup_K J_k = J$. Thus, since each J_k is countable, Theorem 0.16 gives

$$|J| = \left|\bigcup_{k \in K} J_k\right| \leq \aleph_0 |K| = |K|$$

By symmetry, we also have $|K| \leq |J|$, and so the Schröder-Bernstein theorem implies that $|J| = |K|$. ∎

Theorem 13.28 Two Hilbert spaces are isometrically isomorphic if and only if they have the same Hilbert dimension.

Proof. Suppose that $hdim(\mathrm{H}_1) = hdim(\mathrm{H}_2)$. Let $\mathcal{O}_1 = \{\mathbf{u}_k \mid k \in K\}$ be a Hilbert basis for H_1 and $\mathcal{O}_2 = \{\mathbf{v}_k \mid k \in K\}$ be a Hilbert basis for H_2. We may define a map $\tau: \mathrm{H}_1 \to \mathrm{H}_2$ as follows

$$\tau\left(\sum_{k \in K} r_k \mathbf{u}_k\right) = \sum_{k \in K} r_k \mathbf{v}_k$$

We leave it as an exercise to verify that τ is a bijective isometry. The converse is also left as an exercise. ∎

A Characterization of Hilbert Spaces

We have seen that any vector space V is isomorphic to a vector space $(F^B)_0$ of all functions from B to F that have finite support. There is a corresponding result for Hilbert spaces. Let K be any nonempty set, and let

$$\ell^2(K) = \left\{ f:K \to \mathbb{C} \,\middle|\, \sum_{k \in K} |f(k)|^2 < \infty \right\}$$

The functions in $\ell^2(K)$ are referred to as **square summable functions.** (We can also define a real version of this set by replacing $\mathbb{C}$ by $\mathbb{R}$.) We define an inner product on $\ell^2(K)$ by

$$\langle f,g \rangle = \sum_{k \in K} f(k)\overline{g(k)}$$

The proof that $\ell^2(K)$ is a Hilbert space is quite similar to the proof that $\ell^2 = \ell^2(\mathbb{N})$ is a Hilbert space, and the details are left to the reader. If we define $\delta_k \in \ell^2(K)$ by

$$\delta_k(j) = \delta_{k,j} = \begin{cases} 1 & \text{if } j = k \\ 0 & \text{if } j \neq k \end{cases}$$

then the collection

$$\mathcal{O} = \{\delta_k \mid k \in K\}$$

is a Hilbert basis for $\ell^2(K)$, of cardinality $|K|$. To see this, observe that

$$\langle \delta_i, \delta_j \rangle = \sum_{k \in K} \delta_i(k)\overline{\delta_j(k)} = \delta_{i,j}$$

and so $\mathcal{O}$ is orthonormal. Moreover, if $f \in \ell^2(K)$, then $f(k) \neq 0$ for only a countable number of $k \in K$, say $\{k_1, k_2, \ldots\}$. If we define f' by

$$f' = \sum_{i=1}^{\infty} f(k_i)\delta_{k_i}$$

then $f' \in cspan(\mathcal{O})$ and $f'(j) = f(j)$ for all $j \in K$, which implies that $f = f'$. This shows that $\ell^2(K) = cspan(\mathcal{O})$, and so $\mathcal{O}$ is a total orthonormal set, that is, a Hilbert basis for $\ell^2(K)$.

Now let H be a Hilbert space, with Hilbert basis $\mathcal{B} = \{\mathbf{u}_k \mid k \in K\}$. We define a map $\phi: H \to \ell^2(K)$ as follows. Since $\mathcal{B}$ is a Hilbert basis, any $\mathbf{x} \in H$ has the form

$$\mathbf{x} = \sum_{k \in K} \langle \mathbf{x}, \mathbf{u}_k \rangle \mathbf{u}_k$$

Since the series on the right converges, Theorem 13.22 implies that the series

$$\sum_{k \in K} |\langle \mathbf{x}, \mathbf{u}_k \rangle|^2$$

converges. Hence, another application of Theorem 13.22 implies that the following series converges, and so we may set

$$\phi(\mathbf{x}) = \sum_{k \in K} \langle \mathbf{x}, \mathbf{u}_k \rangle \delta_k$$

It follows from Theorem 13.19 that ϕ is linear, and it is not hard to see that it is also bijective. Notice that $\phi(\mathbf{u}_k) = \delta_k$, and so ϕ takes the Hilbert basis $\mathfrak{B}$ for H to the Hilbert basis $\mathfrak{O}$ for $\ell^2(K)$.

Notice also that

$$\| \phi(\mathbf{x}) \|^2 = \langle \phi(\mathbf{x}), \phi(\mathbf{x}) \rangle = \sum_{k \in K} | \langle \mathbf{x}, \mathbf{u}_k \rangle |^2 = \left\| \sum_{k \in K} \langle \mathbf{x}, \mathbf{u}_k \rangle \mathbf{u}_k \right\|^2 = \| \mathbf{x} \|^2$$

and so ϕ is an isometric isomorphism. We have proved the following theorem.

Theorem 13.29 If H is a Hilbert space of Hilbert dimension κ, and if K is any set of cardinality κ, then H is isometrically isomorphic to $\ell^2(K)$. ∎

The Riesz Representation Theorem

We conclude our discussion of Hilbert spaces by discussing the Riesz representation theorem. As it happens, not all linear functionals on a Hilbert space have the form "take the inner product with...," as in the finite dimensional case. To see this, observe that if $\mathbf{y} \in H$, then the function

$$f_{\mathbf{y}}(\mathbf{x}) = \langle \mathbf{x}, \mathbf{y} \rangle$$

is certainly a linear functional on H. However, it has a special property. In particular, the Cauchy-Schwarz inequality gives, for all $\mathbf{x} \in H$

$$| f_{\mathbf{y}}(\mathbf{x}) | = | \langle \mathbf{x}, \mathbf{y} \rangle | \leq \| \mathbf{x} \| \, \| \mathbf{y} \|$$

or, for all $\mathbf{x} \neq \mathbf{0}$,

$$\frac{| f_{\mathbf{y}}(\mathbf{x}) |}{\| \mathbf{x} \|} \leq \| \mathbf{y} \|$$

Noticing that equality holds if $\mathbf{x} = \mathbf{y}$, we have

$$\sup_{\mathbf{x} \neq \mathbf{0}} \frac{| f_{\mathbf{y}}(\mathbf{x}) |}{\| \mathbf{x} \|} = \| \mathbf{y} \|$$

This prompts us to make the following definition, which we do for linear transformations between Hilbert spaces (this covers the case of linear functionals).

Definition Let $\tau:H_1 \to H_2$ be a linear transformation from H_1 to H_2. Then τ is said to be **bounded** if

$$\sup_{\mathbf{x} \neq \mathbf{0}} \frac{\| \tau(\mathbf{x}) \|}{\| \mathbf{x} \|} < \infty$$

If the supremum on the left is finite, we denote it by $\| \tau \|$ and call it the **norm** of τ. □

Of course, if $f:H \to F$ is a bounded linear functional on H, then

$$\| f \| = \sup_{\mathbf{x} \neq \mathbf{0}} \frac{| f(\mathbf{x}) |}{\| \mathbf{x} \|}$$

The set of all *bounded* linear functionals on a Hilbert space H is called the **continuous dual space**, or **conjugate space**, of H, and denoted by H^*. Note that this differs from the *algebraic* dual of H, which is the set of *all* linear functionals on H. In the finite dimensional case, however, since all linear functionals are bounded (exercise), the two concepts agree. (Unfortunately, there is no universal agreement on the notation for the algebraic dual versus the continuous dual. Since we will discuss only the continuous dual in this section, no confusion should arise.)

The following theorem gives some simple reformulations of the definition of norm.

Theorem 13.30 Let $\tau:H_1 \to H_2$ be a bounded linear transformation.

1) $\| \tau \| = \sup_{\| \mathbf{x} \| = 1} \| \tau(\mathbf{x}) \|$
2) $\| \tau \| = \sup_{\| \mathbf{x} \| \leq 1} \| \tau(\mathbf{x}) \|$
3) $\| \tau \| = \inf\{c \in \mathbb{R} \mid \| \tau(\mathbf{x}) \| \leq c \| \mathbf{x} \| \text{ for all } \mathbf{x} \in H\}$ ∎

The following theorem explains the importance of bounded linear transformations.

Theorem 13.31 Let $\tau:H_1 \to H_2$ be a linear transformation. The following are equivalent.

1) τ is bounded
2) τ is continuous at any point $\mathbf{x}_0 \in H$
3) τ is continuous.

Proof. Suppose that τ is bounded. Then

$$\| \tau(\mathbf{x}) - \tau(\mathbf{x}_0) \| = \| \tau(\mathbf{x} - \mathbf{x}_0) \| \leq \| \tau \| \, \| \mathbf{x} - \mathbf{x}_0 \| \to 0$$

as $\mathbf{x}\to\mathbf{x}_0$. Hence, τ is continuous at $\mathbf{x}_0$. Thus, (1) implies (2). If (2) holds, then for any $\mathbf{y}\in H$, we have

$$\|\tau(\mathbf{x})-\tau(\mathbf{y})\| = \|\tau(\mathbf{x}-\mathbf{y}+\mathbf{x}_0)-\tau(\mathbf{x}_0)\|\to 0$$

as $\mathbf{x}\to\mathbf{y}$, since τ is continuous at $\mathbf{x}_0$, and $\mathbf{x}-\mathbf{y}+\mathbf{x}_0\to\mathbf{x}_0$ as $\mathbf{y}\to\mathbf{x}$. Hence, τ is continuous at any $\mathbf{y}\in H$, and (3) holds. Finally, suppose that (3) holds. Thus, τ is continuous at $\mathbf{0}$, and so there exists a $\delta>0$ such that

$$\|\mathbf{x}\|\le\delta \;\Rightarrow\; \|\tau(\mathbf{x})\|\le 1$$

In particular,

$$\|\mathbf{x}\|=\delta \;\Rightarrow\; \frac{\|\tau(\mathbf{x})\|}{\|\mathbf{x}\|}\le\frac{1}{\delta}$$

and so

$$\|\mathbf{x}\|=1\Rightarrow\|\delta\mathbf{x}\|=\delta\Rightarrow\frac{\|\tau(\delta\mathbf{x})\|}{\|\delta\mathbf{x}\|}\le\frac{1}{\delta}\Rightarrow\frac{\|\tau(\mathbf{x})\|}{\|\mathbf{x}\|}\le\frac{1}{\delta}$$

Thus, τ is bounded. ∎

Now we can state and prove the Riesz representation theorem.

Theorem 13.32 **(The Riesz representation theorem)** Let H be a Hilbert space. For any *bounded* linear functional f on H, there is a unique $\mathbf{z}_0\in H$ such that

$$f(\mathbf{x})=\langle\mathbf{x},\mathbf{z}_0\rangle$$

for all $\mathbf{x}\in H$. Moreover, $\|\mathbf{z}_0\| = \|f\|$.

Proof. If $f=0$, we may take $\mathbf{z}_0=\mathbf{0}$, so let us assume that $f\neq 0$. Hence, $K=ker(f)\neq H$, and since f is continuous, K is closed. Thus

$$H=K\oplus K^{\perp}$$

Now, the first isomorphism theorem, applied to the linear functional $f{:}H\to F$, implies that $H/K\approx F$ (as vector spaces). In addition, Theorem 3.5 implies that $H/K\approx K^{\perp}$, and so $K^{\perp}\approx F$. In particular, $dim(K^{\perp})=1$.

For any $\mathbf{z}\in K^{\perp}$, we have

$$\mathbf{x}\in K \;\Rightarrow\; f(\mathbf{x})=0=\langle\mathbf{x},\mathbf{z}\rangle$$

Since $dim(K^{\perp})=1$, all we need do is find a $\mathbf{0}\neq\mathbf{z}\in K^{\perp}$ for which

$$f(\mathbf{z})=\langle\mathbf{z},\mathbf{z}\rangle$$

for then $f(r\mathbf{z})=rf(\mathbf{z})=r\langle\mathbf{z},\mathbf{z}\rangle=\langle r\mathbf{z},\mathbf{z}\rangle$ for all $r\in F$, showing that $f(\mathbf{x})=\langle\mathbf{x},\mathbf{z}\rangle$ for $\mathbf{x}\in K$ as well.

But if $\mathbf{0}\neq\mathbf{z}\in K^{\perp}$, then

$$\mathbf{z}_0 = \frac{\overline{f(\mathbf{z})}}{\langle \mathbf{z},\mathbf{z}\rangle}\mathbf{z}$$

has this property, as can be easily checked. The fact that $\|\mathbf{z}_0\| = \|f\|$ has already been established. ▮

EXERCISES

1. Prove that the sup metric on the metric space $C[a,b]$ of continuous functions on $[a,b]$ does not come from an inner product. *Hint*: let $f(t) = 1$ and $g(t) = (t-a)/(b-a)$, and consider the parallelogram law.
2. Prove that any Cauchy sequence that has a convergent subsequence must itself converge.
3. Let V be an inner product space, and let A and B be subsets of V. Show that
   a) $A \subset B \Rightarrow B^{\perp} \subset A^{\perp}$
   b) $A^{\perp}$ is a closed sub*space* of V
   c) $[cspan(A)]^{\perp} = A^{\perp}$
4. Let V be an inner product space and $S \subset V$. Under what conditions is $S^{\perp\perp\perp} = S^{\perp}$?
5. Prove that a subspace S of a Hilbert space H is closed if and only if $S = S^{\perp\perp}$.
6. Let V be the subspace of ℓ^2 consisting of all sequences of real numbers, with the property that each sequence has only a finite number of nonzero terms. Thus, V is an inner product space. Let K be the subspace of V consisting of all sequences $\mathbf{x} = (x_n)$ in V with the property that $\Sigma x_n/n = 0$. Show that K is closed, but that $K^{\perp\perp} \neq K$. *Hint*: For the latter, show that $K^{\perp} = \{\mathbf{0}\}$ by considering the sequences $\mathbf{u} = (1,\ldots,-n,\ldots)$, where the term $-n$ is in the nth coordinate position.
7. Let $\mathcal{O} = \{\mathbf{u}_1,\mathbf{u}_2,\ldots\}$ be an orthonormal set in H. If $\mathbf{x} = \Sigma r_k \mathbf{u}_k$ converges, show that
$$\|\mathbf{x}\|^2 = \sum_{k=1}^{\infty} |r_k|^2$$
8. Prove that if an infinite series
$$\sum_{k=1}^{\infty} \mathbf{x}_k$$
converges absolutely in a Hilbert space H, then it also converges in the sense of the "net" definition given in this section.
9. Let $\{r_k \mid k \in K\}$ be a collection of nonnegative real numbers. If the sum on the left below converges, show that
$$\sum_{k \in K} r_k = \sup_{\substack{J \text{ finite} \\ J \subset K}} \sum_{k \in J} r_k$$

10. Find a countably infinite sum of real numbers that converges in the sense of partial sums, but not in the sense of nets.
11. Prove that if a Hilbert space H has infinite Hilbert dimension, then no Hilbert basis for H is a Hamel basis.
12. Prove that $\ell^2(K)$ is a Hilbert space for any nonempty set K.
13. Prove that any linear transformation between finite dimensional Hilbert spaces is bounded.
14. Prove that if $f \in H^*$, then $ker(f)$ is a closed subspace of H.
15. Prove that a Hilbert space is separable if an only if $hdim(H) \le \aleph_0$.
16. Can a Hilbert space have countably infinite Hamel dimension?
17. What is the Hamel dimension of $\ell^2(\mathbb{N})$?
18. Let τ and σ be bounded linear operators on H. Verify the following.
    a) $\| r\tau \| = |r| \, \| \tau \|$
    b) $\| \tau + \sigma \| \le \| \tau \| + \| \sigma \|$
    c) $\| \tau\sigma \| \le \| \tau \| \, \| \sigma \|$
19. Use the Riesz representation theorem to show that $H^* \approx H$ for any Hilbert space H.

CHAPTER 14

Tensor Products

Contents: *Free Vector Spaces. Another Look at the Direct Sum. Bilinear Maps and Tensor Products. Properties of the Tensor Product. The Tensor Product of Linear Transformations. Change of Base Field. Multilinear Maps and Iterated Tensor Products. Alternating Maps and Exterior Products. Exercises.*

In the preceding chapters, we have seen several ways to construct new vector spaces from old ones. Two of the most important such constructions are the direct sum $U \oplus V$ and the set $\mathcal{L}(U,V)$ of all linear transformations from U to V. In this chapter, we consider another construction, known as the tensor product.

There are several ways to define the tensor product but, unfortunately, they are all a bit less perspicuous than one might like. Therefore, in order to provide some motivation, we will first recast the definition of the familiar external direct sum. In order to do this (and to define tensor products) we need the concept of a free vector space.

Free Vector Spaces

Let F be a field. Given any nonempty set X, we may construct a vector space $\mathcal{F}_X$ over F with X as basis, simply by taking $\mathcal{F}_X$ to be the set of all formal finite linear combinations of elements of X

$$\mathcal{F}_X = \left\{ \sum_{\text{finite}} r_i x_i \,\middle|\, x_i \in X,\ r_i \in F \right\}$$

where the operations are as expected – combine like terms using the rules

$$rx_i + sx_i = (r+s)x_i$$

and

$$r(sx_i) = (rs)x_i$$

The vector space $\mathcal{F}_X$ is called the **free vector space** on X. The term *free* is meant to connote the fact that there is no relationship between the elements of X.

In fact, any vector space V is the free vector space on any basis for V. Thus, in some sense, we have introduced nothing new. However, the concept of free object occurs in many other contexts, as we have seen with regard to modules, where not all modules are free. Moreover, even in the context of vector spaces, it gives us a new viewpoint from which to develop new ideas.

We may characterize the free vector space $\mathcal{F}_X$ as the set $(F^X)_0$ of all functions from X to F that have *finite support.* Recall that the support of a function $f:X \to F$ is defined by

$$supp(F) = \{x \in X \mid f(x) \neq 0\}$$

It is easy to see that a function $f:X \to F$ with finite support corresponds to a finite sum of elements of X, via

$$f \leftrightarrow \sum f(x_i)x_i$$

and therefore that the two constructions of $\mathcal{F}_X$ are equivalent. We will feel free to use either construction.

We can express the concept of freeness in a much more general way as follows. Consider the map $j:X \to \mathcal{F}_X$ defined by $j(x) = x$, and called the **canonical injection** of X into $\mathcal{F}_X$. The pair $(\mathcal{F}_X, j)$ has a very special property. Referring to Figure 14.1, if $f:X \to V$ is any map from X to any vector space V, then there is a *unique linear transformation* τ from $\mathcal{F}_X$ to V for which $\tau \circ j = f$.

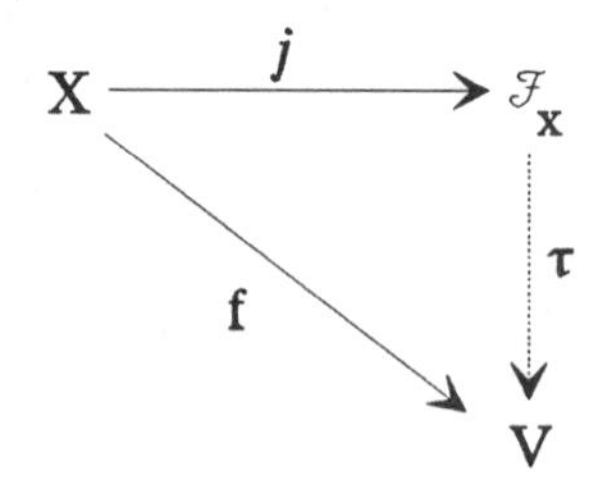

Figure 14.1

For if $f:X \to V$, then we can define a linear transformation $\tau:\mathcal{F}_X \to V$ by setting $\tau(x) = f(x)$ and extending by linearity to $\mathcal{F}_X$. This is legitimate since X is a basis for $\mathcal{F}_X$. The uniqueness of τ also follows from the fact that X is a basis for $\mathcal{F}_X$.

When any two paths in a diagram, such as Figure 14.1, that begin and end at the same locations describe equal functions, we say that the diagram **commutes.** Thus, saying that $\tau \circ j = f$ is the same as saying that the diagram in Figure 14.1 commutes. We can also describe this situation by saying that any function $f:X \to V$ can be *factored through* the canonical injection j.

Now, it so happens that the commutativity of Figure 14.1, and the uniqueness of τ, completely determine the pair $(\mathcal{F}_X, j)$. More specifically, we have the following, known as the *universal property* of the free vector space $\mathcal{F}_X$.

Theorem 14.1 (The Universal Property of Free Vector Spaces) Let X be a nonempty set. Suppose that $\mathcal{F}$ is a vector space over F, and $k:X \to \mathcal{F}$ is a function, and that the pair $(\mathcal{F}, k)$ has the following property. Referring to Figure 14.2, for any function $f:X \to V$, where V is a vector space over F, there exists a unique linear transformation $\tau:\mathcal{F} \to V$ for which $\tau \circ k = f$, that is, for which the diagram in Figure 14.2 commutes. Then $\mathcal{F}$ is isomorphic to the free vector space $\mathcal{F}_X$.

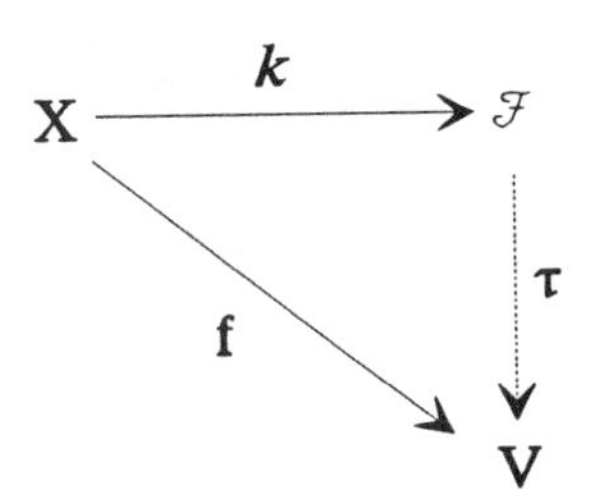

Figure 14.2

Proof. Consider the diagrams in Figure 14.3. The first diagram reflects the fact that we may put $V = \mathcal{F}$ in Figure 14.1. Since this diagram commutes, we have

$$\tau \circ j = k$$

The second diagram reflects the fact that we may set $V = \mathcal{F}_X$ in Figure 14.2. Since this diagram commutes, we have

$$\sigma \circ k = j$$

Making the appropriate substitutions gives

$$\tau \circ \sigma \circ k = k \quad \text{and} \quad \sigma \circ \tau \circ j = j$$

But, the third commutative diagram in Figure 14.3 indicates that the identity is the *unique* linear transformation for which $\iota \circ k = k$, and so

$\tau \circ \sigma = \iota$. Similarly, by drawing the appropriate commutative diagram, we deduce that $\sigma \circ \tau = \iota$. Thus, τ is an *isomorphism* from $\mathcal{F}_X$ to $\mathcal{F}$. ∎

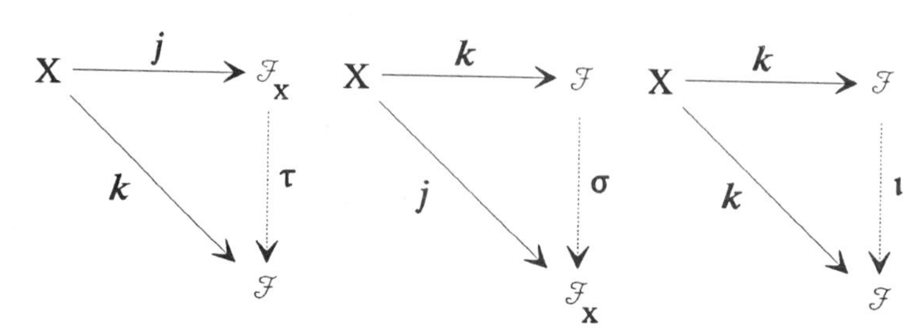

Figure 14.3

Another Look at the Direct Sum

By way of motivation for defining tensor products, let us take another look at the external direct sum construction. Our plan is to characterize this sum in three difference ways.

First, we have the definition. Suppose that U and V are vector spaces over the same field F. The external direct sum $U \boxplus V$ is the vector space of all ordered pairs

$$U \boxplus V = \{(\mathbf{u},\mathbf{v}) \mid \mathbf{u} \in U,\ \mathbf{v} \in V\}$$

with coordinatewise operations

$$(\mathbf{u},\mathbf{v}) + (\mathbf{u}',\mathbf{v}') = (\mathbf{u}+\mathbf{u}',\mathbf{v}+\mathbf{v}')$$

and

$$r(\mathbf{u},\mathbf{v}) = (r\mathbf{u},r\mathbf{v})$$

For the second characterization, we begin by considering the **Cartesian product** $U \times V$, which is simply the set of all ordered pairs

$$U \times V = \{(\mathbf{u},\mathbf{v}) \mid \mathbf{u} \in U,\ \mathbf{v} \in V\}$$

with *no* algebraic structure. Let $\mathcal{F}_{U \times V}$ be the free vector space on $U \times V$. Thus,

$$\mathcal{F}_{U \times V} = \left\{ \sum_{\text{finite}} r_i(\mathbf{u}_i,\mathbf{v}_i) \,\middle|\, (\mathbf{u}_i,\mathbf{v}_i) \in U \times V,\ r_i \in F \right\} \tag{14.1}$$

It is important to keep in mind that we allow no manipulations of the coordinates of the ordered pairs in $\mathcal{F}_{U \times V}$. For instance, we cannot replace $r(\mathbf{u},\mathbf{v})$ by $(r\mathbf{u},r\mathbf{v})$ nor $(\mathbf{u},\mathbf{v}) + (\mathbf{u}',\mathbf{v}')$ by $(\mathbf{u}+\mathbf{u}',\mathbf{v}+\mathbf{v}')$. In a sense, the ordered pairs in (14.1) act simply as "placekeepers" to separate the coefficients r_i.

In fact, the difference between $\mathcal{F}_{U \times V}$ and $U \boxplus V$ is that, in $U \boxplus V$, we do have

$$r(\mathbf{u},\mathbf{v}) - (r\mathbf{u},r\mathbf{v}) = \mathbf{0}$$

and

$$(\mathbf{u},\mathbf{v}) + (\mathbf{u}',\mathbf{v}') - (\mathbf{u}+\mathbf{u}',\mathbf{v}+\mathbf{v}') = \mathbf{0}$$

for all $r \in F$, $\mathbf{u} \in U$ and $\mathbf{v} \in V$.

Let us define S to be the subspace of $\mathcal{F}_{U \times V}$ generated by all vectors of the form

$$r(\mathbf{u},\mathbf{v}) - (r\mathbf{u},r\mathbf{v})$$

and

$$(\mathbf{u},\mathbf{v}) + (\mathbf{u}',\mathbf{v}') - (\mathbf{u}+\mathbf{u}',\mathbf{v}+\mathbf{v}')$$

for all $r \in F$, $\mathbf{u} \in U$ and $\mathbf{v} \in V$. It seems reasonable that the quotient space $\mathcal{F}_{U \times V}/S$ should be isomorphic to the direct sum $U \boxplus V$.

To prove this, consider the map $\tau:\mathcal{F}_{U \times V}/S \to U \boxplus V$ defined by

$$\tau\Big(\sum r_i(\mathbf{u}_i,\mathbf{v}_i)+S\Big) = \sum r_i(\mathbf{u}_i,\mathbf{v}_i)$$

This map is well-defined, since if

$$\sum r_i(\mathbf{u}_i,\mathbf{v}_i)+S = \sum s_i(\mathbf{x}_i,\mathbf{y}_i)+S$$

then

$$\sum r_i(\mathbf{u}_i,\mathbf{v}_i) - \sum s_i(\mathbf{x}_i,\mathbf{y}_i) \in S$$

But any element of S is equal to the zero vector in $U \boxplus V$, and so the vectors $\Sigma r_i(\mathbf{u}_i,\mathbf{v}_i)$ and $\Sigma s_i(\mathbf{x}_i,\mathbf{y}_i)$ are equal in $U \boxplus V$. Hence,

$$\tau\Big(\sum r_i(\mathbf{u}_i,\mathbf{v}_i)+S\Big) = \tau\Big(\sum s_i(\mathbf{x}_i,\mathbf{y}_i)+S\Big) \tag{14.2}$$

Furthermore, τ is linear, and surjective. To see that τ is injective, we must show that if

$$\sum r_i(\mathbf{u}_i,\mathbf{v}_i) = \mathbf{0} \text{ in } U \boxplus V \tag{14.3}$$

then

$$\sum r_i(\mathbf{u}_i,\mathbf{v}_i) \in S$$

To this end, observe that, as formal sums, $\Sigma r_i(\mathbf{u}_i,\mathbf{v}_i) \in S$ if and only if the sum that results by replacing any terms, using the rules

$$r(\mathbf{u},\mathbf{v}) \to (r\mathbf{u},r\mathbf{v}),\ (r\mathbf{u},r\mathbf{v}) \to r(\mathbf{u},\mathbf{v})$$

or

$$(\mathbf{u},\mathbf{v}) + (\mathbf{u}',\mathbf{v}') \to (\mathbf{u}+\mathbf{u}',\mathbf{v}+\mathbf{v}'),\ (\mathbf{u}+\mathbf{u}',\mathbf{v}+\mathbf{v}') \to (\mathbf{u},\mathbf{v}) + (\mathbf{u}',\mathbf{v}')$$

is also in S. Hence, since (14.3) simply says that, by performing such replacements, we may reduce $\Sigma r_i(\mathbf{u}_i,\mathbf{v}_i)$ to $\mathbf{0}$, which is in S, the sum $\Sigma r_i(\mathbf{u}_i,\mathbf{v}_i)$ must be in S. Thus, τ is an isomorphism from $\mathcal{F}_{U\times V}/S$ to $U \boxplus V$.

As can be seen from the previous paragraph, it can be a bit awkward to describe $U \boxplus V$ as a quotient space. However, we do have another characterization, in terms of commutative diagrams. Associated with the direct sum $U \boxplus V$ are the two projections $\rho_1:U \boxplus V \to U$ and $\rho_2:U \boxplus V \to V$ defined by

$$\rho_1((\mathbf{u},\mathbf{v})) = \mathbf{u} \quad \text{and} \quad \rho_2((\mathbf{u},\mathbf{v})) = \mathbf{v}$$

Let us consider the triple $(U \boxplus V,\rho_1,\rho_2)$. Referring to Figure 14.4, if W is any vector space over F, with linear maps $f_1:W \to U$ and $f_2:W \to V$, then there exists a unique linear transformation $\tau:W \to U \boxplus V$ for which the diagram commutes, that is, for which

$$\rho_1\tau = f_1 \quad \text{and} \quad \rho_2\tau = f_2$$

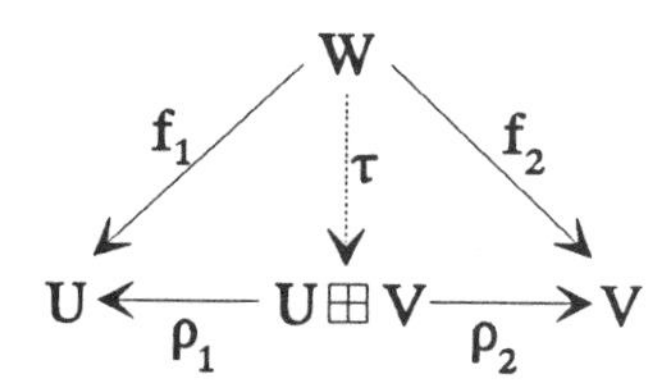

Figure 14.4

To see this, observe that, if such a τ were to exist, then we would have

$$\rho_1(\tau(\mathbf{w})) = f_1(\mathbf{w}) \quad \text{and} \quad \rho_2(\tau(\mathbf{w})) = f_2(\mathbf{w})$$

and so we must have

(14.4) $$\tau(\mathbf{w}) = (f_1(\mathbf{w}),f_2(\mathbf{w}))$$

We leave it to the reader to show that this actually defines a unique linear transformation τ from W to $U \boxplus V$. The following theorem shows that this property characterizes the direct sum. The proof is very similar to that of Theorem 14.1.

Theorem 14.2 (The universal property of external direct sums) Let U and V be vector spaces over F. Let D be a vector space over F, and let $\sigma_1:D \to U$ and $\sigma_2:D \to W$ be linear transformations, as in Figure 14.5. Suppose that the triple (D,σ_1,σ_2) has the following property. If W is any vector space over F, and if $f_1:W \to U$ and $f_2:W \to V$ are linear transformations, then there exists a unique linear transformation $\tau:W\to D$ that makes the diagram commute, that is, or which

$$\sigma_1\tau = f_1 \quad \text{and} \quad \sigma_2\tau = f_2$$

Then D is isomorphic to the external direct sum $U \boxplus V$. ∎

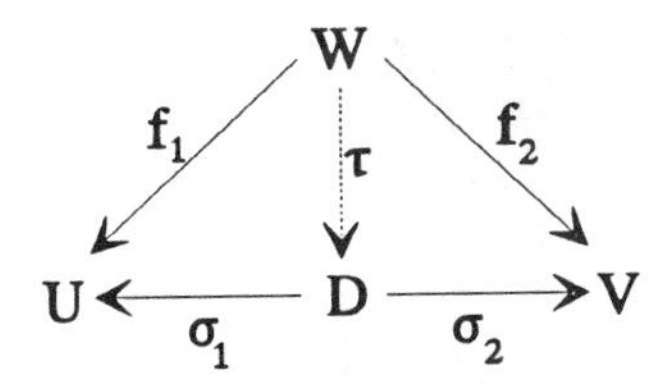

Figure 14.5

In summary, we have three equivalent characterizations of the external direct sum $U \boxplus V$

1) The definition: $U \boxplus V = \{(\mathbf{u},\mathbf{v}) \mid \mathbf{u} \in U,\ \mathbf{v} \in V\}$

2) The quotient space

$$\frac{\mathcal{F}_{U\times V}}{S}$$

where $\mathcal{F}_{U\times V}$ is the free vector space on $U \times V$ and

$$S = span\{r(\mathbf{u},\mathbf{v}) - (r\mathbf{u},r\mathbf{v}),\ (\mathbf{u},\mathbf{v}) + (\mathbf{u}',\mathbf{v}') - (\mathbf{u}+\mathbf{u}',\mathbf{v}+\mathbf{v}')\}$$

3) By the universal property of external direct sums given in Theorem 14.2.

Bilinear Maps and Tensor Products

Before defining tensor products, we need a preliminary definition.

Definition Let U, V and W be vector spaces over F. A function $f:U\times V\to W$ is **bilinear** if it is linear in both variables separately, that is,

$$f(r\mathbf{u} + s\mathbf{u}',\mathbf{v}) = rf(\mathbf{u},\mathbf{v}) + sf(\mathbf{u}',\mathbf{v})$$

and

$$f(\mathbf{u},r\mathbf{v} + s\mathbf{v}') = rf(\mathbf{u},\mathbf{v}) + sf(\mathbf{u},\mathbf{v}')$$

The set of all bilinear functions from $U \times V$ to W is denoted by $\mathcal{B}(U,V;W)$. A bilinear function $f:U\times V\to F$, with values in the base field F, is called a **bilinear form** on $U \times V$. □

Example 14.1

1) A *real* inner product $\langle , \rangle : V \times V \to \mathbb{R}$ is a bilinear form on $V \times V$.
2) If A is an algebra, the product map $\mu : A \times A \to A$ defined by $\mu(a,b) = ab$ is bilinear. In short, multiplication is linear in each variable. □

If V is a vector space, we have two classes of functions from $V \times V$ to W, the linear maps $\mathcal{L}(V \times V, W)$ and the bilinear maps $\mathcal{B}(V, V; W)$. We leave it as an exercise to show that these two classes of maps have only the zero map in common. In other words, the only map that is both linear and bilinear is the zero map.

Now we can define the tensor product of two vector spaces.

Definition Let U and V be vector spaces over F, and let T be the subspace of the free vector space $\mathcal{F}_{U \times V}$ generated by all vectors of the form

$$r(\mathbf{u},\mathbf{v}) + s(\mathbf{u}',\mathbf{v}) - (r\mathbf{u} + s\mathbf{u}',\mathbf{v}) \tag{14.5}$$

and

$$r(\mathbf{u},\mathbf{v}) + s(\mathbf{u},\mathbf{v}') - (\mathbf{u},r\mathbf{v} + s\mathbf{v}') \tag{14.6}$$

for all $r,s \in F$, $\mathbf{u},\mathbf{u}' \in U$ and $\mathbf{v},\mathbf{v}' \in V$. The quotient space $\mathcal{F}_{U \times V}/T$ is called the **tensor product** of U and V and is denoted by $U \otimes V$. □

Note that in the case of the tensor product, we divide by the space spanned by all vectors in $U \times V$ that would be zero if the vector space operations were linear in each coordinate separately. According to this definition, an element of $U \otimes V$ has the form

$$\sum r_i(\mathbf{u}_i,\mathbf{v}_i)+T$$

It is customary to denote the coset $(\mathbf{u},\mathbf{v})+T$ by $\mathbf{u} \otimes \mathbf{v}$, and therefore any element of $U \otimes V$ has the form

$$\sum \mathbf{u}_i \otimes \mathbf{v}_i$$

where

$$r(\mathbf{u} \otimes \mathbf{v}) + s(\mathbf{u}' \otimes \mathbf{v}) = (r\mathbf{u} + s\mathbf{u}') \otimes \mathbf{v} \tag{14.7}$$

and

$$r(\mathbf{u} \otimes \mathbf{v}) + s(\mathbf{u} \otimes \mathbf{v}') = \mathbf{u} \otimes (r\mathbf{v} + s\mathbf{v}') \tag{14.8}$$

Thus,

$$\sum \mathbf{u}_i \otimes \mathbf{v}_i = \sum \mathbf{x}_i \otimes \mathbf{y}_i$$

if and only if we can obtain one expression from the other by a finite number of replacements using (14.7) and (14.8).

As with the external direct sum, this definition, while intuitively pleasing, can be a bit difficult to work with, so we turn to a characterization via a universal property.

Theorem 14.3 (The universal property of tensor products) Let U and V be vector spaces over the same field F. The pair $(U\otimes V,t)$, where $t:U\times V\to U\otimes V$ is the bilinear map defined by

$$t(\mathbf{u},\mathbf{v})=\mathbf{u}\otimes\mathbf{v}$$

has the following property. Referring to Figure 14.6, if $f:U\times V\to W$ is any bilinear function from $U\times V$ to a vector space W over F, then there is a unique *linear* transformation $\tau:U\otimes V\to W$ that makes the diagram in Figure 14.6 commute, that is, for which

$$\tau\circ t=f$$

Moreover, $U\otimes V$ is unique, in the sense that if a pair (X,s) also has this property, then X is isomorphic to $U\otimes V$.

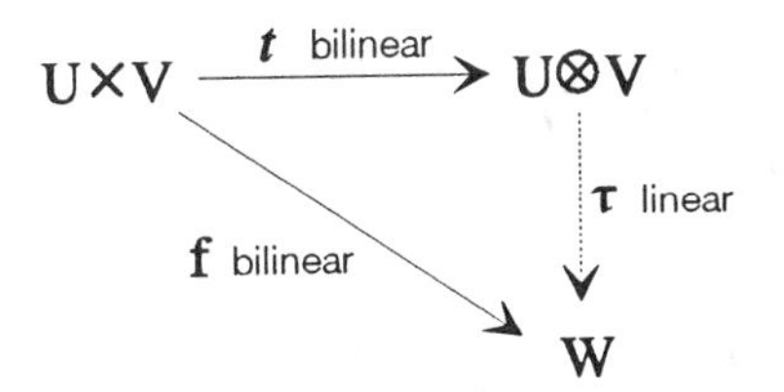

Figure 14.6

Proof. To prove that $(U\otimes V,t)$ has the desired property, consider the diagram in Figure 14.7.

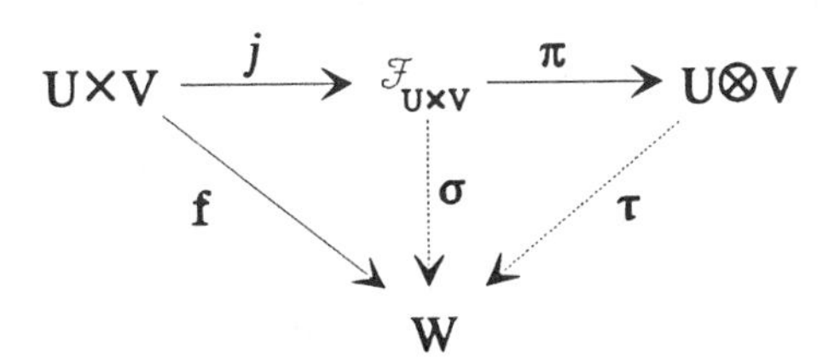

Figure 14.7

Since $t(\mathbf{u},\mathbf{v})=\mathbf{u}\otimes\mathbf{v}=(\mathbf{u},\mathbf{v})+T$, the map $t:U\times V\to U\otimes V$ is just the composition of the canonical injection $j:U\times V\to\mathcal{F}_{U\times V}$ followed by the canonical projection $\pi:\mathcal{F}_{U\times V}\to U\otimes V=\mathcal{F}_{U\times V}/T$. That is,

$$t=\pi\circ j$$

Now, the universal property of free vector spaces implies that there is a *unique* linear transformation $\sigma:\mathcal{F}_{U \times V} \to W$ for which

$$\sigma \circ j = f$$

Note that, since f is bilinear, it sends any of the vectors (14.5) and (14.6) that generate T to the zero vector, so $T \subset ker(\sigma)$. Hence, we may apply Theorem 3.3, to deduce the existence of a *unique* linear transformation $\tau: U \otimes V \to W$ for which

$$\tau \circ \pi = \sigma$$

Hence,

$$\tau \circ t = \tau \circ \pi \circ j = \sigma \circ j = f$$

Moreover, if $\tau' \circ t = f$, then $\sigma' = \tau' \circ \pi: \mathcal{F}_{U \times V} \to W$ is a linear transformation for which

$$\sigma' \circ j(\mathbf{u},\mathbf{v}) = \tau' \circ \pi \circ j(\mathbf{u},\mathbf{v}) = \tau' \circ t(\mathbf{u},\mathbf{v}) = f(\mathbf{u},\mathbf{v}) = \sigma \circ j(\mathbf{u},\mathbf{v})$$

and so $\sigma' \circ j = \sigma \circ j$, implying that $\sigma' = \sigma$, which in turn implies that $\tau' = \tau$. Hence, τ is unique. We leave proof of the uniqueness of $U \otimes V$ as an exercise. ∎

Theorem 14.3 says that to each *bilinear* function $f: U \times V \to W$, there corresponds a unique *linear* function $\tau: U \otimes V \to W$, through which f can be factored (that is, $f = \tau \circ t$). This establishes a map $\phi: \mathcal{B}(U,V;W) \to \mathcal{L}(U \otimes V, W)$ given by $\phi(f) = \tau$. In other words, $\phi(f)$ is the unique linear map for which

$$\phi(f): U \otimes V \to W \qquad \phi(f)(\mathbf{u} \otimes \mathbf{v}) = f(\mathbf{u},\mathbf{v})$$

Observe that ϕ is linear, since if $f,g \in \mathcal{B}(U,V;W)$, then

$$[r\phi(f) + s\phi(g)](\mathbf{u} \otimes \mathbf{v}) = rf(\mathbf{u},\mathbf{v}) + sg(\mathbf{u},\mathbf{v}) = (rf + sg)(\mathbf{u},\mathbf{v})$$

and so the uniqueness part of the universal property implies that

$$r\phi(f) + s\phi(g) = \phi(rf + sg)$$

Also, ϕ is surjective, since if $\tau: U \otimes V \to W$ is any linear map, then $f = \tau \circ t: U \times V \to W$ is bilinear, and by the uniqueness part of the universal property, we have $\phi(f) = \tau$. Finally, ϕ is injective, for if $\phi(f) = 0$, then $f = \phi(f) \circ t = 0$. We have established the following result.

Theorem 14.4 Let U, V and W be vector spaces over F. Then the map $\phi: \mathcal{B}(U,V;W) \to \mathcal{L}(U \otimes V, W)$ defined by the fact that $\phi(f)$ is the unique linear map for which $f = \phi(f) \circ t$, is an isomorphism. Thus,

$$\mathcal{B}(U,V;W) \approx \mathcal{L}(U \otimes V, W)$$

∎

Properties of the Tensor Product

Armed with the definition and the universal property, we can now discuss some of the basic properties of tensor products.

Theorem 14.5 If $\{\mathbf{u}_1,\ldots,\mathbf{u}_n\}$ are linearly independent vectors in U, and $\{\mathbf{v}_1,\ldots,\mathbf{v}_n\}$ are arbitrary vectors in V, then

$$\sum \mathbf{u}_i \otimes \mathbf{v}_i = \mathbf{0} \Rightarrow \mathbf{v}_i = \mathbf{0} \text{ for all } i$$

Proof. Let us consider the dual vectors $\delta_i \in U^*$ to the vectors $\mathbf{u}_i$. Thus, $\delta_i(\mathbf{u}_j) = \delta_{i,j}$. For any linear functionals $\epsilon_i:V \to F$, we define a bilinear form $f:U \times V \to F$ by

$$f(u,v) \to \sum_{j=1}^{n} \delta_j(x)\epsilon_j(y)$$

Then, by the universal property of tensor products, there exists a unique linear functional $\tau:U \otimes V \to F$ for which $\tau \circ t = f$. Hence,

$$\begin{aligned} 0 &= \tau\Big(\sum_i \mathbf{u}_i \otimes \mathbf{v}_i\Big) = \sum_i \tau \circ t(\mathbf{u}_i,\mathbf{v}_i) \\ &= \sum_i f(\mathbf{u}_i,\mathbf{v}_i) = \sum_i \sum_j \delta_j(\mathbf{u}_i)\epsilon_j(\mathbf{v}_i) = \sum_i \epsilon_i(\mathbf{v}_i) \end{aligned}$$

Since the ϵ_i's are arbitrary, we deduce that $\mathbf{v}_i = \mathbf{0}$ for all i. ∎

Corollary 14.6 If $\mathbf{u} \neq \mathbf{0}$ and $\mathbf{v} \neq \mathbf{0}$, then $\mathbf{u} \otimes \mathbf{v} \neq \mathbf{0}$. ∎

Theorem 14.7 Let $\mathfrak{B} = \{\mathbf{e}_i \mid i \in I\}$ be a basis for U and $\mathfrak{C} = \{\mathbf{f}_j \mid j \in J\}$ be a basis for V. Then the set $\mathfrak{D} = \{\mathbf{e}_i \otimes \mathbf{f}_j \mid i \in I, j \in J\}$ is a basis for $U \otimes V$.

Proof. To see that the $\mathfrak{D}$ is linearly independent, suppose that

$$\sum_{i,j} r_{i,j}(\mathbf{e}_i \otimes \mathbf{f}_j) = \mathbf{0}$$

This can be written

$$\sum_i \mathbf{e}_i \otimes \Big(\sum_j r_{i,j}\mathbf{f}_j\Big) = \mathbf{0}$$

and so, by Theorem 14.5, we must have

$$\sum_j r_{i,j}\mathbf{f}_j = \mathbf{0}$$

for all i, and hence $r_{i,j} = 0$ for all i and j. To see that $\mathfrak{D}$ spans $U \otimes V$, let $\mathbf{u} \otimes \mathbf{v} \in U \otimes V$. Since $\mathbf{u} = \sum_i r_i\mathbf{e}_i$, and $\mathbf{v} = \sum_j s_j\mathbf{f}_j$, we have

$$\mathbf{u} \otimes \mathbf{v} = \sum_i r_i\mathbf{e}_i \otimes \sum_j s_j\mathbf{f}_j = \sum_j s_j\Big(\sum_i r_i\mathbf{e}_i \otimes \mathbf{f}_j\Big)$$

$$= \sum_j s_j\Big(\sum_i r_i(\mathbf{e}_i \otimes \mathbf{f}_j)\Big) = \sum_{i,j} r_i s_j(\mathbf{e}_i \otimes \mathbf{f}_j)$$

Since any vector in $U \otimes V$ is a finite sum of vectors $\mathbf{u} \otimes \mathbf{v}$, we deduce that $\mathfrak{D}$ spans $U \otimes V$. ∎

Corollary 14.8 For finite dimensional vector spaces,

$$dim(U \otimes V) = dim(U) \cdot dim(V)$$ ∎

Theorem 14.9 Let U and V be finite dimensional vector spaces. Then

$$U^* \otimes V^* \approx (U \otimes V)^*$$

via the isomorphism $\tau: U^* \otimes V^* \to (U \otimes V)^*$ defined by

$$\tau(\alpha \otimes \beta)(\mathbf{u} \otimes \mathbf{v}) = \alpha(\mathbf{u})\beta(\mathbf{v})$$

Proof. We must show that τ is an isomorphism. Let us first fix $\alpha \in U^*$ and $\beta \in V^*$, and consider the map $\sigma_{\alpha,\beta}: U \times V \to F$ defined by

$$\sigma_{\alpha,\beta}(\mathbf{u},\mathbf{v}) = \alpha(\mathbf{u})\beta(\mathbf{v})$$

This map is bilinear, and so the universal property of tensor products implies that there exists a unique linear map $\widehat{\sigma}_{\alpha,\beta}: U \otimes V \to F$ for which

$$\widehat{\sigma}_{\alpha,\beta}(\mathbf{u} \otimes \mathbf{v}) = \sigma_{\alpha,\beta}(\mathbf{u},\mathbf{v}) = \alpha(\mathbf{u})\beta(\mathbf{v})$$

Thus, $\widehat{\sigma}_{\alpha,\beta} \in (U \otimes V)^*$. Now we define a map $\sigma: U^* \times V^* \to (U \otimes V)^*$ by

$$\sigma(\alpha,\beta) = \widehat{\sigma}_{\alpha,\beta}$$

This map is also bilinear. For instance,

$$\begin{aligned}
\sigma(r\alpha + s\beta,\gamma)(\mathbf{u} \otimes \mathbf{v}) &= (r\alpha + s\beta)(\mathbf{u})\gamma(\mathbf{v}) \\
&= r\alpha(\mathbf{u})\gamma(\mathbf{v}) + s\beta(\mathbf{u})\gamma(\mathbf{v}) \\
&= r\sigma(\alpha,\gamma)(\mathbf{u},\mathbf{v}) + s\sigma(\beta,\gamma)(\mathbf{u},\mathbf{v}) \\
&= [r\sigma(\alpha,\gamma) + s\sigma(\beta,\gamma)](\mathbf{u},\mathbf{v})
\end{aligned}$$

and so

$$\sigma(r\alpha + s\beta,\gamma) = r\sigma(\alpha,\gamma) + s\sigma(\beta,\gamma)$$

which shows that σ is linear in its first coordinate. Hence, the universal property implies that there exists a unique linear map $\tau: U^* \otimes V^* \to (U \otimes V)^*$ for which

$$\tau(\alpha \otimes \beta) = \sigma(\alpha,\beta)$$

that is,

$$\tau(\alpha \otimes \beta)(\mathbf{u} \otimes \mathbf{v}) = \sigma(\alpha,\beta)(\mathbf{u} \otimes \mathbf{v}) = \widehat{\sigma}_{\alpha,\beta}(\mathbf{u} \otimes \mathbf{v}) = \alpha(\mathbf{u})\beta(\mathbf{v})$$

To show that τ is an isomorphism, let $\mathfrak{B} = \{\mathbf{b}_i\}$ be a basis for U, with dual basis $\mathfrak{B}' = \{\beta_i\}$, and let $\mathfrak{C} = \{\mathbf{c}_i\}$ be a basis for V, with dual basis $\mathfrak{C}' = \{\gamma_i\}$. Then

$$\tau(\beta_i \otimes \gamma_j)(\mathbf{b}_u \otimes \mathbf{c}_v) = \beta_i(\mathbf{b}_u)\gamma_j(\mathbf{c}_v) = \delta_{i,u}\delta_{j,v} = \delta_{(i,j),(u,v)}$$

and so $\tau(\beta_i \otimes \gamma_j) \in (U \otimes V)^*$ is a dual basis vector to the basis $\{\mathbf{b}_u \otimes \mathbf{c}_v\}$ for $U \otimes V$. Thus, τ takes the basis $\{\beta_i \otimes \gamma_j\}$ for $U^* \otimes V^*$ to the basis $\{\tau(\beta_i \otimes \gamma_j)\}$ Hence, τ is an isomorphism. ∎

Combining the isomorphisms of Theorem 14.4 and Theorem 14.9, we have, for finite dimensional vector spaces U and V,

$$U^* \otimes V^* \approx (U \otimes V)^* \approx \mathfrak{B}(U,V;F)$$

The Tensor Product of Linear Transformations

Let $\tau{:}V \to V'$ and $\sigma{:}W \to W'$ be linear transformations. Then there is a unique linear transformation $(\tau \odot \sigma){:}V \otimes W \to V' \otimes W'$ satisfying

$$(\tau \odot \sigma)(\mathbf{v} \otimes \mathbf{w}) = \tau(\mathbf{v}) \otimes \sigma(\mathbf{w}) \tag{14.9}$$

To see this, observe that the function $f{:}V \times W \to V' \otimes W'$ defined by $f(\mathbf{v},\mathbf{w}) = \tau(\mathbf{v}) \otimes \sigma(\mathbf{w})$ is bilinear, and so by the universal property of tensor products, there exists a unique linear transformation $\tau \odot \sigma$ for which (14.9) holds. The map $\tau \odot \sigma$ is called the **tensor product** of τ and σ.

Thus, we have a map $\phi{:}\mathfrak{L}(V,W) \times \mathfrak{L}(V',W') \to \mathfrak{L}(V \otimes W, V' \otimes W')$ defined by

$$\phi(\tau,\sigma) = \tau \odot \sigma \tag{14.10}$$

This map is bilinear and so there is a unique linear transformation

$$\theta{:}\mathfrak{L}(V,W) \otimes \mathfrak{L}(V',W') \to \mathfrak{L}(V \otimes W, V' \otimes W')$$

satisfying $\theta(\tau \otimes \sigma) = \tau \odot \sigma$.

We propose to show that θ is injective. Observe that any nonzero vector $\xi \in \mathfrak{L}(V,W) \otimes \mathfrak{L}(V',W')$ has the form

$$\xi = \sum_{i=1}^{n} \tau_i \otimes \sigma_i$$

where the τ_i's are linearly independent, and the σ_i's are linearly independent. To show that $ker(\theta) = \{\mathbf{0}\}$, suppose that

$$\theta(\xi) = \theta\Big(\sum_{i=1}^{n} \tau_i \otimes \sigma_i\Big) = 0$$

Then

$$\sum_{i=1}^{n} \tau_i(\mathbf{v}) \otimes \sigma_i(\mathbf{w}) = 0 \tag{14.11}$$

for all $\mathbf{v} \in V$ and $\mathbf{w} \in W$. Let us choose $\mathbf{v} \in V$ so that $\tau_1(\mathbf{v}) \neq \mathbf{0}$, and suppose (by renumbering if necessary) that $\tau_1(\mathbf{v}),\ldots,\tau_k(\mathbf{v})$ is a maximal linearly independent set among $\tau_1(\mathbf{v}),\ldots,\tau_n(\mathbf{v})$. Thus,

$$\tau_u(\mathbf{v}) = \sum_{j=1}^{k} r_{u,j}\tau_j(\mathbf{v})$$

for $u = k+1,\ldots,n$. Hence, (14.11) gives

$$\begin{aligned} 0 &= \sum_{i=1}^{k} \tau_i(\mathbf{v}) \otimes \sigma_i(\mathbf{w}) + \sum_{u=k+1}^{n} \Big(\sum_{j=1}^{k} r_{u,j}\tau_j(\mathbf{v})\Big) \otimes \sigma_u(\mathbf{w}) \\ &= \sum_{i=1}^{k} \tau_i(\mathbf{v}) \otimes \sigma_i(\mathbf{w}) + \sum_{j=1}^{k} \tau_j(\mathbf{v}) \otimes \Big(\sum_{u=k+1}^{n} r_{u,j}\sigma_u(\mathbf{w})\Big) \\ &= \sum_{i=1}^{k} \tau_i(\mathbf{v}) \otimes \Big(\sigma_i(\mathbf{w}) + \sum_{u=k+1}^{n} r_{u,j}\sigma_u(\mathbf{w})\Big) \end{aligned}$$

and since $\tau_1(\mathbf{v}),\ldots,\tau_k(\mathbf{v})$ are linearly independent, we must have

$$\sigma_i(\mathbf{w}) + \sum_{u=k+1}^{n} r_{u,j}\sigma_u(\mathbf{w}) = 0$$

for all $i = 1,\ldots,k$, and all $\mathbf{w} \in W$. Hence,

$$\sigma_i + \sum_{u=k+1}^{n} r_{u,j}\sigma_u = 0$$

which is in contradiction to the fact that the σ_i's are linearly independent. Hence, $\theta(\xi) \neq 0$ and so θ is injective.

Note that if all vector spaces are finite dimensional, then θ is also surjective, and hence is an isomorphism. In any case, the fact that $\theta{:}\tau \otimes \sigma \mapsto \tau \odot \sigma$ is injective motivates the commonly used notation $\tau \otimes \sigma$ for the tensor product $\tau \odot \sigma$. Let us summarize.

Theorem 14.10 Let $\tau \in \mathcal{L}(V,V')$ and $\sigma \in \mathcal{L}(W,W')$. There is a unique linear transformation $\tau \odot \sigma \in \mathcal{L}(V \otimes W, V' \otimes W')$, called the tensor product of τ and σ, satisfying

$$(\tau \odot \sigma)(\mathbf{v} \otimes \mathbf{w}) = \tau(\mathbf{v}) \otimes \sigma(\mathbf{w})$$

Moreover, there is a (unique) injective linear transformation

$$\theta:\mathcal{L}(V,W)\otimes\mathcal{L}(V',W')\to\mathcal{L}(V\otimes W,V'\otimes W')$$

satisfying $\theta(\tau\otimes\sigma)=\tau\odot\sigma$. In case all vector spaces are finite dimensional, θ is an isomorphism. ∎

Change of Base Field

We have seen in earlier chapters that a linear operator τ, defined on a *real* n-dimensional vector space V, may not have n eigenvalues (counting multiplicity), since its characteristic polynomial may not split over $\mathbb{R}$. On the other hand, a linear operator over the *complex* n-dimensional inner product space does have n eigenvalues. This leads us to wonder whether we can extend a real vector space to a complex vector space, and correspondingly extend a real operator to a complex operator.

Let us approach this question in more generality. For convenience, we refer to a vector space over a field F as an **F-space.** There are several approaches to "upgrading" the base field of a vector space. For instance, suppose that V is an F-space, and that F′ is an extension field of F, that is, $F'\supset F$. If $\{\mathbf{b}_i\}$ is a basis for V, then every element $\mathbf{x}$ of V has the form

$$\mathbf{x}=\sum r_i\mathbf{b}_i$$

where $r_i\in F$. We can define an F′-space V′ simply by taking all formal linear combinations of the form

$$\mathbf{x}=\sum r_i'\mathbf{b}_i$$

where $r_i'\in F'$. In other words, V′ is the free F′-space on the set $\{\mathbf{b}_i\}$. Note that the dimension of V′ as an F′-space is the same as the dimension of V as an F-space. Also, V′ is an F-space (just restrict the scalars to F), and as such, the inclusion map $j:V\to V'$ sending $\mathbf{x}\in V$ to $j(\mathbf{x})=\mathbf{x}\in V'$, is an F-monomorphism.

The approach described in the previous paragraph uses an arbitrarily chosen basis for V, and is therefore not coordinate free. However, we can give a coordinate-free approach using tensor products as follows. If V is an F-space, let

$$V'=F'\otimes_F V$$

It is customary to include the subscript F on $\otimes_F$ to denote the fact that the tensor product is taken with respect to the base field F. (All relevant maps are F-bilinear and F-linear.) However, since we will not take tensor products with respect to any other field, we will not always use this notation.

The vector space V′ is an F-space by definition of tensor product, but we may make it into an F'-space as follows. Fix an $s' \in F'$, and consider the map $f_{s'}:(F' \times V) \to (F' \otimes_F V)$ by

$$f_{s'}(r',\mathbf{v}) = s'r' \otimes \mathbf{v}$$

Since $f_{s'}$ is bilinear, the universality property of tensor products implies that there is a unique F-linear map $\tau_{s'}:(F' \otimes_F V) \to (F' \otimes_F V)$ for which

$$\tau_{s'}(r' \otimes \mathbf{v}) = s'r' \otimes \mathbf{v}$$

This map is intended to be multiplication by the scalar $s' \in F'$. Note that, since $\tau_{s'}$ is F-linear, it is additive, and so

$$\tau_{s'}(r' \otimes \mathbf{v} + u' \otimes \mathbf{w}) = \tau_{s'}(r' \otimes \mathbf{v}) + \tau_{s'}(u' \otimes \mathbf{w})$$

that is,

$$s'(r' \otimes \mathbf{v} + u' \otimes \mathbf{w}) = s'(r' \otimes \mathbf{v}) + s'(u' \otimes \mathbf{w})$$

Since all of the defining properties of scalar multiplication are satisfied, V' is indeed an F'-space.

It is not hard to see that if $\{\mathbf{b}_i\}$ is a basis for the F-space V, then $\{1 \otimes \mathbf{b}_i\}$ is a basis for the F'-space V', and so the dimension of the F'-space V' is equal to the dimension of the F-space V.

The map $\upsilon:V \to V'$ defined by $\upsilon(\mathbf{v}) = 1 \otimes \mathbf{v}$ is easily seen to be an F-monomorphism, and so the F-space V' contains an isomorphic copy of V. The F-linear monomorphism υ is sometimes called the **F'-extension map** of V. This map has a universal property of its own, as described in the next theorem.

Theorem 14.11 Let $\upsilon:V \to V' = F' \otimes_F V$ be the F'-extension map of an F-space V. Then υ has the following universal property. For any F-linear map $f:V \to W'$, where W' is any F'-space, there exists a unique F'-linear map $\tau:V' \to W'$ for which the diagram in Figure 14.8 is commutative, that is,

$$\tau \circ \upsilon = f$$

Proof. If such a map $\tau:F' \otimes_F V \to W'$ is to exist, then it must satisfy

$$\tau(r' \otimes \mathbf{v}) = r'\tau(1 \otimes \mathbf{v}) = r'f(\mathbf{v}) \tag{14.12}$$

This shows that, if τ exists, it is uniquely determined by f. To see that τ exists, consider the map $g:(F' \times V) \to W'$ defined by

$$g(r',\mathbf{v}) = r'f(\mathbf{v})$$

Since this is bilinear, there exists a unique F-linear map τ for which (14.12) holds. It is easy to see that τ is also F'-linear, since

$$\tau[s'(r' \otimes \mathbf{v})] = \tau(s'r' \otimes \mathbf{v}) = s'r'f(\mathbf{v}) = s'\tau(r' \otimes \mathbf{v}) \qquad ∎$$

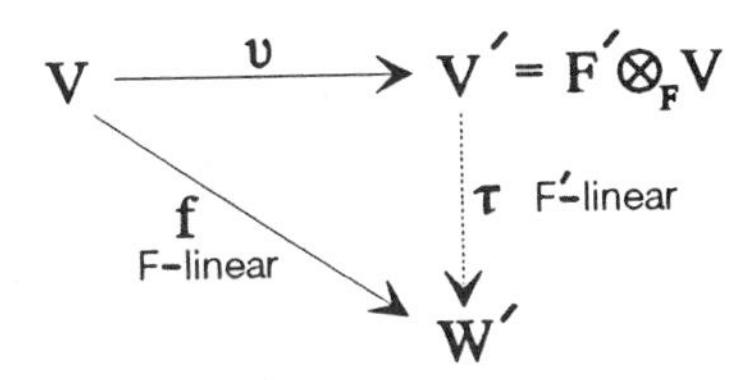

Figure 14.8

Theorem 14.11 is the key to describing how to extend an F-linear map to an F′-linear map.

Theorem 14.12 Let V and W be F-spaces, with F′-extension maps ν and μ, respectively. (See Figure 14.9.) Then for any F-linear map $\tau{:}V\to W$, the map $\tau' = \iota_{F'} \otimes \tau{:}V'\to W'$ is the unique F′-linear map that makes the diagram in Figure 14.9 commutative, that is, for which

$$\mu \circ \tau = \tau' \circ \upsilon$$

Proof. The map $\mu \circ \tau$ is an F-linear map from the F-space V to the F′-space W′. Hence, Theorem 14.11 shows that there is a unique F′-linear map $\tau'{:}V'\to W'$ such that

$$\mu \circ \tau = \tau' \circ \upsilon$$

To see that $\tau' = \iota_{F'} \otimes \tau$, observe that

$$\begin{aligned}\tau'(r' \otimes \mathbf{v}) &= r'\tau'(1 \otimes \mathbf{v}) = r'(\tau' \circ \nu)(\mathbf{v}) = r'(\mu \circ \tau)(\mathbf{v}) \\ &= r'(1 \otimes \tau(\mathbf{v})) = \iota_{F'}(r') \otimes \tau(\mathbf{v}) = (\iota_{F'} \otimes \tau)(r' \otimes \mathbf{v})\end{aligned}$$ ∎

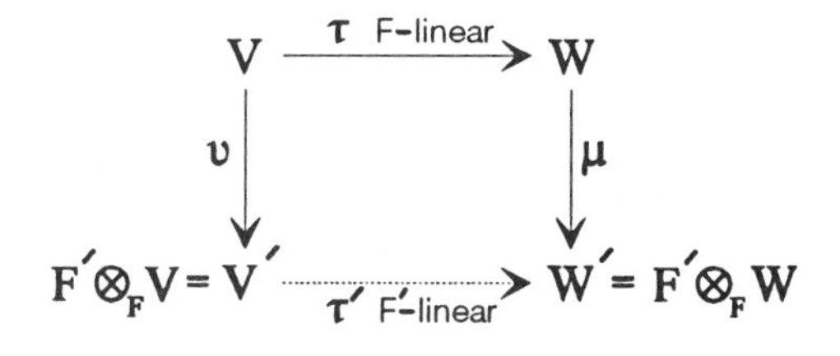

Figure 14.9

Multilinear Maps and Iterated Tensor Products

The tensor product operation can easily be extended to more than two vector spaces. We begin with the extension of the concept of bilinearity.

Definition If $V_1,\ldots,V_n$ and W are vector spaces over F, a function $f:V_1\times\cdots\times V_n\to W$ is said to be **multilinear** if it is linear in each variable separately, that is, if

$$f(\mathbf{u}_1,\ldots,\mathbf{u}_{k-1},r\mathbf{v}+s\mathbf{v}',\mathbf{u}_{k+1},\ldots,\mathbf{u}_n) = \\ rf(\mathbf{u}_1,\ldots,\mathbf{u}_{k-1},\mathbf{v},\mathbf{u}_{k+1},\ldots,\mathbf{u}_n)+sf(\mathbf{u}_1,\ldots,\mathbf{u}_{k-1},\mathbf{v}',\mathbf{u}_{k+1},\ldots,\mathbf{u}_n)$$

for all $k=1,\ldots,n$. A multilinear function of n variables is also referred to as an **n-linear function.** The set of all multilinear functions will be denoted by $\mathrm{Mul}(V_1,\ldots,V_n;W)$. A multilinear function from $V_1\times\cdots\times V_n$ to the base field F is called a **multilinear form** (or **n-form**). □

Example 14.2

1) If A is an algebra then the product map $\mu:A\times\cdots\times A\to A$ defined by $\mu(a_1,\ldots,a_n)=a_1\cdots a_n$ is n-linear.
2) The determinant function $det:\mathcal{M}_n\to F$ is an n-linear form on the columns of the matrices in $\mathcal{M}_n$. □

Definition Let $V_1,\ldots,V_n$ be vector spaces over F, and let T be the subspace of the free vector space $\mathcal{F}$ on $V_1\times\cdots\times V_n$ generated by all vectors of the form

$$r(\mathbf{v}_1,\ldots,\mathbf{v}_{k-1},\mathbf{u},\mathbf{v}_{k+1},\ldots,\mathbf{v}_n)+s(\mathbf{v}_1,\ldots,\mathbf{v}_{k-1},\mathbf{u}',\mathbf{v}_{k+1},\ldots,\mathbf{v}_n) \\ -(\mathbf{v}_1,\ldots,\mathbf{v}_{k-1},r\mathbf{u}+s\mathbf{u}',\mathbf{v}_{k+1},\ldots,\mathbf{v}_n)$$

for all $r,s\in F$, $\mathbf{u},\mathbf{u}'\in U$ and $\mathbf{v}_1,\ldots,\mathbf{v}_n\in V$. The quotient space $\mathcal{F}/T$ is called the **tensor product** of $V_1,\ldots,V_n$, and denoted by $V_1\otimes\cdots\otimes V_n$. □

As before, we denote the coset $(\mathbf{v}_1,\ldots,\mathbf{v}_n)+T$ by $\mathbf{v}_1\otimes\cdots\otimes\mathbf{v}_n$, and so any element of $V_1\otimes\cdots\otimes V_n$ has the form

$$\sum \mathbf{v}_{i_1}\otimes\cdots\otimes\mathbf{v}_{i_n}$$

where the vector space operations are linear in each variable.

The tensor product can also be characterized by a universal property.

Theorem 14.13 **(The universal property of tensor products)** Let $V_1,\ldots,V_n$ be vector spaces over the field F. The pair $(V_1\otimes\cdots\otimes V_n,t)$, where $t:V_1\times\cdots\times V_n\to V_1\otimes\cdots\otimes V_n$ is the multilinear map defined by

$$t(\mathbf{v}_1,\ldots,\mathbf{v}_n)=\mathbf{v}_1\otimes\cdots\otimes\mathbf{v}_n$$

has the following property. Referring to Figure 14.10, if $f:V_1 \times \cdots \times V_n \to W$ is any multilinear function from $V_1 \times \cdots \times V_n$ to a vector space W over F, then there is a unique *linear* transformation $\tau:V_1 \otimes \cdots \otimes V_n \to W$ that makes the diagram in Figure 14.10 commute, that is, for which

$$\tau \circ t = f$$

Moreover, $V_1 \otimes \cdots \otimes V_n$ is unique in the sense that if a pair (X,s) also has this property, then X is isomorphic to $V_1 \otimes \cdots \otimes V_n$. ∎

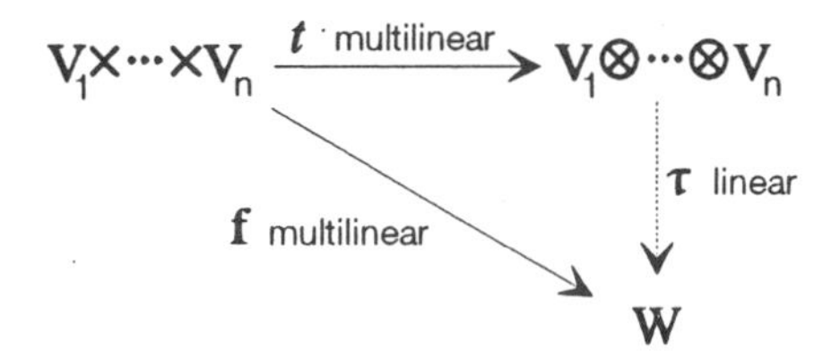

Figure 14.10

Here are some of the basic properties of multiple tensor products.

Theorem 14.14 The tensor product has the following properties. Note that all vector spaces are over the same field F.

1) (**Associativity**) There exists an isomorphism

$$\tau:(V_1 \otimes \cdots \otimes V_n) \otimes (W_1 \otimes \cdots \otimes W_m) \to V_1 \otimes \cdots \otimes V_n \otimes W_1 \otimes \cdots \otimes W_m$$

for which

$$\tau[(\mathbf{v}_1 \otimes \cdots \otimes \mathbf{v}_n) \otimes (\mathbf{w}_1 \otimes \cdots \otimes \mathbf{w}_m)] = \mathbf{v}_1 \otimes \cdots \otimes \mathbf{v}_n \otimes \mathbf{w}_1 \otimes \cdots \otimes \mathbf{w}_m$$

In particular,

$$(U \otimes V) \otimes W \approx U \otimes (V \otimes W) \approx U \otimes V \otimes W$$

2) (**Commutativity**) Let π be any permutation of the indices $\{1,\ldots,n\}$. Then there is an isomorphism

$$\sigma:V_1 \otimes \cdots \otimes V_n \to V_{\pi(1)} \otimes \cdots \otimes V_{\pi(n)}$$

for which

$$\sigma(\mathbf{v}_1 \otimes \cdots \otimes \mathbf{v}_n) = \mathbf{v}_{\pi(1)} \otimes \cdots \otimes \mathbf{v}_{\pi(n)}$$

3) There is an isomorphism $\rho_1:F \otimes V \to V$ for which

$$\rho_1(r \otimes \mathbf{v}) = r\mathbf{v}$$

and similarly, there is an isomorphism $\rho_2:V \otimes F \to V$ for which

$$\rho_2(\mathbf{v} \otimes r) = r\mathbf{v}$$

Hence, $F \otimes V \approx V \approx V \otimes F$. ∎

The analog of Theorem 14.4 is the following.

Theorem 14.15 Let $V_1, \ldots, V_n$ and W be vector spaces over F. Then the map $\phi: Mul(V_1, \ldots, V_n; W) \to \mathcal{L}(V_1 \otimes \cdots \otimes V_n, W)$, defined by the fact that $\phi(f)$ is the unique linear map for which $f = \phi(f) \circ t$, is an isomorphism. Thus,

$$Mul(V_1, \ldots, V_n; W) \approx \mathcal{L}(V_1 \otimes \cdots \otimes V_n, W)$$

Moreover, if all vector spaces are finite dimensional, then

$$dim[Mul(V_1, \ldots, V_n; W)] = dim(V_1) \cdots dim(V_n) \cdot dim(W)$$ ∎

Alternating Maps and Exterior Products

We will use the notation V^n to denote the Cartesian product of V with itself n times, and $\otimes^n V$ to denote the n-fold tensor product. The following definitions describe some special types of multilinear maps.

Definition

1) A multilinear map $f: V^n \to W$ is **symmetric** if

$$f(\mathbf{v}_1, \ldots, \mathbf{v}_i, \ldots, \mathbf{v}_j, \ldots, \mathbf{v}_n) = f(\mathbf{v}_1, \ldots, \mathbf{v}_j, \ldots, \mathbf{v}_i, \ldots, \mathbf{v}_n)$$

for any $i \neq j$.

2) A multilinear map $f: V^n \to W$ is **skew-symmetric** if

$$f(\mathbf{v}_1, \ldots, \mathbf{v}_i, \ldots, \mathbf{v}_j, \ldots, \mathbf{v}_n) = -f(\mathbf{v}_1, \ldots, \mathbf{v}_j, \ldots, \mathbf{v}_i, \ldots, \mathbf{v}_n)$$

for $i \neq j$.

3) A multilinear map $f: V^n \to W$ is **alternating** if

$$f(\mathbf{v}_1, \ldots, \mathbf{v}_n) = 0$$

whenever any two of the vectors $\mathbf{v}_i$ are equal. □

A few remarks about permutations, with which the reader may very well be familiar, are in order. A **permutation** of the set $N = \{1, \ldots, n\}$ is a bijective function $\pi: N \to N$. We denote the set of all such permutations by S_n. This is the *symmetric group* on n symbols. A **cycle** of length k is a permutation of the form $(i_1, i_2, \ldots, i_k)$, that sends i_1 to i_2, i_2 to $i_3, \ldots, i_{k-1}$ to i_k and i_k to i_1. (We assume that $i_u \neq i_v$ for $u \neq v$.) All other elements of N are left fixed. Every permutation is the product (composition) of disjoint cycles.

A **transposition** is a cycle (i,j) of length 2. Every cycle (and therefore every permutation) is the product of transpositions. In general, a permutation can be expressed as a product of transpositions in many ways. However, no matter how one represents a given permutation as such a product, the number of transpositions is either always even or always odd. Therefore, we can define the **parity** of a permutation $\pi \in S_n$ to be the parity of the number of transpositions in any decomposition of π as a product of transpositions. The **sign** of a permutation is defined by

$$sg(\pi) = (-1)^{\text{parity}(\pi)}$$

Thus, $sg(\pi) = 1$ if π is an even permutation, and -1 is π is an odd permutation. The sign of π is often written $(-1)^\pi$.

With these facts in mind, it is apparent that f is symmetric if and only if

$$f(\mathbf{v}_1,\ldots,\mathbf{v}_n) = f(\mathbf{v}_{\pi(1)},\ldots,\mathbf{v}_{\pi(n)})$$

for all permutations $\pi \in S_n$, and that f is alternating if and only if

$$f(\mathbf{v}_1,\ldots,\mathbf{v}_n) = (-1)^\pi f(\mathbf{v}_{\pi(1)},\ldots,\mathbf{v}_{\pi(n)})$$

for all permutations $\pi \in S_n$.

If f is a multilinear function, then

$$\begin{aligned} f(\mathbf{v}_1,&\ldots,\mathbf{v}_i+\mathbf{v}_j,\ldots,\mathbf{v}_i+\mathbf{v}_j,\ldots,\mathbf{v}_n) \\ &= f(\mathbf{v}_1,\ldots,\mathbf{v}_i,\ldots,\mathbf{v}_i,\ldots,\mathbf{v}_n) + f(\mathbf{v}_1,\ldots,\mathbf{v}_i,\ldots,\mathbf{v}_j,\ldots,\mathbf{v}_n) \\ &\quad + f(\mathbf{v}_1,\ldots,\mathbf{v}_j,\ldots,\mathbf{v}_i,\ldots,\mathbf{v}_n) + f(\mathbf{v}_1,\ldots,\mathbf{v}_j,\ldots,\mathbf{v}_j,\ldots,\mathbf{v}_n) \end{aligned}$$

Hence, if f is alternating, then it is also skew-symmetric. On the other hand, if f is skew-symmetric, we have

$$f(\mathbf{v}_1,\ldots,\mathbf{v}_i,\ldots,\mathbf{v}_i,\ldots,\mathbf{v}_n) = -f(\mathbf{v}_1,\ldots,\mathbf{v}_i,\ldots,\mathbf{v}_i,\ldots,\mathbf{v}_n)$$

and so, provided that $\text{char}(F) \neq 2$, this gives

$$f(\mathbf{v}_1,\ldots,\mathbf{v}_i,\ldots,\mathbf{v}_i,\ldots,\mathbf{v}_n) = 0$$

and so f is alternating.

We have discussed symmetric and alternating bilinear functions in Chapter 11. Our intention here is to briefly discuss alternating multilinear functions, which play an especially important role in differential geometry and its applications.

Definition Let V be a vector space over a field F with $\text{char}(F) \neq 2$, and let $\otimes^n V$ be the n-fold tensor product of V with itself. Let U be the subspace of $\otimes^n V$ generated by all elements of the form

$$(\mathbf{v}_1 \otimes \cdots \otimes \mathbf{v}_i \otimes \cdots \otimes \mathbf{v}_j \otimes \cdots \otimes \mathbf{v}_n) + (\mathbf{v}_1 \otimes \cdots \otimes \mathbf{v}_j \otimes \cdots \otimes \mathbf{v}_i \otimes \cdots \otimes \mathbf{v}_n)$$

for all $i<j$. The quotient space $(\otimes^n V)/U$ is called the **nth exterior product space** of V and is denoted by

$$\wedge^n V \quad \text{or} \quad \underbrace{V \wedge \cdots \wedge V}_{\text{n factors}} \qquad \square$$

It is customary to denote the coset $(\mathbf{v}_1 \otimes \cdots \otimes \mathbf{v}_n)+U$ by $\mathbf{v}_1 \wedge \cdots \wedge \mathbf{v}_n$ and refer to $\wedge$ as the **wedge product.** Thus, any element of $V_1 \wedge \cdots \wedge V_n$ has the form

$$\sum \mathbf{v}_{i_1} \wedge \cdots \wedge \mathbf{v}_{i_n}$$

where the vector space operations are linear in each variable, and where the interchange of any two variables introduces a minus sign.

The exterior product can also be characterized by a universal property.

Theorem 14.16 (The universal property of exterior products) Let $V_1,\ldots,V_n$ be vector spaces over a field F with $\text{char}(F) \neq 2$. The pair $(V_1 \wedge \cdots \wedge V_n, a)$, where $a:V_1 \times \cdots \times V_n \to V_1 \wedge \cdots \wedge V_n$ is the alternating multilinear map defined by

$$a(\mathbf{v}_1,\ldots,\mathbf{v}_n) = \mathbf{v}_1 \wedge \cdots \wedge \mathbf{v}_n$$

has the following property. Referring to Figure 14.11, if $f:V_1 \times \cdots \times V_n \to W$ is any *alternating* multilinear function from $V_1 \times \cdots \times V_n$ to a vector space W over F, then there is a unique *linear* transformation $\tau:V_1 \wedge \cdots \wedge V_n \to W$ that makes the diagram in Figure 14.11 commute, that is, for which

$$\tau \circ a = f$$

Moreover, $V_1 \wedge \cdots \wedge V_n$ is unique in the sense that if a pair (X,σ) also has this property, then X is isomorphic to $V_1 \wedge \cdots \wedge V_n$. ∎

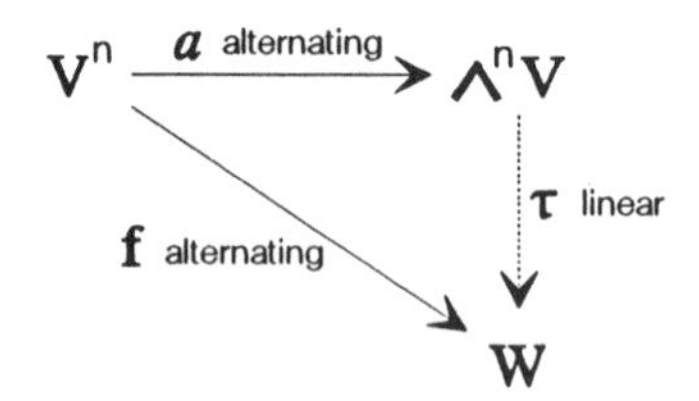

Figure 14.11

EXERCISES

1. Verify that the set $\mathcal{F}_{U \times V}$ is a vector space.
2. Show that if $\tau: W \to X$ is a linear map, and $b: U \times V \to W$ is bilinear, then $\tau \circ b: U \times V \to X$ is bilinear.
3. Show that the only map that is both linear and n-linear (for $n \geq 2$) is the zero map.
4. Find an example of a bilinear map $\tau: V \times V \to W$ whose image $im(\tau) = \{\tau(\mathbf{u},\mathbf{v}) \mid \mathbf{u},\mathbf{v} \in V\}$ is *not* a subspace of W.
5. Prove that $U \otimes V \approx V \otimes U$.
6. Let X and Y be nonempty sets. Use the universal property of tensor products to prove that $\mathcal{F}_{X \times Y} \approx \mathcal{F}_X \otimes \mathcal{F}_Y$.
7. Let $\mathbf{u},\mathbf{u}' \in U$ and $\mathbf{v},\mathbf{v}' \in V$. Assuming that $\mathbf{u} \otimes \mathbf{v} \neq \mathbf{0}$, show that $\mathbf{u} \otimes \mathbf{v} = \mathbf{u}' \otimes \mathbf{v}'$ if and only if $\mathbf{u}' = r\mathbf{u}$ and $\mathbf{v}' = r^{-1}\mathbf{v}$, for $r \neq 0$.
8. Let $\mathcal{B} = \{\mathbf{b}_i\}$ be a basis for U and $\mathcal{C} = \{\mathbf{c}_i\}$ be a basis for V. Show that *any* function $f: U \times V \to W$ can be extended to a linear function $\bar{f}: U \otimes V \to W$. Deduce that the function f can be extended in a unique way to a *bilinear* map $\hat{f}: U \times V \to W$. Show that all bilinear maps are obtained in this way.
9. Let S_1, S_2 be subspaces of U. Show that
$$(S_1 \otimes V) \cap (S_2 \otimes V) \approx (S_1 \cap S_2) \otimes V$$
10. Let $S \subset U$ and $T \subset V$ be subspaces of vector spaces U and V, respectively. Show that
$$(S \otimes V) \cap (U \otimes T) \approx S \otimes T$$
11. Let $S_1, S_2 \subset U$ and $T_1, T_2 \subset V$ be subspaces of U and V, respectively. Show that
$$(S_1 \otimes T_1) \cap (S_2 \otimes T_2) \approx (S_1 \cap S_2) \otimes (T_1 \otimes T_2)$$
12. Find an example of two vector spaces U and V and a nonzero vector $\mathbf{x} \in U \otimes V$ that has at least two distinct (not including order of the terms) representations of the form
$$\mathbf{x} = \sum_{i=1}^{n} \mathbf{u}_i \otimes \mathbf{v}_i$$
where the $\mathbf{u}_i$'s are linearly independent, and so are the $\mathbf{v}_i$'s. However, prove that the number n of terms is the same for all such representations.
13. What is the dimension of the space $\mathcal{B}(U,V;F)$ of all bilinear forms on $U \times V$? (Assume U and V are finite dimensional.)
14. Let ι_X denote the identity operator on a vector space X. Prove that $\iota_V \odot \iota_W = \iota_{V \otimes W}$.
15. Suppose that $\tau_1: U \to V$, $\tau_2: V \to W$, and $\sigma_1: U' \to V'$, $\sigma_2: V' \to W'$. Prove that

$$(\tau_2 \circ \tau_1) \odot (\sigma_2 \circ \sigma_1) = (\tau_2 \odot \sigma_2) \circ (\tau_1 \odot \sigma_1)$$

16. Let V be an F-space, and $F' \supset F$. Prove that if $\{\mathbf{b}_i\}$ is a basis for the F-space V, then $\{1 \otimes \mathbf{b}_i\}$ is a basis for the F'-space V'.
17. Connect the two approaches to extending the base field of an F-space V to F' (at least in the finite dimensional case) by showing that $F^n \otimes_F F' \approx (F')^n$.
18. Prove that any permutation $\pi \in S_n$ is the product of disjoint cycles. Then prove that any cycle is the product of transpositions.
19. Prove that if $\pi \in S_n$, then any decomposition of π as a product of transpositions has the same parity. *Hint:* Consider the polynomial
$$p(x_1,\ldots,x_n) = \prod_{i<j} (x_i - x_j)$$
and let $\pi(p) = p(x_{\pi(1)},\ldots,x_{\pi(n)})$. Show that $\pi(p) = p$ if p is the product of an even number of transpositions, and $\pi(p) = -p$ if π is the product of an odd number of transpositions.

CHAPTER 15

Affine Geometry

Contents: *Affine Geometry. Affine Combinations. Affine Hulls. The Lattice of Flats. Affine Independence. Affine Transformations. Projective Geometry. Exercises.*

In this chapter, we will study the geometry of a finite dimensional vector space V, along with its structure preserving maps. *Throughout this chapter, all vector spaces are assumed to be finite dimensional.*

Affine Geometry

Definition Let V be a vector space. If $\mathbf{v} \in V$ and S is a subspace of V, then the set

$$\mathbf{v}+S=\{\mathbf{v}+\mathbf{s} \mid \mathbf{s} \in S\}$$

is called a **flat**, or **coset** in V. The set $\mathcal{A}(V)$ of all flats in V is called the **affine geometry** of V. The **dimension** $dim(\mathcal{A}(V))$ of $\mathcal{A}(V)$ is defined to be $dim(V)$. □

It is clear that a flat in V is nothing more than a translated subspace of V. We will denote subspaces of V by the letters S,T,... and flats in V by X,Y,.... Here are some of the basic intersection properties of flats.

Theorem 15.1

1) The following are equivalent:
   a) $\mathbf{x}+S=\mathbf{y}+S$ b) $\mathbf{x} \in \mathbf{y}+S$ c) $\mathbf{x}-\mathbf{y} \in S$

Let $X = \mathbf{x}+S$ and $Y = \mathbf{y}+T$ be flats in V. Then

2) $S \subset T \Leftrightarrow \mathbf{v}+X \subset Y$ for some $\mathbf{v} \in V$
3) $S = T \Leftrightarrow \mathbf{v}+X = Y$ for some $\mathbf{v} \in V$
4) $X \cap Y \neq \emptyset, S \subset T \Rightarrow X \subset Y$
5) $X \cap Y \neq \emptyset, S = T \Rightarrow X = Y$

Proof. We leave proof of part (1) as an exercise. To prove (2), observe that $S = -\mathbf{x}+X$ and $T = -\mathbf{y}+T$, and so

$$S \subset T \Leftrightarrow -\mathbf{x}+X \subset -\mathbf{y}+Y \Leftrightarrow (\mathbf{y}-\mathbf{x})+X \subset Y$$

As for (3), we have

$$(\mathbf{y}-\mathbf{x})+X \subset Y \quad \text{and} \quad (\mathbf{x}-\mathbf{y})+Y \subset X$$

and so

$$(\mathbf{y}-\mathbf{x})+X \subset Y \subset (\mathbf{y}-\mathbf{x})+X$$

which implies that $(\mathbf{y}-\mathbf{x})+X = Y$.

To prove (4), let $\mathbf{z} \in X \cap Y$. Then part (2) tells us that $\mathbf{v}+X \subset Y$, and so $\mathbf{v}+\mathbf{z} = \mathbf{y} \in Y$, which implies that $\mathbf{v} = \mathbf{y}-\mathbf{z} \in T$. Hence, $X \subset -\mathbf{v}+Y \subset Y$. Part (5) follows from (4). ∎

Part (1) of the previous theorem says that a flat can be represented in many ways, in the form $\mathbf{x}+S$. When a flat is written $\mathbf{x}+S$, we refer to $\mathbf{x}$ as the **flat representative**, or **coset representative** of the flat. Any element of a flat can be used as a flat representative. On the other hand, part (3) of Theorem 15.1, with $\mathbf{v} = \mathbf{0}$, implies that each flat $\mathbf{x}+S$ is associated with a *unique* subspace S. This allows us to make the following definition.

Definition The **dimension** of a flat $\mathbf{x}+S$ is $dim(S)$. A flat of dimension k is called a **k-flat.** A 0-flat is a **point,** a 1-flat is a **line** and a 2-flat is a **plane.** A flat of dimension $dim(\mathcal{A}(V))-1$ is called a **hyperplane.** □

Definition Two flats $X = \mathbf{x}+S$ and $Y = \mathbf{y}+T$ are said to be **parallel** if $S \subset T$ or $T \subset S$. This is denoted by $X \parallel Y$. □

According to Theorem 15.1, if $X \parallel Y$, then $X \subset Y$, $Y \subset X$ or $X \cap Y = \emptyset$. Moreover, part (2) of Theorem 15.1 says that X and Y are parallel if and only if some translation of one of these flats is contained in the other.

Affine Combinations

If $r_i \in F$ and $r_1+\cdots+r_n = 1$, then the linear combination

$$\mathbf{r}_1\mathbf{x}_1 + \cdots + \mathbf{r}_n\mathbf{x}_n$$

is referred to as an **affine combination** of the vectors $\mathbf{x}_1, \ldots, \mathbf{x}_n$.

Theorem 15.2 If char(F) $\neq 2$, then the following are equivalent for a subset X of V.

1) X is closed under the taking of affine combinations of any two of its points, that is,
$$\mathbf{x},\mathbf{y} \in X \Rightarrow r\mathbf{x} + (1-r)\mathbf{y} \in X$$

2) X is closed under the taking of affine combinations, that is,
$$\mathbf{x}_1, \ldots, \mathbf{x}_n \in X,\ r_1 + \cdots + r_n = 1 \Rightarrow r_1\mathbf{x}_1 + \cdots + r_n\mathbf{x}_n \in X$$

Proof. It is clear that (2) implies (1). For the converse, we proceed by induction. According to (1), for $\mathbf{x}_1, \mathbf{x}_2 \in X$,

$$r_1 + r_2 = 1 \Rightarrow r_1\mathbf{x}_1 + r_2\mathbf{x}_2 \in X$$

Assume for the purposes of induction that for $\mathbf{x}_i \in X$

$$r_1 + \cdots + r_{n-1} = 1 \Rightarrow r_1\mathbf{x}_1 + \cdots + r_{n-1}\mathbf{x}_{n-1} \in X$$

Let $\mathbf{x}_1, \ldots, \mathbf{x}_n \in X$ and $r_1 + \cdots + r_n = 1$, and consider the affine combination

$$\mathbf{z} = r_1\mathbf{x}_1 + \cdots + r_n\mathbf{x}_n$$

If one of r_1 or r_2 is different from 1, say $r_1 \neq 1$, then we may write

$$\mathbf{z} = r_1\mathbf{x}_1 + (1 - r_1)\Big(\frac{r_2}{1-r_1}\mathbf{x}_2 + \cdots + \frac{r_n}{1-r_1}\mathbf{x}_n\Big)$$

and since the sum of the coefficients of the sum inside the large parentheses is 1, the induction hypothesis implies that this sum is in X. Then (1) shows that $\mathbf{z} \in X$. On the other hand, if $r_1 = r_2 = 1$, then since char(F) $\neq 2$, we may write

$$\mathbf{z} = 2[\tfrac{1}{2}\mathbf{x}_1 + \tfrac{1}{2}\mathbf{x}_2] + r_3\mathbf{x}_3 + \cdots + r_n\mathbf{x}_n$$

and since (1) implies that $\frac{1}{2}\mathbf{x}_1 + \frac{1}{2}\mathbf{x}_2 \in X$, we may again deduce from the induction hypothesis that $\mathbf{z} \in X$. In any case, $\mathbf{z} \in X$, and so (2) holds. ∎

Note that the requirement char(F) $\neq 2$ is necessary, for the subset $X = \{(0,0), (1,0), (0,1)\}$ of $(F_2)^2$ satisfies (1), but not (2), in Theorem 15.2. We can now characterize flats.

Theorem 15.3

1) A subset X of V is a flat in V if and only if it is closed under the taking of affine combinations, that is,

$$\mathbf{x}_1,\ldots,\mathbf{x}_n \in X,\ r_1+\cdots+r_n=1 \ \Rightarrow\ r_1\mathbf{x}_1+\cdots+r_n\mathbf{x}_n \in X$$

2) If $\mathrm{char}(F) \neq 2$, a subset X of V is a flat if and only if X contains the line through any two of its points, that is, if and only if

$$\mathbf{x},\mathbf{y} \in X \ \Rightarrow\ r\mathbf{x}+(1-r)\mathbf{y} \in X$$

Proof. Suppose that $X=\mathbf{x}+S$ is a flat, and $\mathbf{x}_1,\ldots,\mathbf{x}_n \in X$. Then $\mathbf{x}_i=\mathbf{x}+\mathbf{s}_i$, for $\mathbf{s}_i \in S$, and so if $\Sigma r_i = 1$, we have

$$\sum r_i\mathbf{x}_i = \sum r_i(\mathbf{x}+\mathbf{s}_i) = \mathbf{x}+\sum r_i\mathbf{s}_i \in \mathbf{x}+S$$

and so X is closed under affine combinations. Conversely, suppose that X is closed under the taking of affine combinations, and let

$$S=\{\mathbf{x}-\mathbf{x}_0 \mid \mathbf{x}\in X\}$$

for some $\mathbf{x}_0 \in X$. If $\mathbf{x}_1-\mathbf{x}_0,\ldots,\mathbf{x}_n-\mathbf{x}_0$ are arbitrary vectors in S, and $r_1,\ldots,r_n \in F$, then

$$\sum r_i(\mathbf{x}_i-\mathbf{x}_0) = r_1\mathbf{x}_1+\cdots+r_n\mathbf{x}_n+(1-r_1-\cdots-r_n)\mathbf{x}_0-\mathbf{x}_0 \in S$$

Thus, S is closed under the taking of linear combinations, and so is a subspace of V. This implies that $X=\mathbf{x}_0+S$ is a flat. Part (2) follows from part (1) and Theorem 15.2. ∎

Affine Hulls

The following definition gives the analog of the subspace spanned by a collection of vectors.

Definition Let C be a nonempty set of vectors in V. The **affine hull** *hull*(C) of C is the smallest flat containing C. We also refer to *hull*(C) as the **flat generated** by C. □

Theorem 15.4 Let C be any nonempty subset of V. The affine hull *hull*(C) is the set of all affine combinations of vectors in C

$$hull(C)=\left\{\sum_{i=1}^{n} r_i\mathbf{x}_i \,\middle|\, n\geq 1,\ \mathbf{x}_1,\ldots,\mathbf{x}_n\in C,\ \sum_{i=1}^{n} r_i=1\right\}$$

Proof. According to Theorem 15.3, any flat containing C must contain all affine combinations of vectors in C. It remains only to show that the set X of all such affine combinations is a flat. To this end, let $\mathbf{y}\in X$, and consider the set

$$S=\{\mathbf{y}_j-\mathbf{y} \mid \mathbf{y}_j\in X\}$$

It suffices to show that S is a subspace of V, for then $X=\mathbf{y}+S$ is indeed a flat. To this end, let

$$\mathbf{y}=\sum_{i=1}^{n} r_{0,i}\mathbf{x}_i, \quad \mathbf{y}_1=\sum_{i=1}^{n} r_{1,i}\mathbf{x}_i \quad \text{and} \quad \mathbf{y}_2=\sum_{i=1}^{n} r_{2,i}\mathbf{x}_i$$

(By including additional zero coefficients if necessary, we may assume that the upper limits of summation are the same.) Hence, any linear combination of $\mathbf{y}_1-\mathbf{y}$ and $\mathbf{y}_2-\mathbf{y}$ has the form

$$\begin{aligned}
\mathbf{z} &= s(\mathbf{y}_1-\mathbf{y})+t(\mathbf{y}_2-\mathbf{y}) \\
&= s\sum_{i=1}^{n} r_{1,i}\mathbf{x}_i + t\sum_{i=1}^{n} r_{2,i}\mathbf{x}_i-(s+t)\mathbf{y} \\
&= \sum_{i=1}^{n}(sr_{1,i}+tr_{2,i})\mathbf{x}_i-(s+t-1)\mathbf{y}-\mathbf{y} \\
&= \sum_{i=1}^{n}\left(sr_{1,i}+tr_{2,i}-(s+t-1)r_{0,i}\right)\mathbf{x}_i-\mathbf{y}
\end{aligned}$$

But,

$$\begin{aligned}
\sum_{i=1}^{n}\left(sr_{1,i}+tr_{2,i}-(s+t-1)r_{0,i}\right) &= s\sum_{i=1}^{n} r_{1,i}+t\sum_{i=1}^{n} r_{2,i}-(s+t-1)\sum_{i=1}^{n} r_{0,i} \\
&= s+t-(s+t-1)=1
\end{aligned}$$

which shows that $\mathbf{z}\in S$. Hence, S is a subspace of V. ∎

The affine hull of a finite set of vectors is denoted by $hull\{\mathbf{x}_1,\ldots,\mathbf{x}_n\}$. We leave it as an exercise to show that

(15.1) $$hull\{\mathbf{x}_1,\ldots,\mathbf{x}_n\}=\mathbf{x}_i+\langle \mathbf{x}_1-\mathbf{x}_i,\ldots,\mathbf{x}_{i-1}-\mathbf{x}_i,\mathbf{x}_{i+1}-\mathbf{x}_i,\ldots,\mathbf{x}_n-\mathbf{x}_i\rangle$$

where $\langle \mathbf{x}_1-\mathbf{x}_i,\ldots,\mathbf{x}_{i-1}-\mathbf{x}_i,\mathbf{x}_{i+1}-\mathbf{x}_i,\ldots,\mathbf{x}_n-\mathbf{x}_i\rangle$ is the subspace spanned by the vectors within the angle brackets. This shows that

$$dim(hull\{\mathbf{x}_1,\ldots,\mathbf{x}_n\})\leq n-1$$

The affine hull of a pair of distinct points is the line through those points, denoted by

$$\overline{\mathbf{xy}}=\{r\mathbf{x}+(1-r)\mathbf{y}\mid r\in F\}=\mathbf{y}+\langle \mathbf{x}-\mathbf{y}\rangle$$

The Lattice of Flats

Since flats are subsets of V, they are partially ordered by set inclusion.

Theorem 15.5 The intersection of a nonempty collection $\mathcal{C} = \{\mathbf{x}_i + S_i \mid i \in K\}$ of flats in V is either empty or is a flat. If the intersection is nonempty, then

$$\bigcap_{i \in K} (\mathbf{x}_i + S_i) = \mathbf{x} + \bigcap_{i \in K} S_i$$

for any vector $\mathbf{x}$ in the intersection.

Proof. If

$$\mathbf{x} \in \bigcap_{i \in K} (\mathbf{x}_i + S_i)$$

then $\mathbf{x}_i + S_i = \mathbf{x} + S_i$ for all $i \in K$, and so

$$\bigcap_{i \in K} (\mathbf{x}_i + S_i) = \bigcap_{i \in K} (\mathbf{x} + S_i) = \mathbf{x} + \bigcap_{i \in K} S_i$$ ∎

Definition The **join** of a nonempty collection $\mathcal{C} = \{\mathbf{x}_i + S_i \mid i \in K\}$ of flats in V is the smallest flat containing all flats in $\mathcal{C}$. We denote the join of the collection $\mathcal{C}$ of flats by $\vee \mathcal{C}$, or by

$$\bigvee_{i \in K} \{\mathbf{x}_i + S_i\}$$

The join of two flats is denoted by $(\mathbf{x} + S) \vee (\mathbf{y} + T)$. □

Theorem 15.6 Let $\mathcal{C} = \{\mathbf{x}_i + S_i \mid i \in K\}$ be a nonempty collection of flats in V.

1) $\vee \mathcal{C}$ is the intersection of all flats that contain all flats in $\mathcal{C}$.
2) $\vee \mathcal{C}$ is *hull*(C), where C is the union of all flats in $\mathcal{C}$. ∎

Theorem 15.7 For any two flats in V,

$$(\mathbf{x} + S) \vee (\mathbf{y} + T) = \mathbf{x} + [\langle \mathbf{x} - \mathbf{y} \rangle + S + T]$$

Proof. Since $\mathbf{x}, \mathbf{y} \in (\mathbf{x} + S) \vee (\mathbf{y} + T)$, we have

$$(\mathbf{x} + S) \vee (\mathbf{y} + T) = \mathbf{x} + U = \mathbf{y} + U$$

for some subspace U of V. Hence, $\mathbf{x} - \mathbf{y} \in U$, and so $\langle \mathbf{x} - \mathbf{y} \rangle \subset U$. Moreover, $\mathbf{x} + S \subset \mathbf{x} + U$ implies that $S \subset U$, and similarly $T \subset U$. Hence,

$$\mathbf{x} + [\langle \mathbf{x} - \mathbf{y} \rangle + S + T] \subset \mathbf{x} + U$$

Since $\mathbf{x} + S$ and $\mathbf{y} + T$ are both contained in $\mathbf{x} + [\langle \mathbf{x} - \mathbf{y} \rangle + S + T]$, we deduce that $\mathbf{x} + U \subset \mathbf{x} + \langle \mathbf{x} - \mathbf{y} \rangle + S + T$. The result follows. ∎

We can now describe the dimension of the join of two flats.

Theorem 15.8 Let $X = \mathbf{x} + S$ and $Y = \mathbf{y} + T$ be flats in V.

1) If $X \cap Y \neq \emptyset$ then
   a) $X \vee Y = \mathbf{x} + S + T$
   b) $dim(X \vee Y) = dim(S+T) = dim(X) + dim(Y) - dim(X \cap Y)$
2) If $X \cap Y = \emptyset$ then
$$dim(X \vee Y) = dim(S+T) + 1$$

Proof. Using Theorem 15.7, we have

$$(\mathbf{x}+S) \cap (\mathbf{y}+T) \neq \emptyset \Leftrightarrow \exists\, \mathbf{s} \in S,\ \mathbf{t} \in T \text{ s.t. } \mathbf{x}+\mathbf{s} = \mathbf{y}+\mathbf{t}$$

$$\Leftrightarrow \mathbf{x}-\mathbf{y} \in S+T \Leftrightarrow \langle \mathbf{x}-\mathbf{y}\rangle + S + T = S + T \Leftrightarrow X \vee Y = \mathbf{x} + S + T$$

This establishes (1a) and (2). As to (1b), note that

$$dim(S+T) = dim(S) + dim(T) - dim(S \cap T)$$

and that, if $(\mathbf{x}+S) \cap (\mathbf{y}+T) \neq \emptyset$, then

$$dim(S \cap T) = dim(\mathbf{x} + [S \cap T]) = dim([\mathbf{x}+S] \cap [\mathbf{y}+T]) = dim(X \cap Y) \quad \blacksquare$$

Affine Independence

We now discuss the affine counterpart of linear independence.

Theorem 15.9 Let $\mathbf{x}_1, \ldots, \mathbf{x}_n$ be vectors in V. The following are equivalent.

1) $X = hull\{\mathbf{x}_1, \ldots, \mathbf{x}_n\}$ has dimension $n-1$.
2) $\{\mathbf{x}_1 - \mathbf{x}_i, \ldots, \mathbf{x}_{i-1} - \mathbf{x}_1, \mathbf{x}_{i+1} - \mathbf{x}_i, \ldots, \mathbf{x}_n - \mathbf{x}_i\}$ is linearly independent for all $i = 1, \ldots, n$.
3) $\mathbf{x}_i \notin hull\{\mathbf{x}_1, \ldots, \mathbf{x}_{i-1}, \mathbf{x}_{i+1}, \ldots, \mathbf{x}_n\}$ for all $i = 1, \ldots, n$.
4) If $\Sigma r_j \mathbf{x}_j$ and $\Sigma s_j \mathbf{x}_j$ are affine combinations, then
$$\sum_j r_j \mathbf{x}_j = \sum_j s_j \mathbf{x}_j \Rightarrow r_j = s_j \text{ for all } j$$

Proof. The fact that (1) and (2) are equivalent follows directly from (15.1). If (3) does not hold, we have

$$hull\{\mathbf{x}_1, \ldots, \mathbf{x}_n\} = hull\{\mathbf{x}_1, \ldots, \mathbf{x}_{i-1}, \mathbf{x}_{i+1}, \ldots, \mathbf{x}_n\}$$

where by (15.1), the latter has dimension at most $n-2$. Hence, (1) cannot hold, and so (1) implies (3).

Next we show that (3) implies (4). Suppose that (3) holds, and that $\Sigma r_j \mathbf{x}_j = \Sigma s_j \mathbf{x}_j$. Setting $t_j = r_j - s_j$ gives

$$\sum_j t_j \mathbf{x}_j = \mathbf{0} \quad \text{and} \quad \sum_j t_j = 0$$

But if any of the t_j's are nonzero, say $t_1 \neq 0$, then dividing by t_1 gives

$$\mathbf{x}_1 + \sum_{j>1} (t_j/t_1)\mathbf{x}_j = \mathbf{0}$$

or

$$\mathbf{x}_1 = \sum_{j>1} -(t_j/t_1)\mathbf{x}_j$$

where

$$\sum_{j>1} -(t_j/t_1) = 1$$

Hence, $\mathbf{x}_1 \in \mathit{hull}\{\mathbf{x}_2,\ldots,\mathbf{x}_n\}$. This contradiction implies that $t_j = 0$ for all j, that is, $r_j = s_j$ for all j. Thus, (3) implies (4).

Finally, we show that (4) implies (2). For concreteness, let us show that (4) implies that $\{\mathbf{x}_2 - \mathbf{x}_1,\ldots,\mathbf{x}_n - \mathbf{x}_1\}$ is linearly independent. Indeed, if $\alpha_2,\ldots,\alpha_n \in F$ and $\Sigma\alpha_j = \alpha$, then

$$\sum_{j\geq 2} \alpha_j(\mathbf{x}_j - \mathbf{x}_1) = \mathbf{0} \;\Rightarrow\; \sum_{j\geq 2} \alpha_j\mathbf{x}_j = \alpha\mathbf{x}_1 \;\Rightarrow\; (1-\alpha)\mathbf{x}_1 + \sum_{j\geq 2} \alpha_j\mathbf{x}_j = \mathbf{x}_1$$

But the latter is an equality between two affine combinations, and so corresponding coefficients must be equal, which implies that $\alpha_j = 0$ for all $j = 2,\ldots,n$. This shows that (4) implies (2). ∎

Definition The vectors $\mathbf{x}_1,\ldots,\mathbf{x}_n$ are **affinely independent** if they satisfy any (and hence all) of the conditions of Theorem 15.9. □

Theorem 15.10 If X is a flat of dimension n, then there exist n+1 vectors $\mathbf{x}_1,\ldots,\mathbf{x}_{n+1}$ for which every vector $\mathbf{x} \in X$ has a *unique* expression as an affine combination

$$\mathbf{x} = r_1\mathbf{x}_1 + \cdots + r_{n+1}\mathbf{x}_{n+1}$$

The coefficients r_i are called the **barycentric coordinates** of $\mathbf{x}$ with respect to the vectors $\mathbf{x}_1,\ldots,\mathbf{x}_{n+1}$. ∎

Affine Transformations

Now let us discuss some properties of maps that preserve affine structure.

Definition A function $f:V\to V$ that preserves affine combinations, that is, for which

$$\sum_i r_i = 1 \;\Rightarrow\; f\Big(\sum_i r_i\mathbf{x}_i\Big) = \sum_i r_i f(\mathbf{x}_i)$$

is called an **affine transformation** (or **affine map**, or **affinity**). □

We should mention that some authors require that f be bijective in order to be an affine map. The following theorem is the analog of Theorem 15.2.

Theorem 15.11 If $\operatorname{char}(F) \neq 2$, then the following are equivalent for a function $f:V \to V$.

1) f preserves affine combinations of any two of its points, that is,

$$f(r\mathbf{x} + (1-r)\mathbf{y}) = rf(\mathbf{x}) + (1-r)f(\mathbf{y})$$

2) f preserves affine combinations, that is,

$$\sum_i r_i = 1 \Rightarrow f\Big(\sum_i r_i \mathbf{x}_i\Big) = \sum_i r_i f(\mathbf{x}_i) \quad \blacksquare$$

Thus, if $\operatorname{char}(F) \neq 2$, then a map f is an affine transformation if and only if it sends the line through $\mathbf{x}$ and $\mathbf{y}$ to the line through $f(\mathbf{x})$ and $f(\mathbf{y})$. It is clear that linear transformations are affine transformations. So are the following maps.

Definition Let $\mathbf{v} \in V$. The affine map $T_\mathbf{v}:V \to V$ defined by

$$T_\mathbf{v}(\mathbf{x}) = \mathbf{x} + \mathbf{v}$$

for all $\mathbf{x} \in V$, is called **translation** by $\mathbf{v}$. $\square$

It is not hard to see that any map of the form $T_\mathbf{v} \circ \tau$, where $\tau \in \mathcal{L}(V)$, is affine. Conversely, any affine map must have this form.

Theorem 15.12 A function $f:V \to V$ is an affine transformation if and only if $f = T_\mathbf{v} \circ \tau$, where $\mathbf{v} \in V$ and $\tau \in \mathcal{L}(V)$.

Proof. We leave proof that $T_\mathbf{v} \circ \tau$ is an affine transformation to the reader. Conversely, suppose that f is an affine map. Then

$$f(r\mathbf{x} + s\mathbf{y}) = f(r\mathbf{x} + s\mathbf{y} + (1-r-s)\mathbf{0}) = rf(\mathbf{x}) + sf(\mathbf{y}) + (1-r-s)f(\mathbf{0})$$

Rearranging gives

$$f(r\mathbf{x} + s\mathbf{y}) - f(\mathbf{0}) = r[f(\mathbf{x}) - f(\mathbf{0})] + s[f(\mathbf{y}) - f(\mathbf{0})]$$

which is equivalent to

$$(T_{-f(\mathbf{0})} \circ f)(r\mathbf{x} + s\mathbf{y}) = r(T_{-f(\mathbf{0})} \circ f)(\mathbf{x}) + s(T_{-f(\mathbf{0})} \circ f)(\mathbf{y})$$

and so $\tau = T_{-f(\mathbf{0})} \circ f$ is linear. Thus, $f = T_{f(\mathbf{0})} \circ \tau$. $\blacksquare$

Corollary 15.13

1) The composition of two affine transformations is an affine transformation.

2) An affine transformation $f = T_{\mathbf{v}} \circ \tau$ is bijective if and only if τ is bijective.
3) The set $Aff(V)$ of all *bijective* affine transformations on V is a group under composition of maps, called the **affine group** of V. ∎

Let us make a few remarks for those familiar with the basics of group theory. The set $Trans(V)$ of all translations of V is a subgroup of $Aff(V)$. We can define a function $\phi: Aff(V) \to \mathcal{L}(V)$ by

$$\phi(T_{\mathbf{v}} \circ \tau) = \tau$$

It is not hard to see that ϕ is a well-defined group homomorphism from $Aff(V)$ onto $\mathcal{L}(V)$, with kernel $Trans(V)$. Hence, $Trans(V)$ is a normal subgroup of $Aff(V)$ and

$$\frac{Aff(V)}{Trans(V)} \approx \mathcal{L}(V)$$

Projective Geometry

If $dim(V) = 2$, then the join of any two distinct points in V is a line. On the other hand, it is not the case that the intersection of any two lines is a point. Thus, we see a certain asymmetry between the concepts of points and lines in V. This asymmetry can be removed by constructing the so-called *projective plane.* Our plan here is to very briefly describe one possible construction of projective geometries of all dimensions.

By way of motivation, let us consider Figure 15.1.

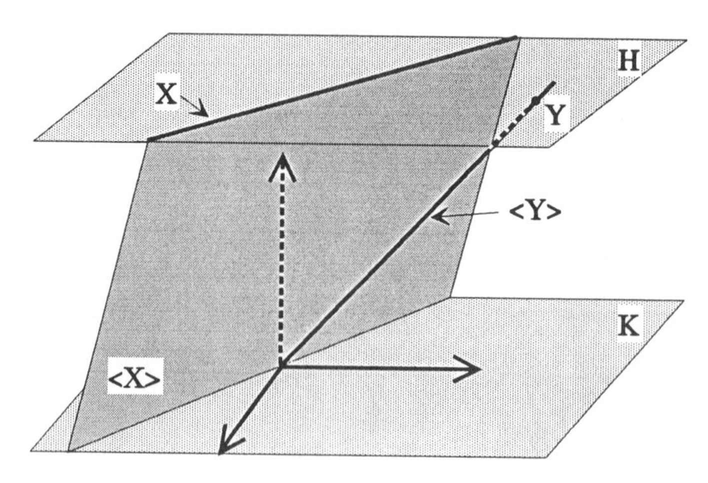

Figure 15.1

Note that H is a hyperplane in a 3-dimensional vector space V and that $0 \notin H$. Now, the set $\mathcal{A}(H)$ of all flats of V that lie in H is an affine geometry of dimension 2. (According to our definition of affine geometry, H must be a vector space in order to define $\mathcal{A}(H)$. However, we hereby extend the definition of affine geometry to include the collection of all flats contained in a *flat* of V.)

To each flat X in H, we associate the subspace $\langle X \rangle$ of V generated by X. This defines a function

$$P:\mathcal{A}(H) \to \mathcal{S}(V), \qquad P(X) = \langle X \rangle$$

where $\mathcal{S}(V)$ is the set of all subspaces of V. Note that P is not onto $\mathcal{S}(V)$, but *only* because $im(P)$ does not contain any subspaces of the subspace K that contains the origin and is parallel to H. Figure 15.1 shows a one-dimensional flat X, and its image $P(X) = \langle X \rangle$, as well as a zero-dimensional flat Y, and its image $\langle Y \rangle$. Note that, for any flat X in H, we have $dim(P(X)) = dim(X) + 1$.

Note also that if L_1 and L_2 are *any* two distinct lines in H, the corresponding planes $P(L_1)$ and $P(L_2)$ have the property that their intersection is a line through the origin. We are now ready to define projective geometries.

Definition Let V be a vector space. The set $\mathcal{P}(V)$ of all subspaces of V is called the **projective geometry** of V. If S is a subspace of V, its **projective dimension**, denoted by $pdim(S)$ is equal to $dim(S) - 1$. The **projective dimension** of $\mathcal{P}(V)$ is defined to be $pdim(V) = dim(V) - 1$. A subspace of projective dimension 0, 1 or 2 is called a **projective point, projective line**, or **projective plane**, respectively. □

Thus, referring to Figure 15.1, a projective point is a line through the origin and, provided that it is not contained in the plane K described earlier, it meets H in an (affine) point. Similarly, a projective line is a plane through the origin and, provided that it is not K, it will meet H in a line. (This holds in higher dimensions as well.) In short,

$$P(\text{point}) = \text{projective point}, \; P(\text{line}) = \text{projective line}$$

$$P(\text{plane}) = \text{projective plane}$$

and so on.

Given a vector space V of any dimension, and any hyperplane H in V not containing the origin, we can define the function $P:\mathcal{A}(H) \to \mathcal{P}(V)$, as shown in Figure 15.1 for $dim(V) = 3$. It is also clear from this figure that a projective geometry of projective dimension n is an "extension" of an affine geometry of (affine) dimension n, formed in such a way that all "objects" intersect. More specifically, the map $P:\mathcal{A}(H)\to\mathcal{P}(V)$ satisfies the properties described in the following theorem.

Theorem 15.14 The map $P:\mathcal{A}(H)\to\mathcal{P}(V)$ from the affine geometry $\mathcal{A}(H)$ to the projective geometry $\mathcal{P}(V)$ satisfies the following.

1) P is injective, with inverse given by
$$P^{-1}(U) = U \cap H$$
2) $im(P)$ is the set of all subspaces of V that are not contained in the subspace K parallel to H
3) $X \subset Y$ if and only if $P(X) \subset P(Y)$
4) If X_i are flats in H with nonempty intersection, then
$$P\Big(\bigcap_{i\in K} X_i\Big) = \bigcap_{i\in K} P(X_i)$$
5) For any collection of flats in H,
$$P(\bigvee_{i\in K} X_i) = \bigoplus_{i\in K} P(X_i)$$
6) P preserves dimension, in the sense that
$$pdim(P(X)) = dim(X)$$
7) $X \parallel Y$ if and only if one of $P(X)\cap K$ and $P(Y)\cap K$ is contained in the other.

Proof. To prove part (1), let $\mathbf{x}+S$ be a flat in H. Then $\mathbf{x}\in H$, and so $H = \mathbf{x}+K$, which implies that $S\subset K$. Note also that $P(\mathbf{x}+S) = \langle \mathbf{x}\rangle + S$, and
$$\mathbf{z}\in P(\mathbf{x}+S)\cap H = (\langle\mathbf{x}\rangle + S)\cap(\mathbf{x}+K) \ \Rightarrow\ \mathbf{z} = r\mathbf{x}+\mathbf{s} = \mathbf{x}+\mathbf{k}$$
for some $\mathbf{s}\in S$, $\mathbf{k}\in K$ and $r\in F$. This implies that $(1-r)\mathbf{x}\in K$, which implies that either $\mathbf{x}\in K$ or $r=1$. But $\mathbf{x}\in H$ implies $\mathbf{x}\notin K$, and so $r=1$, which implies that $\mathbf{z}=\mathbf{x}+\mathbf{s}\in\mathbf{x}+S$. In other words,
$$P(\mathbf{x}+S)\cap H \subset \mathbf{x}+S$$
Since the reverse inclusion is clear, we have
$$P(\mathbf{x}+S)\cap H = \mathbf{x}+S$$
This establishes (1).

To prove (2), let U be a subspace of V that is not contained in K. We wish to show that U is in the image of $\mathcal{P}$. Note first that since $U\not\subset K$, and $dim(K) = dim(V)-1$, we have $U+K=V$, and so
$$dim(U\cap K) = dim(U)+dim(K)-dim(U+K) = dim(U)-1$$
Now, let $\mathbf{0}\neq\mathbf{x}\in U-K$. Then

$$\begin{aligned}\mathbf{x} \notin \mathrm{K} &\Rightarrow \langle \mathbf{x} \rangle + \mathrm{K} = \mathrm{V} \\ &\Rightarrow \mathrm{r}\mathbf{x} + \mathbf{k} \in \mathrm{H} \text{ for some } 0 \neq \mathrm{r} \in \mathrm{F}, \mathbf{k} \in \mathrm{K} \Rightarrow \mathrm{r}\mathbf{x} \in \mathrm{H}\end{aligned}$$

Thus, $\mathrm{r}\mathbf{x} \in \mathrm{U} \cap \mathrm{H}$ for some $0 \neq \mathrm{r} \in \mathrm{F}$. Hence, the flat $\mathrm{r}\mathbf{x} + (\mathrm{U} \cap \mathrm{K})$ lies in H, and

$$dim(\mathrm{r}\mathbf{x} + (\mathrm{U} \cap \mathrm{K})) = dim(\mathrm{U} \cap \mathrm{K}) = dim(\mathrm{U}) - 1$$

which implies that $\mathrm{P}(\mathrm{r}\mathbf{x} + (\mathrm{U} \cap \mathrm{K})) = \langle \mathrm{r}\mathbf{x} \rangle + (\mathrm{U} \cap \mathrm{K})$ lies in U, and has the same dimension as U. In other words,

$$\mathrm{P}(\mathrm{r}\mathbf{x} + (\mathrm{U} \cap \mathrm{K})) = \langle \mathrm{r}\mathbf{x} \rangle + (\mathrm{U} \cap \mathrm{K}) = \mathrm{U}$$

We leave proof of the remaining parts of the theorem as exercises. ∎

EXERCISES

1. Show that if $\mathbf{x}_1, \ldots, \mathbf{x}_n \in \mathrm{V}$, then the set $\mathrm{S} = \{\Sigma \mathrm{r}_i \mathbf{x}_i \mid \Sigma \mathrm{r}_i = 0\}$ is a subspace of V.
2. Prove that $hull\{\mathbf{x}_1, \ldots, \mathbf{x}_n\} = \mathbf{x}_1 + \langle \mathbf{x}_2 - \mathbf{x}_1, \ldots, \mathbf{x}_n - \mathbf{x}_1 \rangle$.
3. Prove that the set $\mathrm{X} = \{(0,0), (1,0), (0,1)\}$ in $(\mathrm{F}_2)^2$ is closed under the formation of lines, but not affine hulls.
4. Prove that a flat contains the origin $\mathbf{0}$ if and only if it is a subspace.
5. Prove that a flat X is a subspace if and only if for some $\mathbf{x} \in \mathrm{X}$ we have $\mathrm{r}\mathbf{x} \in \mathrm{X}$ for some $1 \neq \mathrm{r} \in \mathrm{F}$.
6. Show that the join of a collection $\mathcal{C} = \{\mathbf{x}_i + \mathrm{S}_i \mid i \in \mathrm{K}\}$ of flats in V is the intersection of all flats that contain all flats in $\mathcal{C}$.
7. Is the collection of all flats in V a lattice under set inclusion? If not, how can you "fix" this?
8. Prove that if $dim(\mathrm{X}) = dim(\mathrm{Y})$ and $\mathrm{X} \parallel \mathrm{Y}$ then $\mathrm{S} = \mathrm{T}$, where $\mathrm{X} = \mathbf{x} + \mathrm{S}$ and $\mathrm{Y} = \mathbf{y} + \mathrm{T}$.
9. Suppose that $\mathrm{X} = \mathbf{x} + \mathrm{S}$ and $\mathrm{Y} = \mathbf{y} + \mathrm{T}$ are disjoint hyperplanes in V. Show that $\mathrm{S} = \mathrm{T}$.
10. (*The parallel postulate*) Let X be a flat in V, and $\mathbf{v} \notin \mathrm{X}$. Show that there is exactly one flat containing $\mathbf{v}$, parallel to X, and having the same dimension as X.
11. a) Find an example to show that the join $\mathrm{X} \vee \mathrm{Y}$ of two flats may not be the set of all lines connecting all points in the union of these flats.
    b) Show that if X and Y are flats with $\mathrm{X} \cap \mathrm{Y} \neq \emptyset$, then $\mathrm{X} \vee \mathrm{Y}$ is the union of all lines $\overline{\mathbf{xy}}$ where $\mathbf{x} \in \mathrm{X}$ and $\mathbf{y} \in \mathrm{Y}$.
12. Show that if $\mathrm{X} \parallel \mathrm{Y}$ and $\mathrm{X} \cap \mathrm{Y} = \emptyset$ then $dim(\mathrm{X} \vee \mathrm{Y}) = \max\{dim(\mathrm{X}), dim(\mathrm{Y})\} + 1$.
13. Let $dim(\mathrm{V}) = 2$. Prove the following.
    a) The join of any two distinct points is a line.

b) The intersection of any two nonparallel lines is a point.

14. Let $dim(V) = 3$. Prove the following.
 a) The join of any two distinct points is a line.
 b) The intersection of any two nonparallel planes is a line.
 c) The join of any two lines whose intersection is a point is a plane.
 d) The intersection of two coplanar nonparallel lines is a point.
 e) The join of any two distinct parallel lines is a plane.
 f) The join of a line and a point not on that line is a plane.
 g) The intersection of a plane and a line not on that plane is a point.
15. Prove that $f V \to V$ is an affine transformation if and only if $f = \tau \circ T_{\mathbf{w}}$ for some $\mathbf{w} \in V$ and $\tau \in \mathcal{L}(V)$.
16. Verify the group-theoretic remarks about the group homomorphism $\phi: Aff(V) \to \mathcal{L}(V)$, and the subgroup $Trans(V)$ of $Aff(V)$.

CHAPTER 16

The Umbral Calculus

Contents: *Formal Power Series. The Umbral Algebra. Formal Power Series as Linear Operators. Sheffer Sequences. Examples of Sheffer Sequences. Umbral Operators and Umbral Shifts. Continuous Operators on the Umbral Algebra. Operator Adjoints. Automorphisms of the Umbral Algebra. Derivations of the Umbral Algebra. Exercises.*

In this chapter, we give a brief introduction to a relatively new subject, called the umbral calculus. This is an algebraic theory used to study certain types of polynomial functions that play an important role in applied mathematics. We give only a brief introduction to the subject – emphasizing the algebraic aspects rather than the applications. For more on the umbral calculus, we suggest *The Umbral Calculus*, by Roman [1984].

Formal Power Series

We begin with a few remarks concerning formal power series. Let $\mathcal{F}$ denote the algebra of formal power series in the variable t, with complex coefficients. Thus, $\mathcal{F}$ is the set of all formal sums of the form

$$f(t) = \sum_{k=0}^{\infty} a_k t^k \tag{16.1}$$

where $a_k \in \mathbb{C}$. Addition and multiplication are purely formal

$$\sum_{k=0}^{\infty} a_k t^k + \sum_{k=0}^{\infty} b_k t^k = \sum_{k=0}^{\infty} (a_k + b_k) t^k$$

and

$$\left(\sum_{k=0}^{\infty} a_k t^k\right)\left(\sum_{k=0}^{\infty} b_k t^k\right) = \sum_{k=0}^{\infty}\left(\sum_{j=0}^{k} a_j b_{k-j}\right) t^k$$

The **order** $o(f)$ of f is the *smallest* exponent of t that appears with a nonzero coefficient. The order of the zero series is $+\infty$. A series f has a multiplicative inverse, denoted by f^{-1}, if and only if $o(f) = 0$. We leave it to the reader to show that

$$o(fg) = o(f) + o(g)$$

and

$$o(f+g) \geq \min\{o(f), o(g)\}$$

If f_k is a sequence in $\mathcal{F}$ with $o(f_k) \to \infty$ as $k \to 0$, then for any series

$$g(t) = \sum_{k=0}^{\infty} b_k t^k$$

we may form the series

$$h(t) = \sum_{k=0}^{\infty} b_k f_k(t)$$

This sum is well-defined since the coefficient of each power of t is a finite sum. In particular, if $o(f) \geq 1$, then $o(f^k) \to \infty$, and so the **composition**

$$(g \circ f)(t) = g(f(t)) = \sum_{k=0}^{\infty} b_k f^k(t)$$

is well-defined. It is easy to see that $o(g \circ f) = o(g)o(f)$.

If $o(f) = 1$, then f has a compositional inverse, denoted by $\bar{f}$ and satisfying $(f \circ \bar{f})(t) = (\bar{f} \circ f)(t) = t$. A series f with $o(f) = 1$ is called a **delta series.**

The sequence of powers f^k of a delta series f forms a **pseudobasis** for $\mathcal{F}$, in the sense that for any $g \in \mathcal{F}$, there exists a unique sequence of constants a_k for which

$$g(t) = \sum_{k=0}^{\infty} a_k f^k(t)$$

Finally, we note that the formal derivative of the series (16.1) is given by

$$\partial_t f(t) = f'(t) = \sum_{k=1}^{\infty} k a_k t^{k-1}$$

The operator ∂_t is a *derivation*, that is,

$$\partial_t(fg) = \partial_t(f)g + f\partial_t(g)$$

The Umbral Algebra

Let $\mathcal{P} = \mathbb{C}[x]$ denote the algebra of polynomials in a single variable x over the complex field. One of the starting points of the umbral calculus is the fact that any formal power series in $\mathcal{F}$ can play three different roles – as a formal power series, as a linear functional on $\mathcal{P}$, and as a linear operator on $\mathcal{P}$. Let us first explore the connection between formal power series and linear functionals.

Let $\mathcal{P}^*$ denote the vector space of *all* linear functionals on $\mathcal{P}$. Note that $\mathcal{P}^*$ is the *algebraic* dual space of $\mathcal{P}$, as defined in Chapter 2. It will be convenient to denote the action of $L \in \mathcal{P}^*$ on $p(x) \in \mathcal{P}$ by

$$\langle L \mid p(x) \rangle$$

The vector space operations on $\mathcal{P}^*$ then take the form

$$\langle L + M \mid p(x) \rangle = \langle L \mid p(x) \rangle + \langle M \mid p(x) \rangle$$

and

$$\langle rL \mid p(x) \rangle = r\langle L \mid p(x) \rangle, \quad r \in \mathbb{C}$$

Note also that since any linear functional on $\mathcal{P}$ is uniquely determined by its values on a basis for $\mathcal{P}$, $L \in \mathcal{P}^*$ is uniquely determined by the values $\langle L \mid x^n \rangle$ for $n \geq 0$.

Now, any formal series in $\mathcal{F}$ can be written in the form

$$f(t) = \sum_{k=0}^{\infty} \frac{a_k}{k!} t^k$$

and we can use this to define a linear functional $f(t)$ by setting

$$\langle f(t) \mid x^n \rangle = a_n$$

for $n \geq 0$. In other words, the *linear functional* $f(t)$ is defined by the condition

$$f(t) = \sum_{k=0}^{\infty} \frac{\langle f(t) \mid x^k \rangle}{k!} t^k$$

Note in particular that

$$\langle t^k \mid x^n \rangle = n!\delta_{n,k}$$

where $\delta_{n,k}$ is the Kronecker delta function. This implies that

$$\langle t^k \mid p(x) \rangle = p^{(k)}(0)$$

and so t^k is the functional "kth derivative at 0." Also, t^0 is evaluation at 0.

As it happens, any linear functional $L \in \mathcal{P}^*$ has the form $f(t)$. To see this, we simply note that if

$$f_L(t) = \sum_{k=0}^{\infty} \frac{\langle L \mid x^k \rangle}{k!} t^k$$

then

$$\langle f_L(t) \mid x^n \rangle = \langle L \mid x^n \rangle$$

for all $n \geq 0$, and so *as linear functionals*, $L = f_L(t)$.

Thus, we can define a map $\phi: \mathcal{P}^* \to \mathcal{F}$ by $\phi(L) = f_L(t)$.

Theorem 16.1 The map $\phi: \mathcal{P}^* \to \mathcal{F}$ defined by $\phi(L) = f_L(t)$ is a vector space isomorphism from $\mathcal{P}^*$ onto $\mathcal{F}$.

Proof. To see that ϕ is injective, note that

$$f_L(t) = f_M(t) \Rightarrow \langle L \mid x^n \rangle = \langle M \mid x^n \rangle \text{ for all } n \geq 0 \Rightarrow L = M$$

Moreover, the map ϕ is surjective, since for any $f \in \mathcal{F}$, the linear functional $L = f(t)$ has the property that $\phi(L) = f_L(t) = f(t)$. Finally,

$$\begin{aligned}\phi(rL + sM) &= \sum_{k=0}^{\infty} \frac{\langle rL + sM \mid x^k \rangle}{k!} t^k \\ &= r\sum_{k=0}^{\infty} \frac{\langle L \mid x^k \rangle}{k!} t^k + s\sum_{k=0}^{\infty} \frac{\langle M \mid x^k \rangle}{k!} t^k = r\phi(L) + s\phi(M)\end{aligned}$$

∎

From now on, we shall identify the vector space $\mathcal{P}^*$ with the vector space $\mathcal{F}$, using the isomorphism $\phi: \mathcal{P}^* \to \mathcal{F}$. Thus, *we think of linear functionals on* $\mathcal{P}$ *simply as formal power series.* The advantage of this approach is that $\mathcal{F}$ is more than just a vector space – it is an algebra. Hence, we have automatically defined a multiplication of linear functionals, namely, the product of formal power series. The algebra $\mathcal{F}$, when thought of as both the algebra of formal power series and the algebra of linear functionals on $\mathcal{P}$, is called the **umbral algebra.**

Let us consider an example.

Example 16.1 For $a \in \mathbb{C}$, the **evaluation functional** $\epsilon_a \in \mathcal{P}^*$ is defined by

$$\langle \epsilon_a \mid p(x) \rangle = p(a)$$

In particular, $\langle \epsilon_a \mid x^n \rangle = a^n$, and so the formal power series representation for this functional is

$$f_{\epsilon_a}(t) = \sum_{k=0}^{\infty} \frac{\langle \epsilon_a \mid x^k \rangle}{k!} t^k = \sum_{k=0}^{\infty} \frac{a^k}{k!} t^k = e^{at}$$

which is the exponential series. If e^{bt} is evaluation at b, then

$$e^{at} e^{bt} = e^{(a+b)t}$$

and so the product of evaluation at a and evaluation at b is evaluation at $a+b$. ▯

When we are thinking of a delta series $f \in \mathcal{F}$ as a linear functional, we refer to it as a **delta functional.** Similarly, an invertible series $f \in \mathcal{F}$ is referred to as an **invertible functional.** Here are some simple consequences of the development so far.

Theorem 16.2

1) For any $f \in \mathcal{F}$,
$$f(t) = \sum_{k=0}^{\infty} \frac{\langle f(t) \mid x^k \rangle}{k!} t^k$$
2) For any $p \in \mathcal{P}$,
$$p(x) = \sum_{k \geq 0} \frac{\langle t^k \mid p(x) \rangle}{k!} x^k$$
3) For any $f,g \in \mathcal{F}$,
$$\langle f(t)g(t) \mid x^n \rangle = \sum_{k=0}^{n} \binom{n}{k} \langle f(t) \mid x^k \rangle \langle g(t) \mid x^{n-k} \rangle t^k$$
4) $o(f(t)) > \deg p(x) \;\Rightarrow\; \langle f(t) \mid p(x) \rangle = 0$
5) If $o(f_k) = k$ for all $k \geq 0$, then
$$\left\langle \sum_{k=0}^{\infty} a_k f_k(t) \,\middle|\, p(x) \right\rangle = \sum_{k \geq 0} a_k \langle f_k(t) \mid p(x) \rangle$$
where the sum on the right is a finite one.
6) If $o(f_k) = k$ for all $k \geq 0$, then
$$\langle f_k(t) \mid p(x) \rangle = \langle f_k(t) \mid q(x) \rangle \text{ for all } k \geq 0 \;\Rightarrow\; p(x) = q(x)$$
7) If $\deg p_k(x) = k$ for all $k \geq 0$, then
$$\langle f(t) \mid p_k(x) \rangle = \langle g(t) \mid p_k(x) \rangle \text{ for all } k \geq 0 \;\Rightarrow\; f(t) = g(t)$$

Proof. We prove only part (3). Let
$$f(t) = \sum_{k=0}^{\infty} \frac{a_k}{k!} t^k \quad \text{and} \quad g(t) = \sum_{j=0}^{\infty} \frac{b_j}{j!} t^j$$
Then
$$f(t)g(t) = \sum_{m=0}^{\infty} \left(\frac{1}{m!} \sum_{k=0}^{m} \binom{m}{k} a_k b_{m-k} \right) t^m$$
and applying both sides of this (as linear functionals) to x^n gives
$$\langle f(t)g(t) \mid x^n \rangle = \sum_{k=0}^{n} \binom{n}{k} a_k b_{n-k}$$
The result now follows from the fact that part (1) implies $a_k = \langle f(t) \mid x^k \rangle$ and $b_{n-k} = \langle g(t) \mid x^{n-k} \rangle$. ∎

We can now present our first "umbral" result.

Theorem 16.3 For any $f(t) \in \mathcal{F}$ and $p(x) \in \mathcal{P}$,

$$\langle f(t) \mid xp(x) \rangle = \langle \partial_t f(t) \mid p(x) \rangle$$

Proof. By linearity, we need only establish this for $p(x) = x^n$. But, if

$$f(t) = \sum_{k=0}^{\infty} \frac{a_k}{k!} t^k$$

then

$$\langle \partial_t f(t) \mid x^n \rangle = \left\langle \sum_{k=1}^{\infty} \frac{a_k}{(k-1)!} t^{k-1} \,\middle|\, x^n \right\rangle$$

$$= \sum_{k=1}^{\infty} \frac{a_k}{(k-1)!} \delta_{k-1,n} = a_{n+1} = \langle f(t) \mid x^{n+1} \rangle \qquad \blacksquare$$

Let us consider a few examples of important linear functionals and their power series representations.

Example 16.2

1) We have already encountered the evaluation functional e^{at}, satisfying
$$\langle e^{at} \mid p(x) \rangle = p(a)$$
2) The **forward difference functional** is the delta functional $e^{at} - 1$, satisfying
$$\langle e^{at} - 1 \mid p(x) \rangle = p(a) - p(0)$$
3) The **Abel functional** is the delta functional te^{at}, satisfying
$$\langle te^{at} \mid p(x) \rangle = p'(a)$$
4) The invertible functional $(1-t)^{-1}$ satisfies
$$\langle (1-t)^{-1} \mid p(x) \rangle = \int_0^{\infty} p(u) e^{-u} du$$
as can be seen by setting $p(x) = x^n$, and expanding the expression $(1-t)^{-1}$.
5) To determine the linear functional f satisfying
$$\langle f(t) \mid p(x) \rangle = \int_0^a p(u)\, du$$
we observe that
$$f(t) = \sum_{k=0}^{\infty} \frac{\langle f(t) \mid x^k \rangle}{k!} t^k = \sum_{k=0}^{\infty} \frac{a^{k+1}}{(k+1)!} t^k = \frac{e^{at} - 1}{t}$$

The inverse $t/(e^{at}-1)$ of this functional is associated with the so-called *Bernoulli polynomials*, which play a very important role in mathematics and its applications. In fact, the numbers

$$B_n = \left\langle \frac{t}{e^{at}-1} \,\middle|\, x^n \right\rangle$$

are known as the **Bernoulli numbers.** □

Formal Power Series as Linear Operators

We now turn to the connection between formal power series and linear operators on $\mathcal{P}$. Let us denote the k-th derivative operator on $\mathcal{P}$ by t^k. Thus,

$$t^k p(x) = p^{(k)}(x)$$

We can then extend this to formal series in t

$$f(t) = \sum_{k=0}^{\infty} \frac{a_k}{k!} t^k \tag{16.2}$$

by defining the linear operator $f(t):\mathcal{P} \to \mathcal{P}$ by

$$f(t)p(x) = \sum_{k=0}^{\infty} \frac{a_k}{k!}[t^k p(x)] = \sum_{k \geq 0} \frac{a_k}{k!}\, p^{(k)}(x)$$

the latter sum being a finite one. Note in particular that

$$f(t)x^n = \sum_{k=0}^{n} \binom{n}{k} a_k x^{n-k} \tag{16.3}$$

With this definition, we see that each formal power series $f \in \mathcal{F}$ plays three roles in the umbral calculus, namely, as a formal power series, as a linear functional, and as a linear operator. The differing notations $\langle f(t) \mid p(x) \rangle$ and $f(t)p(x)$ will make it clear whether we are thinking of f as a functional or as an operator.

It is important to note that $f = g$ in $\mathcal{F}$ if and only if $f = g$ as linear functionals, which holds if and only if $f = g$ as linear operators. It is also worth noting that

$$[f(t)g(t)]p(x) = f(t)[g(t)p(x)]$$

and so we may write $f(t)g(t)p(x)$ without ambiguity. In addition,

$$f(t)g(t)p(x) = g(t)f(t)p(x)$$

for all $f,g \in \mathcal{F}$ and $p \in \mathcal{P}$.

When we are thinking of a delta series f as an operator, we call it a **delta operator**. The following theorem describes the key relationship between linear functionals and linear operators of the form $f(t)$.

Theorem 16.4 If $f,g \in \mathcal{F}$, then

$$\langle f(t)g(t) \mid p(x)\rangle = \langle f(t) \mid g(t)p(x)\rangle$$

for all polynomials $p(x) \in \mathcal{P}$.

Proof. If f has the form (16.2), then by (16.3),

$$\langle t^0 \mid f(t)x^n\rangle = \left\langle t^0 \,\middle|\, \sum_{k=0}^{n} \binom{n}{k} a_k x^{n-k} \right\rangle = a_n = \langle f(t) \mid x^n\rangle \tag{16.4}$$

By linearity, this holds for x^n replaced by any polynomial $p(x)$. Hence, applying this to the product fg gives

$$\begin{aligned}\langle f(t)g(t) \mid p(x)\rangle &= \langle t^0 \mid f(t)g(t)p(x)\rangle \\ &= \langle t^0 \mid f(t)[g(t)p(x)]\rangle = \langle f(t) \mid g(t)p(x)\rangle\end{aligned}$$

∎

Equation (16.4) shows that applying the linear functional $f(t)$ is equivalent to applying the operator $f(t)$, and then following by evaluation at $x = 0$.

Here are the operator versions of the functionals in Example 16.2.

Example 16.3

1) The operator e^{at} satisfies

$$e^{at}x^n = \sum_{k=0}^{\infty} \frac{a^k}{k!} t^k x^n = \sum_{k=0}^{n} \binom{n}{k} a^k x^{n-k} = (x+a)^n$$

and so

$$e^{at}p(x) = p(x+a)$$

for all $p \in \mathcal{P}$. Thus e^{at} is a **translation operator.**

2) The **forward difference operator** is the delta operator $e^{at} - 1$, where

$$(e^{at} - 1)p(x) = p(x+a) - p(a)$$

3) The **Abel operator** is the delta operator te^{at}, where

$$te^{at}p(x) = p'(x+a)$$

4) The invertible operator $(1-t)^{-1}$ satisfies

$$(1-t)^{-1}p(x) = \int_0^{\infty} p(x+u)e^{-u}du$$

5) The operator $(e^{at} - 1)/t$ is easily seen to satisfy

$$\frac{e^{at}-1}{t}\, p(x) = \int_x^{x+a} p(u)\, du$$

□

We have seen that *all* linear functionals on $\mathcal{P}$ have the form $f(t)$, for $f \in \mathcal{F}$. However, not all linear operators on $\mathcal{P}$ have this form. To see this, observe that

$$\deg [f(t)p(x)] \leq \deg p(x)$$

but the linear operator $\phi:\mathcal{P}\to\mathcal{P}$ defined by $\phi(p(x)) = xp(x)$ does not have this property. Proof of the following characterization of operators that do have the form $f(t)$ can be found in Roman [1984].

Theorem 16.5 The following are equivalent for a linear operator $\tau:\mathcal{P}\to\mathcal{P}$.

1) τ has the form $f(t)$, that is, there exists an $f \in \mathcal{F}$ for which $\tau = f(t)$, as linear operators.
2) τ commutes with the derivative operator, that is, $\tau t = t\tau$.
3) τ commutes with any delta operator $g(t)$, that is, $\tau g(t) = g(t)\tau$.
4) τ commutes with any translation operator, that is, $\tau e^{at} = e^{at}\tau$. ∎

Sheffer Sequences

We can now define the principal object of study in the umbral calculus. When referring to a *sequence* $s_n(x)$ in $\mathcal{P}$, we shall always imply that $\deg s_n(x) = n$ for all $n \geq 0$. The proof of the following result is straightforward, but in the interest of space, it will be omitted.

Theorem 16.6 Let f be a delta series, let g be an invertible series, and consider the geometric sequence

$$g, gf, gf^2, gf^3, \ldots$$

in $\mathcal{F}$. Then there is a unique sequence $s_n(x)$ in $\mathcal{P}$ satisfying the *orthogonality conditions*

$$\langle g(t)f^k(t) \mid s_n(x)\rangle = n!\delta_{n,k} \tag{16.5}$$

for all $n,k \geq 0$. ∎

Definition The sequence $s_n(x)$ in (16.5) is called the **Sheffer sequence** for the ordered pair $(g(t),f(t))$. We shorten this by saying that $s_n(x)$ is *Sheffer* for $(g(t),f(t))$. □

Two special types of Sheffer sequences deserve explicit mention.

Definition The Sheffer sequence for a pair of the form $(1,f(t))$ is called the **associated sequence** for $f(t)$. The Sheffer sequence for a pair of the form $(g(t),t)$ is called the **Appell sequence** for $g(t)$. □

Before considering examples, we wish to describe several characterizations of Sheffer sequences. First, we require a key result.

Theorem 16.7 **(The expansion theorems)** Let $s_n(x)$ be Sheffer for $(g(t),f(t))$.

1) For any $h \in \mathcal{F}$,

$$h(t) = \sum_{k=0}^{\infty} \frac{\langle h(t) \mid s_k(x)\rangle}{k!} g(t)f^k(t)$$

2) For any $p \in \mathcal{P}$,

$$p(x) = \sum_{k \geq 0} \frac{\langle g(t)f^k(t) \mid p(x)\rangle}{k!} s_k(x)$$

Proof. Part (1) follows from parts (5) and (7) of Theorem 16.2, since

$$\left\langle \sum_{k=0}^{\infty} \frac{\langle h(t) \mid s_k(x)\rangle}{k!} g(t)f^k(t) \,\middle|\, s_n(x) \right\rangle = \sum_{k=0}^{\infty} \frac{\langle h(t) \mid s_k(x)\rangle}{k!} n!\delta_{n,k}$$

$$= \langle h(t) \mid s_n(x)\rangle$$

Part (2) follows in a similar way from part (6) of Theorem 16.2. ∎

We can now begin our characterization of Sheffer sequences, starting with the generating function. The idea of a generating function is quite simple. If $r_n(x)$ is a sequence of polynomials, we may define a formal power series of the form

$$g(t,x) = \sum_{k=0}^{\infty} \frac{r_k(x)}{k!} t^k$$

This is referred to as the **(exponential) generating function** for the sequence $r_n(x)$. (The term *exponential* refers to the presence of $k!$ in this series. When this is not present, we have an *ordinary* generating function.) Since the series is a formal one, knowing $g(t)$ is equivalent (in theory, if not always in practice) to knowing the polynomials $r_n(x)$. Moreover, a knowledge of the generating function of a sequence of polynomials can often lead to a deeper understanding of the sequence itself, that might not be otherwise easily accessible. For this reason, generating functions are studied quite extensively.

For the proofs of the following characterizations, we refer the reader to Roman [1984].

Theorem 16.8 **(Generating function)**

1) Let $p_n(x)$ be the associated sequence for $f(t)$. The generating function of $p_n(x)$ is

$$e^{y\bar{f}(t)} = \sum_{k=0}^{\infty} \frac{p_k(y)}{k!} t^k$$

where $\bar{f}(t)$ is the compositional inverse of $f(t)$.

2) Let $s_n(x)$ be Sheffer for $(g(t), f(t))$. The generating function of $s_n(x)$ is

$$\frac{1}{g(\bar{f}(t))} e^{y\bar{f}(t)} = \sum_{k=0}^{\infty} \frac{s_k(y)}{k!} t^k$$ ∎

Theorem 16.9 **(Conjugate representation)**

1) A sequence $p_n(x)$ is the associated sequence for $f(t)$ if and only if

$$p_n(x) = \sum_{k=0}^{n} \frac{1}{k!} \langle \bar{f}(t)^k \mid x^n \rangle x^k$$

2) A sequence $s_n(x)$ is Sheffer for $(g(t),f(t))$ if and only if

$$s_n(x) = \sum_{k=0}^{n} \frac{1}{k!} \langle g(\bar{f}(t))^{-1} \bar{f}(t)^k \mid x^n \rangle x^k$$ ∎

Theorem 16.10 **(Operator characterization)**

1) A sequence $p_n(x)$ is the associated sequence for $f(t)$ if and only if

   a) $p_n(0) = \delta_{n,0}$

   b) $f(t)p_n(x) = np_{n-1}(x)$ for $n \geq 0$

2) A sequence $s_n(x)$ is Sheffer for $(g(t),f(t))$, for some $g(t)$, if and only if

$$f(t)s_n(x) = ns_{n-1}(x)$$

for all $n \geq 0$. ∎

Theorem 16.11

1) **(The binomial identity)** A sequence $p_n(x)$ is the associated sequence for a delta series $f(t)$ if and only if it is of **binomial type**, that is, if and only if it satisfies the identity

$$p_n(x+y) = \sum_{k=0}^{n} \binom{n}{k} p_k(y) p_{n-k}(x)$$

for all $y \in \mathbb{C}$.

2) **(The Sheffer identity)** A sequence $s_n(x)$ is Sheffer for $(g(t),f(t))$, for some $g(t)$ if and only if

$$s_n(x+y) = \sum_{k=0}^{n} \binom{n}{k} p_k(y) s_{n-k}(x)$$

for all $y \in \mathbb{C}$, where $p_n(x)$ is the associated sequence for $f(t)$. ∎

Examples of Sheffer Sequences

We can now give some examples of Sheffer sequences. While it is often a relatively straightforward matter to *verify* that a given sequence is Sheffer for a given pair $(g(t),f(t))$, it is quite another matter to *find* the Sheffer sequence for a given pair. The umbral calculus provides two formulas for this purpose, one of which is direct, but requires the usually very difficult computation of the series $(f(t)/t)^{-n}$. The other is a recurrence relation that expresses each $s_n(x)$ in terms of previous terms in the Sheffer sequence. Unfortunately, space does not permit us to discuss these formulas in detail. However, we will discuss the recurrence formula for associated sequences later in this chapter.

Example 16.4 The sequence $p_n(x) = x^n$ is the associated sequence for the delta series $f(t) = t$. The generating function for this sequence is

$$e^{yt} = \sum_{k=0}^{\infty} \frac{y^k}{k!} t^k$$

and the binomial identity is precisely that:

$$(x + y)^n = \sum_{k=0}^{n} \binom{n}{k} x^k y^{n-k}$$

Example 16.5 The **lower factorial polynomials**

$$(x)_n = x(x-1)\cdots(x-n+1)$$

form the associated sequence for the forward difference functional

$$f(t) = e^t - 1$$

discussed in Example 16.2. To see this, we simply compute, using Theorem 16.10. Since $(0)_0$ is defined to be 1, we have $(0)_n = \delta_{n,0}$. Also,

$$\begin{aligned}
(e^t - 1)(x)_n &= (x+1)_n - (x)_n \\
&= (x+1)x(x-1)\cdots(x-n+2) - x(x-1)\cdots(x-n+1) \\
&= x(x-1)\cdots(x-n+2)[(x+1) - (x-n+1)] \\
&= nx(x-1)\cdots(x-n+2) \\
&= n(x)_{n-1}
\end{aligned}$$

The generating function for the lower factorial polynomials is

$$e^{y\log(1+t)} = \sum_{k=0}^{\infty} \frac{(y)_k}{k!} t^k$$

which can be rewritten in the more familiar form

$$(1+t)^y = \sum_{k=0}^{\infty} \binom{y}{k} t^k$$

Of course, this is a formal identity, so there is no need to make any restrictions on t. The binomial identity in this case is

$$(x+y)_n = \sum_{k=0}^{n} \binom{n}{k} (x)_k (y)_{n-k}$$

which can also be written in the form

$$\binom{x+y}{n} = \sum_{k=0}^{n} \binom{x}{k} \binom{y}{n-k}$$

This is known as the **Vandermonde convolution formula.**

Example 16.6 The **Abel polynomials**

$$A_n(x;a) = x(x-an)^{n-1}$$

form the associated sequence for the Abel functional

$$f(t) = te^{at}$$

also discussed in Example 16.2. We leave verification of this to the reader. The generating function for the Abel polynomials is

$$e^{y\bar{f}(t)} = \sum_{k=0}^{\infty} \frac{y(y-ak)^{k-1}}{k!} t^k$$

Taking the formal derivative of this with respect to y gives

$$\bar{f}(t)e^{y\bar{f}(t)} = \sum_{k=0}^{\infty} \frac{k(y-a)(y-ak)^{k-1}}{k!} t^k$$

which, for $y = 0$, gives a formula for the compositional inverse of the series $f(t) = te^{at}$,

$$\bar{f}(t) = \sum_{k=1}^{\infty} \frac{(-a)^k k^{k-1}}{(k-1)!} t^k$$

Example 16.7 The famous **Hermite polynomials** $H_n(x)$ form the Appell sequence for the invertible functional

$$g(t) = e^{t^2/2}$$

We ask the reader to show that $s_n(x)$ is the Appell sequence for $g(t)$ if and only if $s_n(x) = g(t)^{-1}x^n$. Using this fact, we get

$$H_n(x) = e^{-t^2/2}x^n = \sum_{k \geq 0} (-\tfrac{1}{2})^k \frac{(n)_{2k}}{k!} x^{n-k}$$

The generating function for the Hermite polynomials is

$$e^{yt-t^2/2} = \sum_{k=0}^{\infty} \frac{H_k(y)}{k!} t^k$$

and the Sheffer identity is

$$H_n(x+y) = \sum_{k=0}^{n} \binom{n}{k} H_k(x) y^{n-k}$$

We should remark that the Hermite polynomials, as defined in the literature, often differ from our definition by a multiplicative constant. □

Example 16.8 The well-known and important **Laguerre polynomials** $L_n^{(\alpha)}(x)$ of order α form the Sheffer sequence for the pair

$$g(t) = (1-t)^{-\alpha-1}, \quad f(t) = \frac{t}{t-1}$$

It is possible to show (although we will not do so here) that

$$L_n^{(\alpha)}(x) = \sum_{k=0}^{n} \frac{n!}{k!} \binom{\alpha+n}{n-k} (-x)^k$$

The generating function of the Laguerre polynomials is

$$\frac{1}{(1-t)^{\alpha+1}} e^{yt/(t-1)} = \sum_{k=0}^{\infty} \frac{L_k^{(\alpha)}(x)}{k!} t^k$$

As with the Hermite polynomials, some definitions of the Laguerre polynomials differ by a multiplicative constant. □

We presume that the few examples we have given here indicate that the umbral calculus applies to a significant range of important polynomial sequences. In Roman [1984], we discuss approximately 30 different sequences of polynomials that are (or are closely related to) Sheffer sequences.

Umbral Operators and Umbral Shifts

We have now established the basic framework of the umbral calculus. As we have seen, the umbral algebra plays three roles – as the algebra of formal power series in a single variable, as the algebra of all linear functionals on $\mathcal{P}$, and as the algebra of all linear operators on $\mathcal{P}$ that commute with the derivative operator. Moreover, since $\mathcal{F}$ is an algebra, we can consider geometric sequences in $\mathcal{F}$

$$g, gf, gf^2, gf^3, \ldots$$

where $o(g) = 0$ and $o(f) = 1$. We have seen by example that the orthogonality conditions

$$\langle g(t)f^k(t) \mid s_n(x)\rangle = n!\delta_{n,k}$$

define important families of polynomial sequences.

While the machinery that we have developed so far does unify a number of topics from the classical study of polynomial sequences (for example, special cases of the expansion theorem include Taylor's expansion, the Euler-MacLaurin formula and Boole's summation formula), it does not provide much new insight into their study. Our plan now is to take a brief look at some of the deeper results in the umbral calculus, which center around the interplay between operators on $\mathcal{P}$ and their adjoints, which are operators on the umbral algebra $\mathcal{F} = \mathcal{P}^*$.

We begin by defining two important operators on $\mathcal{P}$ associated to each Sheffer sequence.

Definition Let $s_n(x)$ be Sheffer for $(g(t),f(t))$. The linear operator $\lambda_{g,f}:\mathcal{P}\to\mathcal{P}$ defined by

$$\lambda_{g,f}(x^n) = s_n(x)$$

is called the **Sheffer operator** for the pair $(g(t),f(t))$, or for the sequence $s_n(x)$. If $p_n(x)$ is the associated sequence for $f(t)$, the Sheffer operator

$$\lambda_f(x^n) = p_n(x)$$

is called the **umbral operator** for $f(t)$, or for $p_n(x)$. ▯

Definition Let $s_n(x)$ be Sheffer for $(g(t),f(t))$. The linear operator $\theta_{g,f}:\mathcal{P}\to\mathcal{P}$ defined by

$$\theta_{g,f}[s_n(x)] = s_{n+1}(x)$$

is called the **Sheffer shift** for the pair $(g(t),f(t))$, or for the sequence $s_n(x)$. If $p_n(x)$ is the associated sequence for $f(t)$, the Sheffer operator

$$\theta_f[p_n(x)] = p_{n+1}(x)$$

is called the **umbral shift** for $f(t)$, or for $p_n(x)$. ▯

We will confine our attention in this brief introduction to umbral operators and umbral shifts, rather than the more general Sheffer operators and Sheffer shifts. It is clear that each Sheffer sequence uniquely determines a Sheffer operator and vice-versa. Hence, knowing the Sheffer operator of a sequence is equivalent to knowing the sequence.

Continuous Operators on the Umbral Algebra

It is clearly desirable that an operator $T \in \mathcal{L}(\mathcal{F})$ on the umbral algebra pass under infinite sums, that is,

$$T\Big(\sum_{k=0}^{\infty} a_k f_k(t)\Big) = \sum_{k=0}^{\infty} a_k T[f_k(t)] \tag{16.6}$$

whenever the sum on the left is defined, which is precisely when $o(f_k(t)) \to \infty$ as $k \to \infty$. Not all operators on $\mathcal{F}$ have this property, which leads to the following definition.

Definition A linear operator T on the umbral algebra $\mathcal{F}$ is **continuous** if it satisfies (16.6). □

The term continuous can be justified by defining a topology on $\mathcal{F}$. However, since no additional topological concepts will be needed, we will not do so here. Note that in order for (16.6) to make sense, we must have $o(T[f_k(t)]) \to \infty$. It turns out that this condition is also sufficient.

Theorem 16.12 A linear operator T on $\mathcal{F}$ is continuous if and only if

$$o(f_k) \to \infty \ \Rightarrow \ o(T(f_k)) \to \infty \tag{16.7}$$

Proof. The necessity is clear. Suppose that (16.7) holds, and that $o(f_k) \to \infty$. For any $m \geq 0$, we have

$$\Big\langle T\sum_{k=0}^{\infty} a_k f_k(t) \,\Big|\, x^n \Big\rangle = \Big\langle T\sum_{k=0}^{m} a_k f_k(t) \,\Big|\, x^n \Big\rangle + \Big\langle T\sum_{k>m} a_k f_k(t) \,\Big|\, x^n \Big\rangle \tag{16.8}$$

Since $o(\Sigma_{k>m} a_k f_k(t)) \to \infty$, (16.7) implies that we may choose m large enough so that

$$o\Big(T\sum_{k>m} a_k f_k(t)\Big) > n$$

as well as

$$o(T[f_k(t)]) > n \quad \text{for } k > m$$

Hence, (16.8) gives

$$\Big\langle T\sum_{k=0}^{\infty} a_k f_k(t) \,\Big|\, x^n \Big\rangle = \Big\langle T\sum_{k=0}^{m} a_k f_k(t) \,\Big|\, x^n \Big\rangle = \Big\langle \sum_{k=0}^{m} a_k T[f_k(t)] \,\Big|\, x^n \Big\rangle$$

$$= \Big\langle \sum_{k=0}^{\infty} a_k T[f_k(t)] \,\Big|\, x^n \Big\rangle$$

which implies the desired result. ∎

Operator Adjoints

If $\tau:\mathcal{P}\to\mathcal{P}$ is a linear operator on $\mathcal{P}$, then its (operator) adjoint $\tau^{\times}$ is an operator on $\mathcal{P}^* = \mathcal{F}$ defined by

$$\tau^{\times}[h(t)] = h(t) \circ \tau$$

In the symbolism of the umbral calculus, this is

$$\langle \tau^{\times} h(t) \mid p(x) \rangle = \langle h(t) \mid \tau p(x) \rangle$$

(We have reduced the number of parentheses used to aid clarity.)

Let us recall the basic properties of the adjoint from Chapter 3.

Theorem 16.13 For $\tau,\sigma \in \mathcal{L}(\mathcal{P})$,

1) $(\tau+\sigma)^{\times} = \tau^{\times} + \sigma^{\times}$
2) $(r\tau)^{\times} = r\tau^{\times}$ for any $r \in \mathbb{C}$
3) $(\tau\sigma)^{\times} = \sigma^{\times}\tau^{\times}$
4) $(\tau^{-1})^{\times} = (\tau^{\times})^{-1}$ for any invertible $\tau \in \mathcal{L}(\mathcal{P})$ ∎

Thus, the map $\phi:\mathcal{L}(\mathcal{P})\to\mathcal{L}(\mathcal{F})$ that sends $\tau:\mathcal{P}\to\mathcal{P}$ to its adjoint $\tau^{\times}:\mathcal{F}\to\mathcal{F}$ is a linear transformation from $\mathcal{L}(\mathcal{P})$ to $\mathcal{L}(\mathcal{F})$. Moreover, since $\tau^{\times} = 0$ implies that $\langle h(t) \mid \tau p(x) \rangle = 0$ for all $h(t) \in \mathcal{F}$ and $p(x) \in \mathcal{P}$, which in turn implies that $\tau = 0$, we deduce that ϕ is injective. The next theorem describes the range of ϕ.

Theorem 16.14 A linear operator $T \in \mathcal{L}(\mathcal{F})$ is the adjoint of a linear operator $\mathcal{L} \in \mathcal{L}(\mathcal{P})$ if and only if T is continuous.

Proof. First, suppose that $T = \tau^{\times}$ for some $\tau \in \mathcal{L}(\mathcal{P})$. If $o(f_k(t))\to\infty$, then for any $n \geq 0$, there is a k_n for which

$$k > k_n \;\Rightarrow\; o(f_k(t)) > \deg \tau(x^i) \text{ for all } 0 \leq i \leq n$$

Hence,

$$\begin{aligned} k > k_n &\Rightarrow \langle \tau^{\times} f_k(t) \mid x^i \rangle = \langle f_k(t) \mid \tau x^i \rangle = 0 \text{ for all } 0 \leq i \leq n \\ &\Rightarrow o(\tau^{\times} f_k(t)) > n \end{aligned}$$

which shows that $o(\tau^{\times} f_k(t))\to\infty$, and hence that $\tau^{\times}$ is continuous.

For the converse, assume that T is continuous. We can define a linear operator τ on $\mathcal{P}$ by setting

$$\tau x^n = \sum_{k\geq 0} \frac{\langle Tt^k \mid x^n \rangle}{k!} x^k$$

This makes sense since $o(Tt^k)\to\infty$ as $k\to\infty$, and so the sum on the right is a finite one. Then

$$\langle \tau^{\times} t^m \mid x^n \rangle = \langle t^m \mid \tau x^n \rangle = \sum_{k\geq 0} \frac{\langle Tt^k \mid x^n \rangle}{k!} \langle t^m \mid x^k \rangle = \langle Tt^m \mid x^n \rangle$$

which implies that $Tt^m = \tau^\times t^m$ for all $m \geq 0$. Finally, since T and $\tau^\times$ are both continuous, we have $T = \tau^\times$. ∎

Automorphisms of the Umbral Algebra

Figure 16.1 shows the map ϕ, which is an isomorphism from the vector space $\mathcal{L}(\mathcal{P})$ onto the space of all *continuous* linear operators on $\mathcal{F}$. We are interested in determining the images of the set of all umbral operators, and the set of all umbral shifts, under this isomorphism.

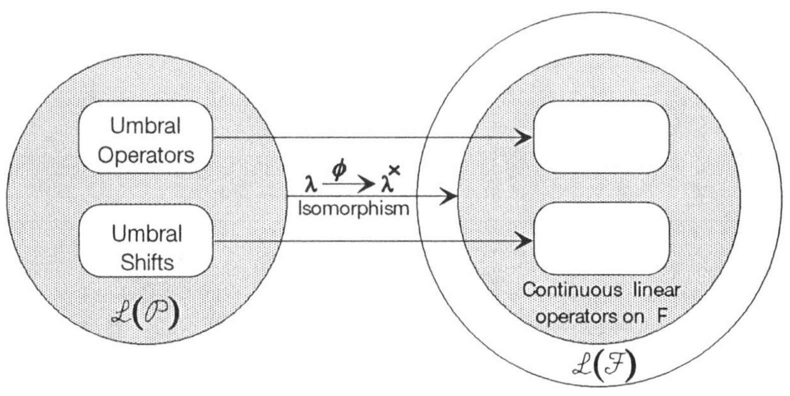

Figure 16.1

Let us begin with umbral operators. Suppose that λ_f is the umbral operator for the associated sequence $p_n(x)$, associated to the delta series $f(t) \in \mathcal{F}$. Then

$$\langle \lambda_f^\times f(t)^k \mid x^n \rangle = \langle f(t)^k \mid \lambda_f x^n \rangle = \langle f(t)^k \mid p_n(x) \rangle = n!\delta_{n,k} = \langle t^k \mid x^n \rangle$$

for all k and n. Hence, $\lambda_f^\times f(t)^k = t^k$, which implies, since $\lambda_f^\times$ is continuous, that

$$\lambda_f^\times t^k = \bar{f}(t)^k$$

More generally, for any $h(t) \in \mathcal{F}$,

$$\lambda_f^\times h(t) = h(\bar{f}(t)) \tag{16.9}$$

In words, $\lambda_f^\times$ is composition by $\bar{f}(t)$.

From (16.9), we deduce that $\lambda_f^\times$ is a vector space isomorphism, and that

$$\lambda_f^\times[g(t)h(t)] = g(\bar{f}(t))h(\bar{f}(t)) = \lambda_f^\times g(t)\lambda_f^\times h(t)$$

Hence, $\lambda_f^\times$ is an *automorphism* of the umbral algebra $\mathcal{F}$. It is a pleasant fact that this characterizes umbral operators. The first step in the proof of this is the following, whose proof is left as an exercise.

Theorem 16.15 If T is an automorphism of the umbral algebra, then T preserves order, that is, $o(Tf(t)) = o(f(t))$. In particular, T is continuous. ∎

Theorem 16.16 A linear operator λ on $\mathcal{P}$ is an umbral operator if and only if its adjoint is an automorphism of the umbral algebra $\mathcal{F}$. Moreover, if λ_f is an umbral operator, then

$$\lambda_f^{\times}h(t) = h(\bar{f}(t))$$

for all $h(t) \in \mathcal{F}$. In particular, $\lambda_f^{\times}f(t) = t$.

Proof. We have already shown that the adjoint of λ_f is an automorphism satisfying (16.9). For the converse, suppose that $\lambda^{\times}$ is an automorphism of $\mathcal{F}$. Theorem 16.15 implies the existence of a unique delta series $f(t)$ for which $\lambda^{\times}f(t) = t$. If $p_n(x)$ is the associated sequence for $f(t)$, then

$$\begin{aligned}\langle f(t)^k \mid \lambda x^n\rangle &= \langle \lambda^{\times}f(t)^k \mid x^n\rangle = \langle [\lambda^{\times}f(t)]^k \mid x^n\rangle \\ &= \langle t^k \mid x^n\rangle = n!\delta_{n,k} = \langle f(t)^k \mid p_n(x)\rangle\end{aligned}$$

and so part (6) of Theorem 16.2 implies that $\lambda x^n = p_n(x)$. Thus, λ is an umbral operator. ∎

Theorem 16.16 allows us to fill in one of the blank boxes on the right side of Figure 16.1, as shown in Figure 16.2.

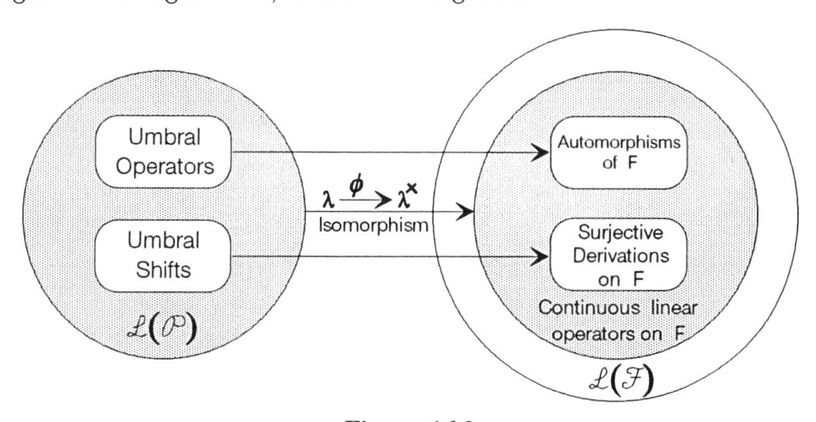

Figure 16.2

Let us see how we might use Theorem 16.16 to advantage in the study of associated sequences. Since the set $Aut(\mathcal{F})$ of all automorphisms of $\mathcal{F}$ is a group under composition, so is the set of umbral operators. More specifically, let

$$\lambda_f{:}x^n \rightarrow p_n(x) \quad \text{and} \quad \lambda_g{:}x^n \rightarrow q_n(x)$$

be umbral operators. Then

$$(\lambda_g \circ \lambda_f)^\times = \lambda_f^\times \circ \lambda_g^\times$$

is an automorphism of $\mathcal{F}$, and so $\lambda_g \circ \lambda_f$ is an umbral operator. In fact, since

$$(\lambda_g \circ \lambda_f)^\times f(g(t)) = \lambda_f^\times \circ \lambda_g^\times f(g(t)) = \lambda_f^\times f(t) = t$$

we deduce that $\lambda_g \circ \lambda_f = \lambda_{f \circ g}$. Also, since

$$\lambda_{\bar{f}} \circ \lambda_f = \lambda_{f \circ \bar{f}} = \lambda_t = \iota$$

we have $\lambda_f^{-1} = \lambda_{\bar{f}}$.

Now, if

$$p_n(x) = \sum_{k=0}^{n} p_{n,k} x^k$$

then $\lambda_g \circ \lambda_f$ is the umbral operator for the associated sequence

$$(\lambda_g \circ \lambda_f) x^n = \lambda_g p_n(x) = \sum_{k=0}^{n} p_{n,k} \lambda_g x^k = \sum_{k=0}^{n} p_{n,k} q_k(x)$$

This sequence, denoted by

$$p_n(\mathbf{q}(x)) = \sum_{k=0}^{n} p_{n,k} q_k(x) \tag{16.10}$$

is called the **umbral composition** of $p_n(x)$ with $q_n(x)$. Let us summarize.

Theorem 16.17 Let $p_n(x)$ and $q_n(x)$ be associated sequences, with umbral operators λ_f and λ_g, respectively.

1) $\lambda_g \circ \lambda_f = \lambda_{f \circ g}$ and $\lambda_f^{-1} = \lambda_{\bar{f}}$
2) The set of associated sequences forms a group under umbral composition, as defined by (16.10). In particular, the umbral composition $p_n(\mathbf{q}(x))$ is the associated sequence for the composition $f \circ g$. The identity is the sequence x^n, and the inverse of $p_n(x)$ is the associated sequence for the compositional inverse $\bar{f}(t)$. ∎

Derivations of the Umbral Algebra

We have seen that an operator on $\mathcal{P}$ is an umbral operator if and only if its adjoint is an automorphism of $\mathcal{F}$. Now suppose that $\theta_f \in \mathcal{L}(\mathcal{P})$ is the umbral shift for the associated sequence $p_n(x)$, associated to the delta series $f(t) \in \mathcal{F}$. Then

$$\begin{aligned}\langle \theta_f^\times f(t)^k \mid p_n(x)\rangle &= \langle f(t)^k \mid \theta_f p_n(x)\rangle = \langle f(t)^k \mid p_{n+1}(x)\rangle \\ &= (n+1)!\delta_{n+1,k} = (n+1)n!\delta_{n,k-1} = \langle k f(t)^{k-1} \mid p_n(x)\rangle\end{aligned}$$

and so

$$(16.11) \qquad \theta_f^\times f(t)^k = kf(t)^{k-1}$$

This implies

$$(16.12) \qquad \theta_f^\times[f(t)^k f(t)^j] = \theta_f^\times[f(t)^k]f(t)^j + f(t)^k\theta_f^\times[f(t)^j]$$

and further, by continuity, that

$$(16.13) \qquad \theta_f^\times[g(t)h(t)] = [\theta_f^\times g(t)]h(t) + g(t)[\theta_f^\times g(t)]$$

Let us pause for a definition.

Definition Let $\mathcal{A}$ be any algebra. A linear operator ∂ on $\mathcal{A}$ is a **derivation** if

$$\partial(ab) = (\partial a)b + a\partial b$$

for all $a,b \in \mathcal{A}$. □

Thus, we have shown that the adjoint of an umbral shift is a derivation of the umbral algebra $\mathcal{F}$. Moreover, the expansion theorem and (16.11) show that $\theta_f^\times$ is surjective. As with umbral operators, this characterizes umbral shifts. First we need a preliminary result on surjective derivations.

Theorem 16.18 Let ∂ be a surjective derivation on the umbral algebra $\mathcal{F}$. Then $\partial c = 0$ for any *constant* $c \in \mathcal{F}$ and $o(\partial f(t)) = o(f(t)) - 1$, if $o(f(t)) \geq 1$. In particular, ∂ is continuous.

Proof. We begin by noting that $\partial 1 = \partial 1^2 = \partial 1 + \partial 1 = 2\partial 1$, and so $\partial c = c\partial 1 = 0$ for all constants $c \in \mathcal{F}$. Since ∂ is surjective, there must exists an $h(t) \in \mathcal{F}$ for which

$$\partial h(t) = 1$$

Writing $h(t) = h_0 + th_1(t)$, we have

$$1 = \partial[h_0 + th_1(t)] = (\partial t)h_1(t) + t\partial h_1(t)$$

which implies that $o(\partial t) = 0$. Finally, if $o(h(t)) = k \geq 1$, then $h(t) = t^k h_1(t)$, where $o(h_1(t)) = 0$, and so

$$o[\partial h(t)] = o[\partial t^k h_1(t)] = o[t^k\partial h(t) + kt^{k-1}h_1(t)\partial t] = k - 1$$ ∎

Theorem 16.19 A linear operator θ on $\mathcal{P}$ is an umbral shift if and only if its adjoint is a surjective derivation of the umbral algebra $\mathcal{F}$. Moreover, if θ_f is an umbral shift, then $\theta_f^\times = \partial_f$ is derivation with respect to $f(t)$, that is,

$$\theta_f^\times f(t)^k = kf(t)^{k-1}$$

for all $k \geq 0$. In particular, $\theta_f^\times f(t) = 1$.

Proof. We have already seen that $\theta_f^\times$ is derivation with respect to $f(t)$. For the converse, suppose that $\theta^\times$ is a surjective derivation. Theorem 16.18 implies that there is a delta functional $f(t)$ such that $\theta^\times f(t) = 1$. If $p_n(x)$ is the associated sequence for $f(t)$, then

$$\begin{aligned}\langle f(t)^k \mid \theta p_n(x)\rangle &= \langle \theta^\times f(t)^k \mid p_n(x)\rangle = \langle kf(t)^{k-1}\theta^\times f(t) \mid p_n(x)\rangle \\ &= \langle kf(t)^{k-1} \mid p_n(x)\rangle = (n+1)!\delta_{n+1,k} = \langle f(t)^k \mid p_{n+1}(x)\rangle\end{aligned}$$

Hence, $\theta p_n(x) = p_{n+1}(x)$, that is, $\theta = \theta_f$ is the umbral shift for $p_n(x)$. ∎

Figure 16.2 is now justified. Let us summarize.

Theorem 16.20 The isomorphism ϕ from $\mathcal{L}(\mathcal{P})$ onto the continuous linear operators on $\mathcal{F}$ is a bijection from the set of all umbral operators to the set of all automorphisms of $\mathcal{F}$, as well as a bijection from the set of all umbral shifts to the set of all surjective derivations on $\mathcal{F}$. ∎

We have seen that the fact that the set of all automorphisms on $\mathcal{F}$ is a group under composition shows that the set of all associated sequences is a group under umbral composition. The set of all surjective derivations on $\mathcal{F}$ does not form a group. However, we do have the chain rule for derivations!

Theorem 16.21 **(The chain rule)** Let ∂_f and ∂_g be surjective derivations on $\mathcal{F}$. Then

$$\partial_g = (\partial_g f(t))\partial_f$$

Proof. This follows from

$$\partial_g f(t)^k = kf(t)^{k-1}\partial_g f(t) = (\partial_g f(t))\partial_f f(t)^k$$

and so continuity implies the result. ∎

The chain rule leads to the following umbral result.

Theorem 16.22 If θ_f and θ_g are umbral shifts, then

$$\theta_f = \theta_g \circ \partial_f g(t)$$

Proof. The chain rule gives

$$\theta_f^\times = (\partial_f g(t))\theta_g^\times$$

and so

$$\begin{aligned}\langle h(t) \mid \theta_f p(x)\rangle &= \langle \theta_f^\times h(t) \mid p(x)\rangle = \langle (\partial_f g(t))\theta_g^\times h(t) \mid p(x)\rangle \\ &= \langle \theta_g^\times h(t) \mid \partial_f g(t) p(x)\rangle = \langle h(t) \mid \theta_g \circ \partial_f g(t) p(x)\rangle\end{aligned}$$

for all $p(x) \in \mathcal{P}$ and all $h(t) \in \mathcal{F}$, which implies the result. ∎

We leave it as an exercise to show that $\partial_g f(t) = [\partial_f g(t)]^{-1}$. Now, by taking $g(t) = t$ in Theorem 16.22, and observing that $\theta_t x^n = x^{n+1}$ and so θ_t is multiplication by x, we get

$$\theta_f = x\partial_f t = x[\partial_t f(t)]^{-1} = x[f'(t)]^{-1}$$

Applying this to the associated sequence $p_n(x)$ for $f(t)$ gives the following important recurrence relation for $p_n(x)$.

Theorem 16.23 **(The recurrence formula)** Let $p_n(x)$ be the associated sequence for $f(t)$. Then

$$p_{n+1}(x) = x[f'(t)]^{-1} p_n(x)$$ ∎

Example 16.9 The recurrence relation can be used to find the associated sequence for the forward difference functional $f(t) = e^t - 1$. Since $f'(t) = e^t$, the recurrence relation is

$$p_{n+1}(x) = xe^{-t} p_n(x) = xp_n(x-1)$$

Using the fact that $p_0(x) = 1$, we have

$$p_1(x) = x, \quad p_2(x) = x(x-1), \quad p_3(x) = x(x-1)(x-2)$$

and so on, leading easily to the lower factorial polynomials

$$p_n(x) = x(x-1)\cdots(x-n+1) = (x)_n$$ □

Example 16.10 Consider the delta functional

$$f(t) = \log(1+t)$$

Since $\bar{f}(t) = e^t - 1$ is the forward difference functional, Theorem 16.17 implies that the associated sequence $\phi_n(x)$ for $f(t)$ is the inverse, under umbral composition, of the lower factorial polynomials. Thus, if we write

$$\phi_n(x) = \sum_{k=0}^{n} S(n,k) x^k$$

then

$$x^n = \sum_{k=0}^{n} S(n,k)(x)_k$$

The coefficients $S(n,k)$ in this equation are known as the **Stirling numbers of the second kind** and have great combinatorial significance.

In fact, $S(n,k)$ is the number of partitions of a set of size n into k blocks. The polynomials $\phi_n(x)$ are called the **exponential polynomials.**

The recurrence relation for the exponential polynomials is

$$\phi_{n+1}(x) = x(1+t)\phi_n(x) = x(\phi_n(x) + \phi_n'(x))$$

Equating coefficients of x^k on both sides of this gives the well-known formula for the Stirling numbers

$$S(n+1,k) = S(n,k-1) + kS(n,k)$$

Many other properties of the Stirling numbers can be derived by umbral means. □

EXERCISES

1. Prove that $o(fg) = o(f) + o(g)$, for any $f,g \in \mathcal{F}$.
2. Prove that $o(f+g) \geq \min\{o(f),o(g)\}$, for any $f,g \in \mathcal{F}$.
3. Show that any delta series has a compositional inverse.
4. Show that for any delta series f, the sequence f^k is a pseudobasis.
5. Prove that ∂_t is a derivation.
6. Show that $f \in \mathcal{F}$ is a delta functional if and only if $\langle f \mid 1\rangle = 0$ and $\langle f \mid x\rangle \neq 0$.
7. Show that $f \in \mathcal{F}$ is invertible if and only if $\langle f \mid 1\rangle \neq 0$.
8. Show that $\langle f(at) \mid p(x)\rangle = \langle f(t) \mid p(ax)\rangle$ for any $a \in \mathbb{C}$, $f \in \mathcal{F}$ and $p \in \mathcal{P}$.
9. Show that $\langle te^{at} \mid p(x)\rangle = p'(a)$ for any polynomial $p(x) \in \mathcal{P}$.
10. Show that $f = g$ in $\mathcal{F}$ if and only if $f = g$ as linear functionals, which holds if and only if $f = g$ as linear operators.
11. Prove that if $s_n(x)$ is Sheffer for $(g(t),f(t))$, then $f(t)s_n(x) = ns_{n-1}(x)$. *Hint:* Apply the functionals $g(t)f^k(t)$ to both sides.
12. Verify that the Abel polynomials form the associated sequence for the Abel functional.
13. Show that a sequence $s_n(x)$ is the Appell sequence for $g(t)$ if and only if $s_n(x) = g(t)^{-1}x^n$.
14. If f is a delta series, show that the adjoint $\lambda_f^\times$ of the umbral operator λ_f is a vector space isomorphism of $\mathcal{F}$.
15. Prove that if T is an automorphism of the umbral algebra, then T preserves order, that is, $o(Tf(t)) = o(f(t))$. In particular, T is continuous.
16. Show that an umbral operator maps associated sequences to associated sequences.
17. Let $p_n(x)$ and $q_n(x)$ be associated sequences. Define a linear operator α by $\alpha{:}p_n(x) \to q_n(x)$. Show that α is an umbral operator.
18. Prove that if ∂_f and ∂_g are surjective derivations on $\mathcal{F}$, then $\partial_g f(t) = [\partial_f g(t)]^{-1}$.

References

Artin, E., *Geometric Algebra*, Interscience Publishers, 1988.
Artin, M., *Algebra*, Prentice-Hall, 1991.
Blyth, T. S., *Module Theory – An Approach to Linear Algebra*, Oxford U. Press, 1990.
Grueb, W., *Linear Algebra*, 4th edition, Springer-Verlag, 1975.
Greub, W., *Multilinear Algebra*, Springer-Verlag, 1978.
Halmos, P., *Finite-Dimensional Vector Spaces*, Van Nostrand, 1958.
Halmos, P., *Naive Set Theory*, Van Nostrand, 1960.
Jacobson, N., *Basic Algebra I*, Freeman, 1985.
Kreyszig, E., *Introductory Functional Analysis with Applications*, John Wiley and Sons, 1978.
MacLane, S. and Birkhoff, G., *Algebra*, Macmillan, 1979.
Roman, S., *Coding and Information Theory*, Springer-Verlag, 1992.
Roman, S., *The Umbral Calculus*, Academic Press, 1984.
Snapper, E. and Troyer, R., *Metric Affine Geometry*, Dover, 1971.

Index of Notation

Index

Graduate Texts in Mathematics

continued from page ii

97 KOBLITZ. Introduction to Elliptic Curves and Modular Forms.
98 BRÖCKER/TOM DIECK. Representations of Compact Lie Groups.
99 GROVE/BENSON. Finite Reflection Groups. 2nd ed.
100 BERG/CHRISTENSEN/RESSEL. Harmonic Analysis on Semigroups: Theory of Positive Definite and Related Functions.
101 EDWARDS. Galois Theory.
102 VARDARAJAN. Lie Groups, Lie Algebras and Their Representations.
103 LANG. Complex Analysis. 2nd ed.
104 DUBROVIN/FOMENKO/NOVIKOV. Modern Geometry--Methods and Applications. Part II.
105 LANG. $SL_2(\mathbf{R})$.
106 SILVERMAN. The Arithmetic of Elliptic Curves.
107 OLVER. Applications of Lie Groups to Differential Equations.
108 RANGE. Holomorphic Functions and Integral Representations in Several Complex Variables.
109 LEHTO. Univalent Functions and Teichmüller Spaces.
110 LANG. Algebraic Number Theory.
111 HUSEMÖLLER. Elliptic Cruves.
112 LANG. Elliptic Functions.
113 KARATZAS/SHREVE. Brownian Motion and Stochastic Calculus. 2nd ed.
114 KOBLITZ. A Course in Number Theory and Cryptography.
115 BERGER/GOSTIAUX. Differential Geometry: Manifolds, Curves, and Surfaces.
116 KELLEY/SRINIVASAN. Measure and Integral. Vol. I.
117 SERRE. Algebraic Groups and Class Fields.
118 PEDERSEN. Analysis Now.
119 ROTMAN. An Introduction to Algebraic Topology.
120 ZIEMER. Weakly Differentiable Functions: Sobolev Spaces and Functions of Bounded Variation.
121 LANG. Cyclotomic Fields I and II. Combined 2nd ed.
122 REMMERT. Theory of Complex Functions.
Readings in Mathematics
123 EBBINGHAUS/HERMES et al. Numbers.
Readings in Mathematics
124 DUBROVIN/FOMENKO/NOVIKOV. Modern Geometry--Methods and Applications. Part III.
125 BERENSTEIN/GAY. Complex Variables: An Introduction.
126 BOREL. Linear Algebraic Groups.
127 MASSEY. A Basic Course in Algebraic Topology.
128 RAUCH. Partial Differential Equations.
129 FULTON/HARRIS. Representation Theory: A First Course.
Readings in Mathematics.
130 DODSON/POSTON. Tensor Geometry.
131 LAM. A First Course in Noncommutative Rings.
132 BEARDON. Iteration of Rational Functions.
133 HARRIS. Algebraic Geometry: A First Course.
134 ROMAN. Coding and Information Theory.
135 ROMAN. Advanced Linear Algebra.